AF812664

Advanced Sciences and Technologies for Security Applications

The series Advanced Sciences and Technologies for Security Applications comprises interdisciplinary research covering the theory, foundations and domain-specific topics pertaining to security. Publications within the series are peer-reviewed monographs and edited works in the areas of:

- biological and chemical threat recognition and detection (e.g., biosensors, aerosols, forensics)
- crisis and disaster management
- terrorism
- cyber security and secure information systems (e.g., encryption, optical and photonic systems)
- traditional and non-traditional security
- energy, food and resource security
- economic security and securitization (including associated infrastructures)
- transnational crime
- human security and health security
- social, political and psychological aspects of security
- recognition and identification (e.g., optical imaging, biometrics, authentication and verification)
- smart surveillance systems
- applications of theoretical frameworks and methodologies (e.g., grounded theory, complexity, network sciences, modelling and simulation)

Together, the high-quality contributions to this series provide a cross-disciplinary overview of forefront research endeavours aiming to make the world a safer place.

More information about this series at http://www.springer.com/series/5540

Anthony J. Masys
Editor

Security by Design

Innovative Perspectives on Complex Problems

Springer

Editor
Anthony J. Masys
Global Disaster Management, Humanitarian Assistance and Homeland Security
University of South Florida
Tampa, FL
USA

ISSN 1613-5113 ISSN 2363-9466 (electronic)
Advanced Sciences and Technologies for Security Applications
ISBN 978-3-030-08602-2 ISBN 978-3-319-78021-4 (eBook)
https://doi.org/10.1007/978-3-319-78021-4

Softcover re-print of the Hardcover 1st edition 2018

Printed on acid-free paper

This Springer imprint is published by the registered company Springer Nature Switzerland AG
The registered company address is: Gewerbestrasse 11, 6330 Cham, Switzerland

Contents

Introduction

Anthony J. Masys

1 Introduction

The current and future security landscape is diverse and complex. It is characterized by such issues as: mass migration and refugee crisis; economic slowdowns in emerging markets; ever-rising numbers of terrorist and cyberattacks; pandemics; active shooters; climate related disasters; global water shortages; energy security; the proliferation of the Internet of Things and its impact on the security of our homes, cities and critical infrastructure. These global risks have been in the headlines in the last year (Global Risks Report) and pose significant security challenges both nationally and globally. In fact, national security is no longer just national. Non-state actors, cyber NGOs, rising powers, pandemics and hybrid wars and crimes in strategic areas pose complex challenges to global security.

Designing security solutions requires understanding the complex interdependencies that reside within the threat landscape to create a plan or action such that '*…if implemented-is intended to bring about a situation with specific desired characteristics without creating unforeseen and undesired side and after effects*' Rittel (1968).

This edited book captures salient global, regional and national security challenges and leveraging key design thinking principles, presents 'design' solutions to address them.

A. J. Masys (✉)
University of South Florida, Tampa, FL, USA
e-mail: tmasys@health.usf.edu

A. J. Masys (ed.), *Security by Design*, Advanced Sciences and Technologies for Security Applications, https://doi.org/10.1007/978-3-319-78021-4_1

2 Designing Security

As part of the Springer book series: Advanced Sciences and Technologies for Security Applications, this edited volume: **Security by Design,** introduces novel perspectives and innovative approaches to how we can design security interventions.

The 17 chapters in this book reflect contributions from various experts and through case studies and research reveal the many perspectives, tools and approaches to support security design. Security design can support risk and threat analysis, risk communication, problem framing and development of intervention strategies.

3 Content

Rama Gheerawo in his chapter '**Design Thinking and Design Doing: describing a process of people-centred innovation**' outlines the benefits of Design Thinking as a creative framework for innovation that can be applied to projects and organisations across government, business and the public sector. This chapter includes a short history of the Design Thinking approach to set context and define the concept. At the heart of the approach is a people-centered focus, and the chapter discusses the role of Inclusive Design and design ethnography in underpinning Design Thinking. It outlines five principles: Involve People, Translate Design Thinking into Design Doing, Create Value and Capture Value, Follow the Arc of Design Thinking and Navigate Complexity. It aims to give an overarching view of Design Thinking, demonstrating the value that it can bring to diverse areas of research and application. Challenges and observations for progressing the approach are also noted.

Craig Perry and Andy Hines in their chapter '**An Integral Futures lens on the future security issues**' argues that conventional analysis of security issues tends to take a disjointed approach through tightly focused tactical lenses. Increasingly, however, security issues are embedded in a wider and diverse range of factors that may escape conventional analysis. An integral futures perspective to exploring security issues is proposed that provides a wider lens by taking an integrated holistic approach that explores individual and collective aspects of emerging issues from both an interior (e.g., motivation, culture) and exterior (e.g., behavior, systems, and infrastructures) perspective. We believe the approach may provide a fresh approach to issue identification as well as more holistic explanatory framework.

Bradley Keating and Anthony Masys in their chapter '**Innovations in Active Shooting Emergency Medical Response and Triage**' argue that with the rise in prevalence of mass shootings, Emergency Medical providers may now find themselves responding to an active shooter event. Traditional triage and victim management techniques leveraged from military experiences are being re-examined. As noted by Smith et al. [15], 'A recent analysis of mass shooting events noted a wounding pattern different from military experience with very few cases of life-threatening extremity hemorrhage; therefore, extrication and transfer to definitive care needs to be a priority in addition to any in-place care provided'. As noted in van Rein et al. [18: 329] 'prehospital trauma triage is essential in providing appropriate care for patients at risk from severe injury to improve their chance of survival'.

This chapter will discuss the requirements for a more 'tactically oriented' medical response in an active shooter incident, and presents novel triage and scene management techniques to facilitate a more rapid and effective response to maximize victim survivability.

Miriam Kaundert and Anthony Masys in their chapter '**Mass Migration, Humanitarian Assistance and Crisis Management: Embracing Social Innovation and Organizational Learning**' argue that the current surge in migration to the European Union (EU) is rapidly becoming the largest and most complex facing Europe since the Second World War. Between January 2015 and February 2016, over 1.1 million people made their way to the EU, escaping conflict and poverty in their countries.

Moore and Westley [8] argue that 'urgent calls for "innovation" have become increasingly frequent as people begin to recognize both the need for fundamental change in human beings' relationships with each other and their environment, and the catastrophic ramifications of inaction or ill-informed decisions as to how such change might be effected'.

This chapter examines the complex causality regarding mass migration of refugees and leverages social innovation and organizational learning as theoretical frameworks to support the design of humanitarian assistance and crisis management.

Ivan Taylor and Anthony Masys in their chapter '**Complexity and Unintended Consequences in a Human Security Crisis: A System Dynamic Model of the Refugee Migration to Europe**' examine the mass migration crisis in Europe. As described by UNHCR [17], 'Europe is living through a maritime refugee crisis of historic proportions. Its evolving response has become one of the continent's defining challenges of the early 21st century, with long-lasting implications for humanitarian practice, regional stability and international public opinion'. UNHCR's figures show that over one million people had reached Europe across the Mediterranean, mainly to Greece and Italy, in 2015. Of these, over 3700 were missing, believed drowned. This represents a regional human security crisis. Sagan [13: 12] argues that 'Things that have never happened before happen all the time'. The mass migration of refugees should not have come as a complete surprise. The mass migration in Europe, North Africa and Middle East derives from a wicked problem space and is linked to humanitarian challenges elsewhere. Thus, the flight to European shores reflected not only the pull of greater long-term security in Europe, but also the failure of the international humanitarian community to meet basic needs in other places [10].

This chapter explores the complexity of the refugee and humanitarian crisis in Europe and North Africa through the application of a system dynamics model. Given the unintended consequences associated with policies and intervention strategies, the system dynamics model examines the volatility of such interventions on migration.

Polinpapilinho F. Katina and Charles B. Keating in their chapter '**Cyber-Physical Systems Governance: A framework for (meta) CyberSecurity Design**' presents a framework for improving security of cyber-physical systems through purposeful design, execution, and evolution of metasystem functions. State actors (i.e., government agencies), non-state actors (i.e., for-profit and non-profit organizations), and their systems operate under highly emergent and complex conditions. Under these conditions, system performance is not always deducible from the constituent systems. Moreover, such systems are often interdependent and dynamically interacting with other systems such that the state of each system is influenced by and is influencing states of the interconnected systems. In these conditions, leaving a system to develop through processes of 'accretion' (ad hoc evolution of a system) or 'self-organization' (totally unconstrained evolution of a system) might increase the probability of missing performance expectations. In contrast, 'purposeful design' is an invitation for a more determined engagement in system development to increase the probability of producing expected and desired performance. In this chapter, emerging research in *Complex System Governance* (CSG) is suggested as an emerging field to direct more purposeful design for systems. Specifically, a CSG enabled framework for security design focused on Cyber-Physical Systems (CPS) is provided. The framework, grounded in Systems Theory and Management Cybernetics, emphasizes more holistic design for integration, coordination, communication, and control for development of CPS.

Erik Hollnagel in his chapter, '**To feel secure or to be secure, that is the question**' focuses on the title of this volume 'Security by design' as an expression of confidence that it is possible to be secure by design. In that case the obvious next step is to think about possible ways of ensuring security by design, starting, for instance, with the solutions that have been developed to solve the seemingly analogous problems of safety or prevention by design. But what if the title had been "Security by design?" or in other words a question rather than a statement? In that case the title would not be taken as an expression of confidence but rather as an admission of uncertainty or even of impotence, a possible doubt of whether security can be ensured by design. The obvious next step would then be to take a closer look at the two central concepts, namely "security" and "design". The current chapter will do just that.

D.B. Skillcorn and C. Leuprecht in their chapter '**Clustering Heterogeneous Semi-Structured Social Science Datasets for Security Applications**' argues that social scientists have begun to collect large datasets that are heterogeneous and semi-structured, but the ability to analyze such data has lagged behind its collection. They design a process to map such datasets to a numerical form, apply singular value decomposition clustering, and explore the impact of individual attributes or fields by overlaying visualizations of the clusters. This provides a new path for understanding such datasets, which we illustrate with three real-world examples: the Global Terrorism Database, which records details of every terrorist attack since 1970; a Chicago police dataset, which records details of every drug-related incident over a period of approximately a month; and a dataset describing members of a Hezbollah crime/terror network in the U.S.

Frederic Petit, Duane Verner, Julia Phillips, and Lawrence Paul Lewis in their chapter '**Critical Infrastructure Protection and Resilience—Integrating Interdependencies**' argues that enhancing the protection and resilience of critical infrastructure has emerged as an urgent goal—a goal made more challenging by the complexity of these systems and their inherent interdependencies. Consideration of critical infrastructure interdependencies, as well as their integration into risk management and business continuity processes, is an important element of *U.S Presidential Policy Directive 21 on Critical Infrastructure Security and Resilience* (PPD–21), the *Canada-United States Action Plan for Critical Infrastructure*, and the *European Program for Critical Infrastructure Protection*. Several international standards also require the consideration of interdependencies between organizations and the effect on their risk management and business continuity practices. All of these documents, from the strategic level through the operational standards, involve the consideration of interdependencies that can exist among infrastructure, how they are managed, and how they affect business continuity, security, and resilience management. A "system of systems" approach can help establish the appropriate scope of an interdependency analysis, as well as the specific assets and/or subsystems for which resilience-related information should be collected.

David Wood in his chapter '**The Application of Warden's Concentric Ring Theory to Applegate's Offensive Cyber Manoeuvre Strategy in Modern Conflicts**' argues that Cyber has become an essential element for governments, militaries and societies. This importance makes it a target when planning a military campaign. The targeting model developed by Colonel John Warden has demonstrated flexibility in that it can be applied to any entity that can be analyzed as a system. Scott Applegate has accurately described three different strategies that can be employed when waging offensive cyber warfare. The targeting model developed by Colonel John Warden can be applied to each of these cyber maneuver strategies to analyse and attack an opponent's cyber capabilities. There is anecdotal evidence that these different cyber strategies have been employed in recent world conflicts. Real world examples accessed from open source literature have been analysed using Warden's theory, to demonstrate that countries have already started basing their attacks against their opponent's cyber capabilities using this model.

Corey Tsang and Anthony Masys in their chapter '**Fentanyl Crisis: a National Security matter**' describe how amongst the global security challenges, synthetic drugs constitute one of the most significant drug problems worldwide [16]. Fentanyl and non-pharmaceutical fentanyls (NPFs) have moved beyond being a concern to emerging as a national and global security matter. In 2015, more than 52,000 Americans (144/day) died from drug overdose. As reported by Chester [4], 63% of overdoses involved opioids whereby they are having '…a considerable impact on public health and public safety in communities across the United States'.

In Canada, the Fentanyl-related deaths has markedly increased across British Columbia, Alberta, Ontario and Quebec, ranging from a doubling to 20-fold increase from 2009 to 2014 [3]. This increasing trend is noted internationally with increasing mortality noted in Australia [12] and the Nordic countries (Denmark, Finland, Iceland, Norway, Sweden) [14].

Richmond (2001:3) argues that 'The way we think is outdated. As a result, the way we act creates problems, and then we are ill-equipped to address them because of the way we think'. His argument stems from the '...fact that we have been applying the same set of non-solutions to the crucial social problems such as drugs, poverty and crime for most of the last fifty years without any positive results'.

This chapter examines the national and global fentanyl crisis and introduces a holistic perspective of the problem space that leverages social innovation and macro-social analysis [2, 7] to examine the public health and criminal dimensions.

Simon Bennett in his chapter '**Risk-based aviation security- designing out terror**' reviews the pros and cons of risk-based aviation security. Advantages include the opportunity to use scarce resources to best affect and—for most airline customers—a speedier passage. Disadvantages include greater operational complexity and additional capital and revenue cost. There is also the matter of political sensitivity: a risk-based aviation security system that is perceived to be discriminatory risks losing the support of those it is designed to protect—with unpredictable consequences. Finally, if the assumptions that inform the computerised vetting algorithms are wrong, the innocent may be inconvenienced, and the malevolent waved through. Referencing the Israeli experience, the author concludes that realising the benefits of risk-based aviation security requires planning, intelligent design and the will and ability to deliver a professional vetting service *in perpetuity*.

Ross Prizzia and Jason Levy in their chapter '**Customizing Web-EOC Crisis Management Software to Facilitate Collaboration and Increase Situational Awareness: Advances in Business Resource Center (BRC) Design for Business Continuity Management**' examines the best design and use of a Business Resource Center (BRC) to increase the disaster management collaboration among the public, private and non-governmental sectors. The visitor industry was identified as the first sector that will be engaged. The project includes five major tasks: creation of a BRC Annex and Standard Operating Guide; the development of a BRC WebEOC portal; the design of small businesses preparedness materials; the identification of emergency communications requirements; and emergency management outreach, training and exercises. This paper focuses on the second task listed above: development of a BRC WebEOC portal for effective disaster management collaboration, information sharing and situational awareness. This involves the design of data fields for the WebEOC crisis management system. Finally, focus groups in Hawaii are examined to learn more about current levels of disaster engagement, gaps in disaster information or services, and opportunities for increased success with disaster management through a central BRC. In general, there was a strong sense of support for the development of a collaborative BRC space for the public and private sectors. Investigations pertaining to business disruptions are undertaken with an emphasis on modeling, understanding and characterizing the complex causality that defines these incidents. In so doing this chapter uncovers creative, timely and important strategies improving business continuity management and disaster resilience in Hawaii.

JiYoung Park, Jason Levy, Minsu Son, Changkeun Park, and Ha Hwang in their chapter '**Advances in Cybersecurity Design: An Integrated Framework to Quantify the Economic Impacts of Cyber-terrorist Behavior**' put forth an integrated framework to quantify the economic impacts of cyber-attacks on the U.S. airport security systems. A cyber-attack on the United States would involve complex strategic behavior by terrorists, who could invade an airport electronic system without any U.S. border or entry point. At the same time, defending entities must consider the complex processes that may cause any turmoil. The possibility of simultaneous threats from cyber-attacks makes it more difficult for defending entities to secure their airports, highlighting the need for improved and integrated inter-governmental collaboration. Collaborative networking requires close inter-governmental coordination to overcome such cyber-terrorism to protect critical infrastructure systems. Constructing a new model for strategic cyber-terror security requires a combination of competitive and cooperative game situations to develop specific strategies against cyber-terrorism. In addition, an airport shutdown would have ripple effects throughout domestic and international economies; these effects must be analyzed with a spatially disaggregate economic model. Combining competitive and cooperative game situations with an economic impact model, this study proposes the Game Theoretic National Interstate Economic Model (G-NIEMO) framework. G-NIEMO identifies which airport may be most vulnerable in the event that an airport electronic system is shut down. Using the probabilistic costs of airport closures, the model identifies the economic importance of cyber security by event location and industry type. From G-NIEMO, equilibrium strategies to protect U.S. airport systems can be developed and a general guideline for the evaluation of resource allocation can be provided to U.S. government agencies.

Anthony Masys in his chapter '**Operationalizing 'Deliverology' for Humanitarian Aid and Disaster Risk Reduction: Accelerating improvement in delivery**' examines recent disasters and humanitarian aid efforts in the Asia/Pacific Region through the lens of the Sendai Framework and 'Deliverology'. Deliverology is a 'systematic process for driving progress and delivering results in government and the public sector' [1]. The key principles of Deliverology, rooted in design thinking and reflective practices can help organizations (humanitarian aid) define and execute their highest-priority objectives so that they have the greatest possible impact. In so doing, the chapter walks the reader through 'operationalizing deliverology and design' to accelerate improvements in delivery of humanitarian aid in support of the 'New Ways of Working (NWOW)' [9]. Deliverology combined with design is more than a process model but is a **mindset** for truly impacting communities through disaster risk reduction.

Suzanne Waldman and Anthony Masys in their chapter '**Nuclear Waste Management: Security and Safety Implications**' argue that given the significant nuclear footprint throughout the Asia/Pacific, issues pertaining to nuclear waste management resonate with Beck's risk discourse on how distinctively modern manufactured risks are both temporally and spatially displaced. Many Asia/Pacific countries, including Japan and South Korea are facing increasingly large nuclear fuel stocks, with safe space for surface storage growing tight. A key event complicating nuclear power waste storage in the region was the Fukushima nuclear accident, which increased the nuclear waste profile while complicating storage siting by elevating public concern and diminishing trust in nuclear industries and regulators [11]. At the same time, equally salient manufactured risks from global climate change due to fossil fuel consumption are contending with nuclear risks, adding urgency to the optimization of cradle-to-grave energy life cycle decisions. All in all, the manufactured risk landscape calls for "reflective practices" that leverage the theoretical frameworks of risk perception and risk communication to shed light on issues challenging the Asia/Pacific region. In this chapter, the authors draw upon theoretical orientations of cultural cognition of risk [6], responsible innovation [5], and design thinking as a lens for reflection and guidance to enable more constructive and conciliatory nuclear waste decision-making within the Asia/Pacific region.

Jason Levy and Ross Prizzia in their chapter '**Building Effective Emergency Management Public-Private Partnerships (PPP) for Information Sharing**' describe how recent disasters have highlighted the need for effective and proactive partnerships between public emergency responders and the "Whole Community" to assist in all phases of disaster management. This paper examines the importance of reliable, credible, and accessible information that can support the private sector and the "Whole Community" in times of disaster It is shown that community resilience is enhanced through a shared, real-time "common operating picture", explicit linkages, and enhanced information sharing. Business Emergency Operations Center (BEOC) designs are compared and contrasted and recommendations are put forth. Optimal Public Private Partnerships for information sharing identified. It is shown that achieving disaster resilience involves government and businesses sharing potential risks, as well as identifying disaster consequences, priorities, information, and capabilities. The conclusions provide guidance that can be used as dual-use roadmap for improving information sharing: on the one hand governments can use the insights of this paper to build new public-private sector programs (or to improve existing programs); on the other hand, the private sector can better understand how to become connected with the information sharing processes of governments.

References

1. Barber M, Moffat A, Kihn P (2011) Deliverology 101: a field guide for educational leaders. Thousand Oaks, California, Corwin
2. Boivin R (2014) Macrosocial network analysis: the case of transnational drug trafficking. In: Masys AJ (ed) Networks and network analysis for defence and security. Springer, Berlin

3. Canadian Community Epidemiology Network on Drug Use (CCENDU) (2015) Deaths involving Fentanyl in Canada, 2009–2014. Ottawa, Ontario
4. Chester KL (2017) Fentanyl: the next wave of the opioid crisis. subcommittee on oversight and investigations committee on energy and commerce United States house of representatives http://docs.house.gov/meetings/IF/IF02/20170321/105739/HHRG-115-IF02-Wstate-ChesterK-20170321.pdf
5. Guston DH, Sarewitz D (2002) Real-time technology assessment. Technol Soc 24(1–2):93–109
6. Kahan D (2015) What is the "science of science communication"? J Sci Commun 14(3):1–10
7. Masys AJ (2016) Disrupting terrorist and criminal networks: crime script analysis through DODAF applications. In: Masys AJ (ed) Exploring the security landscape- non-traditional security challenges. Springer, Berlin
8. Moore M, Westley F (2011) Surmountable chasms: networks and social innovation for resilient systems. Ecology and society 16(1): 5 [online]. http://www.ecologyandsociety.org/vol16/iss1/art5/
9. OCHA (2017) New way of working. https://www.unocha.org/sites/unocha/files/NWOW%20Booklet%20low%20res.002_0.pdf
10. ODI (2016) Special feature: refugees and vulnerable migrants in Europe. Humanitarian Exchange Number 67 September 2016. http://odihpn.org/wp-content/uploads/2016/09/HE-67-FINAL.pdf
11. Rich M (2017) Struggling with Japan's nuclear waste, Six Years After Disaster. New York Times, March 11, 2017. https://www.nytimes.com/2017/03/11/world/asia/struggling-with-japans-nuclear-waste-six-years-after-disaster.html?mwrsm=Email
12. Roxburgh A, Burns L, Drummer OH, Pilgrim J et al (2013) Trends in fentanyl prescriptions and fentanyl-related mortality in Australia. Drug Alcohol Rev 32:269–275
13. Sagan S (1993) The limitation of safety—organizations, accidents and nuclear weapons. Princeton University Press, Princeton, New Jersey
14. Simonsen KW, Edvardsen HM, Thelander G, Ojanpera I et al (2015) Fatal poisoning in drugs addicts in the Nordic countries in 2012. Forensic Sci Int 248:172–180
15. Smith ER, Shapiro G, Sarani B (2016) The profile of wounding in civilian public mass shooting fatalities. J Trauma Acute Care Surg 81(1):86–91
16. UNODC (2017) Global smart update: Fentanyl and its analogous- 50 years on. https://www.unodc.org/documents/scientific/Global_SMART_Update_17_web.pdf
17. UNHCR (2015) UNHCR global trends: forced displacements in 2014. http://www.unhcr.org/556725e69.pdf
18. Van Rein EAJ, Houwert RM, Gunning AC, Lichtveld RA, Leenen LPH, van Heijl M (2017) Accuracy of prehospital triage protocols in selecting severely injured patients: a systematic view. J Trauma Acute Care Surgery 83(2):329–339

Design Thinking and Design Doing: Describing a Process of People-Centred Innovation

Rama Gheerawo

Abstract This chapter outlines the benefits of Design Thinking as a creative framework for innovation that can be applied to projects and organisations across government, business and the public sector. It includes a short history of the Design Thinking approach to set context and define the concept. However, much of the ensuing discussion and evaluation of ideas and methods is based on recent publications, papers and journal articles to give a current view of academic and practitioner activities. At the heart of the approach is a people-centred focus, and the chapter discusses the role of Inclusive Design and design ethnography in underpinning Design Thinking. It outlines five principles: Involve People, Translate Design Thinking into Design Doing, Create Value and Capture Value, Follow the Arc of Design Thinking and Navigate Complexity. It aims to give an overarching view of Design Thinking, demonstrating the value that it can bring to diverse areas of research and application. Challenges and observations for progressing the approach are also noted.

Keywords Design thinking · Inclusive design · Empathy · Innovation People-centred

1 Design Thinking

Design Thinking (DT) has gained significance and importance in using the principles and practice of design to address issues, problems and challenges across a variety of sectors and situations—from healthcare to education, from urban to rural, from local to global. It has proven to be an effective method for harnessing the creative, innovative, and a people-centred approaches enshrined in the design process and applying them across organisations drawn from the private sector,

R. Gheerawo (✉)
The Helen Hamlyn Centre for Design, Royal College of Art,
4 Hester Road, London SW11 4AN, UK
e-mail: rama.gheerawo@rca.ac.uk

A. J. Masys (ed.), *Security by Design*, Advanced Sciences and Technologies for Security Applications, https://doi.org/10.1007/978-3-319-78021-4_2

public sector, government and education. However, DT has had a wider impact and is not simply limited to these organisations, reaching beyond them to impact entrepreneurs, small and medium enterprises as well as social institutions. DT harnesses ideas that have been developed within the design disciplines but makes a unique contribution by presenting disparate concepts and notions in a way that is digestible and implementable to designers and non-designers alike. It has allowed the ideas and creative endeavours of design to be democratised, co-opted and applied to creatively solve critical problems and issues in an innovative, holistic and human-centred way.

The practice of DT predates its modern nomenclature, with some scholars arguing that its ideas and ideals resonate across the centuries and even reach back to the achievements of ancient civilisations. Human history is thought to be full of design thinkers, long before the term was defined and popularised. Leonardo Da Vinci, Isambard Kingdom Brunel and Richard Buckminster Fuller all used or created processes that resonate with a DT approach that modern practitioners would recognise, from a desire to better the human condition using a cross-disciplinary approach, to taking advantage of technological advances to invent and innovate in a way that tackles complex or challenging situations.

Numerous definitions of DT currently exist, with many new ideas being researched, discovered and articulated as the value has spread. However, a widely-accepted positioning was articulated by Tim Brown, Chief Executive Officer of IDEO, a global design and innovation company. Published in Harvard Business Review, this primarily denoted DT as a tool for business. The article described DT as a discipline that uses the designer's sensibility and methods to match people's needs with what is technologically feasible and what a viable business strategy can convert into customer value and market opportunity [4]. It signified an aspiration to shift design from simply being viewed as an aesthetic endeavour to moving upstream in the innovation process and upstairs to the boardroom. Importantly, it put the value of a people-centred approach at the heart of the idea through consultation with users and consumers. The simple strapline of *Thinking like a designer can transform the way you develop products, services, processes—and even strategy*, represented a bold statement of intent, positioning the discipline of design within the 'cut and thrust' of the business environment.

There are several suggestions as to other modern origins of DT as a phrase or a concept, with Mootee [53] and Chae [9] suggesting Herbert Simon's 1969 book *The Sciences of the Artificial* as the initial point at which design was suggested as a way of thinking, highlighting the definition of design in this publication as 'the transformation of existing conditions into preferred ones'. The book proposes seven steps which still relate to current processes of DT as follows: Define, Research, Ideate, Prototype, Choose, Implement, and Learn. Szczepanska [66] talks about the Design Science movement that originated in the US in the 1960s where Buckminster Fuller created multidisciplinary design teams to address complex systemic challenges. He described this as "the effective application of the principles of science to the conscious design of our total environment in order to help make the Earth's finite resources meet the needs of all humanity", a definition that parallels some of the aspirations within the field of DT today.

Peter Rowe's 1987 book titled *Design Thinking* is also notable, though its focus was on delineating a detailed account of the process of designing in architecture and urban planning rather than DT as it is known today. Kleinsmann et al. [42] cite another early use of the term from Bruce Archer in his 1979 article *Whatever became of Design Methodology*, stating that DT originated within the design research community. Archer helped found the Design Research Society in 1967 and established the Department of Design Research at London's Royal College of Art (RCA) where he tackled one of the urgent healthcare issues of the time—the standardisation of the hospital bed. He used the newly established principles of design research to effect widespread consultation, evidence gathering through direct observations, and extensive field trials using mock-ups and prototypes to create a successful design that was written into a British Standard and ultimately taken up by a commercial bed manufacturer [44]. Archer's philosophy shifted the focus of design from being an individual process that could be framed as ego-driven, to one that is more collaborative and observant, noting over 40 years ago that:

> You cannot ignore the nurturing of the material culture and still expect to enjoy its fruits. That is why I invented design research as a back-up to design practice. I don't think it is because I happen to be standing here instead of somewhere else that the world seems to be revolving around this point.

In today's context, DT has become a living, evolving idea that is being adopted by a range of individuals and organisations. It is also being defined and redefined and means different things to different people [39]. Importantly, at the centre of DT is the idea that everyone can access and activate the creative value of design. Cross [13] states that "everyone can—and does—design", noting that humans have had a long history of DT as evidenced by the artefacts and inventions of previous civilisations and the "continuing tradition of vernacular design and traditional craftwork".

Exact definitions of DT vary but there are some commonalities and convergences that emerging published literature and practitioners see developing within the theory and practice. As a starting point, Beverland et al. [2] describe the following characteristics of a DT approach: 'abductive reasoning, iterative thinking, experimentation and human-centeredness', and these are often widely accepted as some distinct benefits of DT. Dorst [21] talks about DT as a 'real alternative to conventional problem-solving strategies', something which DT case studies attest to, whether drawn from industry, public sector or academia. The move to address systemic challenges through DT [47] or create organisational change by establishing a DT culture within an institution or company [6] are ideas that have gained visibility in the last decade.

The link between DT and innovation, whether described as an attitude, a process or a result, is also widely understood within the general understanding of the work. It has even been noted to *promise* innovation inspired by the way designers work [7], hinting that DT when used as a strategy, can actually guarantee results.

The people-centred aspect of DT is also prevalent, with the terms human-centred and user-centred used unilaterally within the field. Fraser [24] talks about 'deep human understanding' based on work at the Rotman School of Management. Curedale [14] notes that design has moved from being a marketing tool to answering human need. This people-centred aspect of DT is further discussed later on in this chapter.

In recent years, academic literature has focused on progressing the application of DT in different contexts and looking at the different aspects and influences. As DT becomes increasingly global, one area that would benefit from further investigation would be how different cultural influences affect DT practice and process. Small-scale pilot studies also indicate the need for further research in this space, with Clemmensen et al. [11] noting that 'the core DT methods of induction, deduction .... are affected by the ongoing appearance and disappearance of cultural difference among design team members'. They viewed their study as a starting point for additional research in this area. This represents one of many areas of further study that will need to be conducted over the coming years including the important aspect of delineating and implementing tools for evaluation that capture the outcomes effects and performance of DT in a variety of contexts, sectors and situations. This will need to be done in a manner that is both qualitative and quantitative to ensure a true representation of DT value-adds as well as challenges. As Kimbell [41] simply notes, more research needs to be done into critical understanding and evaluation of DT.

1.1 Creating a Bridge Between Business and Design

DT has enabled design to enter the business sector with a sense of purpose and legitimacy. Design Councils across the globe urge businesses to work with design to ensure a heightened level of creativity and to ensure longevity through a better understanding of their customers and through value creation. Antonelli [1] goes even further, urging business people to become designers, not just aim to understand them better. Recent financial trauma across the globe has fuelled a need for alternative thinking, and a drive for business to think differently, and this is where DT, with its ability to create and co-create non-traditional and alternative outputs has gained traction.

DT can influence innovation management within a business context, replacing outdated ideas such as strategic planning powers strategic innovation. Applied DT is strategic innovation with management of the process being more about imagination, creativity and mobilising as opposed to planning, reducing risk or control [53]. The limitations of current management and organising paradigms can be improved with the integrative practice of DT and the value-based approach that it affords [65]. Professional guidance on project management remains rooted in a mechanistic paradigm of control and cannot therefore address changes in environment or business needs [48]. They go on to delineate three areas that DT can help with in evolving management approaches: managing the explorative phase, managing stakeholder involvement, and managing the project in relation to the strategizing process of the organisation. Together these make for an important addition to innovative outcomes that may be achieved by DT as they point to impact on process, management and strategy.

A definitive move that strengthens the relationship between DT and business is the development of DT methods, tools, teaching curricula and practice at business schools, a notable example being the d.school at Stanford University founded in 2005 and recognised as one thought leader in human-centred design, and a leading teaching institute for design and experiential learning [15]. Drawing on different design capabilities and a focus on addressing complexity rather than ignoring it, the d.school outlines eight 'core abilities' [16]:

1. Navigate Ambiguity
2. Learn from Others (People and Contexts)
3. Synthesize Information
4. Experiment Rapidly
5. Move Between Concrete and Abstract
6. Build and Craft Intentionally
7. Communicate Deliberately
8. Design your Design Work.

As large business organisations adopt DT methods and begin to train staff in the theory and application of these ideas, a number of books have appeared that describe and capture best practice. A popular publication is *Design Thinking Pocket Guide* by Curedale [14], a DT practitioner and professor who presents a brief introduction to DT and a collection of methods and tools used in the process, from insight generation, synthesis, ideation and prototyping. The book champions DT as the reason for intelligent business change, defining DT as "a human-centred approach which is driven by creative and analytical thinking, customer empathy and iterative learning".

Curedale suggests that DT is most applicable when you have a poorly defined problem, a lack of information and a changing context or environment, something many business organisations face. The book presents a selection of methods and tools collected over a number of years from a range of disciplines such as the social sciences, business, engineering and design. These are commonly used throughout the DT process, and are outlined as:

1. Define intent
2. Through ethnographic research develop empathy for the point of view of the user
3. Synthesise the research
4. Frame insights
5. Explore Concepts
6. Synthesize the concepts generated
7. Prototype the favoured ideas

8. Test the prototypes with users
9. Incorporate changes
10. Iterate prototype and testing till a workable design is reached
11. Implement
12. Deliver Offering.

An empirical interview study of six large organisations by Carlgren et al. [7] lead to five themes that characterise DT. These are: User focus, Problem framing, Visualisation, Experimentation and Diversity. *User focus* is about 'deep empathy building' as well as understanding and involving people in the generation and making of ideas. *Problem framing* looks at 'widening, challenging and reframing' a problem statement instead of simply trying to solve it, and avoiding the trap of narrowing down to a solution too quickly. *Visualisation* aims to use 'visual representations' to show ideas in either two or three dimensions to gain consensus and share ideas rather than simply being about prototyping and delivering a solution. *Experimentation* refers to iteratively developing and testing ideas in ways that are convergent and divergent, working on multiple solutions to maximise the creative value of process and outcome. *Diversity* was seen as a cross-cutting theme applicable to ensuring a range of opinions and perspectives as well as a diversity of team members. This type of detailed study is essential in establishing an evidence-base of the benefits and challenges of DT practice within business organisations and can lead to more empirical measures of the value that it can bring. Understanding this within a business context can lead to insights applicable to other areas and sectors where DT is being implemented, particularly the public sector.

1.2 Developing Focus Beyond Business

Although DT has been articulated primarily with a business focus, the ideology and application has developed to accommodate and impact a more social context. It has been described as an iterative and co-creative approach that can support innovation in both the economy but also importantly within society [17]. This effectively reflects the three aspects of sustainability namely, environmental, economic and social [61]. All three have to be considered in order to create truly sustainable solutions and DT can be instrumental in achieving social, as well as economic sustainability.

DT builds on a number of different movements and practices that take a socially-focused attitude to design. In Scandinavia, Co-operative Design emerged as a powerful antidote to 'top-down' problem-solving, by involving everyone in decisions and processes that the designer facilitated [22]. Participatory Design has also risen in popularity and prominence, involving people as authors, instigators and designers of their own solutions, and ideas [63]. Far from devolving or diminishing the role of the designer, this is being seen as a powerful tool for engaging non-designers within the design process in a way that parallels the ambitions of DT. People-centered design encourages attention to, and consideration of the individual as a central part of any creative process and terms such as 'co-design' [64] and 'co-creation' are now actively being used across education, business and government. Professional designers have also co-opted these terms with most understanding that engagement with their end users can provide innovation benefits across the design process.

A recent publication of significance in the field, *Design Thinking for the Greater Good* [46], explores and emphasises how DT has moved from a business focus and into the social space. It outlines how DT is being used by organisations as diverse as charitable foundations, social innovation start-ups, national governments, and elementary schools to solve difficult social challenges. This builds on the inherent business focus that influenced DT at its inception and helps to evolve the ideas. DT is presented in this book as the perfect tool to tackle increasingly complex problems such as:

- Stakeholders who cannot even agree on the problem, much less the solution;
- Employees who are reluctant to change behaviours and take risks, who are often rewarded for compliance rather than performance;
- Decision makers who have too much data, but little of the kind they need;
- Leaders who are more likely to have short tenures and whose every move is scrutinised by funders, politicians, bureaucrats and the media; and
- Users of their services such as students, patients, customers, citizens—whose expectations are sometimes rising as fast as resources to meet them are declining.

The publication argues that DT can tackle these issues as it is a unique problem-solving approach; one that is human-centered, possibility driven, option focused, and iterative. It has developed as a new democratized form of innovation, moving away from designers as experts and external stakeholders, to participatory design with strategic partnerships. As with other DT definitions, they emphasise the importance of reframing the problem as a key evolution from traditional approaches to innovation. "*The definition of a problem is a hypothesis to be tested, as are its solutions... breakthroughs come with the redefinition of the problem itself*'. DT moves away from the traditional 'one big solution' approach, towards multiple solutions quickly tested because numerous answers are both possible and desirable. Whilst the definitions of DT provide a useful background, the novelty of the publication is in how DT can advance from the business world to be applied to the social sector, where issues can be more challenging, and solving them can matter even more.

The idea of 'designing for' people, which gained prominence in the last century is now evolving towards 'designing with' people and subsequently incorporating processes around 'designing by' people [26]. DT reflects these movements helping organisations to understand and relate to their customers and users. This democratization of design practice enables a move towards more people-centered and equitable processes that are increasingly relevant today.

DT is being applied in progressively diverse areas and with typically high levels of success. A few instances are mentioned in this chapter, but the list is by no means exhaustive. Healthcare has seen an influx of DT approaches partially driven by the rise of technology applications and ‘health tech’. As economist Peter Drucker noted, healthcare institutions are amongst the most complex in human history, and DT is seen as a competent tool for addressing this type of complexity. Even in critical emerging areas such as e-pharmacy, DT has a role on defining and supporting the collection of software requirements [8], and Peterson et al. [59] found in their study that “DT processes used in the development and implementation of the mobile heath app were crucial to creating value for user”.

Education has also benefitted from DT consideration. Hernández-Leo et al. [36] noted that when dealing with the complexities of designing learning experiences for students, DT supported the range of activities from “designing stimulating and engaging tasks, selecting and creating appropriate resources, and deciding how best to support students to successfully complete the tasks”. Even at primary school level, DT education interventions lead to a “paradigm shift” [57] and that it would be beneficial to expose children to the values of design and DT. When looking at higher education, Leonard et al. [45] argue that DT can enable sustainable curriculum development “through processes of rapid reconnaissance, mapping the territory and emergent modelling”, signalling the value that can result when addressing urgent, emerging issues.

DT can also work in areas such as public services, engineering and urban planning but has seen significant results when applied in technological contexts which often lack people-centred consideration [30]. Merging traditional HCI methods with DT can lead to a seamless co-existence of methodologies such as user-centered design and rapid prototyping [54]. Even in a focussed technology context such as data storage, DT can have an effect. During a process to apply DT to an innovative data storage service, the methods were felt to be vague, blurry and very broad [28], however, the collaborative and people-centered emphasis of the method was appreciated, even by the technical participants.

DT processes can be used to ‘design’ creative forms of research in scientific and clinical areas. They were co-opted to create a framework of methods to study cognitive biases that can appear within teams involved in strategic decision-making [43], though the actual DT tool did not have the expected influence on mitigating confirmation bias, signalling that more research needs to be done in translating DT approaches from a primarily creative arena into more technical, technological and scientific areas. Other fields of study could also benefit from a DT approach. Ideas around social entrepreneurship and social business are on the rise, but whilst DT has established itself in business practice, it has received scant attention in entrepreneurship pedagogy [38].

2 Principles of Design Thinking

There are numerous frameworks that aim to codify and summarise DT intention and activity that are delineated by academics and practitioners alike. This section outlines five principles drawn from the Helen Hamlyn Centre for Design (HHCD) at the RCA. This design centre, lead by the author, has completed over 260 projects with organisations, corporations and communities at an international level. Much of the work is based on DT methodologies developed and applied in the field and tested by academics and practitioners alike. The HHCD's approach is inclusive and interdisciplinary and work is organised in three research spaces:

- Age & Diversity, focusing on design for a more inclusive society irrespective of a person's age or ability
- Healthcare, looking creating safer and better products and services in the hospital and home
- Social & Global using people-centred design methods to deliver research projects that have social and global impact for marginalised communities.

The five DT principles developed at HHCD and outlined on the following pages are: Involve People, Translate Design Thinking into Design Doing, Create Value and Capture Value, Follow the Arc of Design Thinking and Navigate Complexity.

2.1 Involve People

Human-centeredness and empathy are widely accepted components of DT as noted by several researchers and practitioners [2, 46, 53, 56], sometimes framed as meeting customer latent needs [62] or an empathic look at users or consumers [31]. The discipline of design, which DT draws its processes from, is also changing, becoming less about making people want things, and more about answering human need [14]. DT can enhance co-creation and facilitate respectful, creative interaction, thereby ensuring that both a project output and the process are people-centred, and providing benefit for the recipients of a product, service or system as well as the creators.

Empathy is fundamental to DT and can range from an observational stance to complete immersion. Immersive techniques draw on some radical sociological experiments. In 1979, aged twenty-six, designer Patricia Moore dressed up as different eighty-year old women and travelled across the US to document society's treatment of older people firsthand [52]. Nearly two decades earlier, John Howard Griffin, a journalist living in Dallas darkened his face using medication to immerse himself in African American communities and experience the difficulties they encountered in some Southern US states at that time [33]. Although these types of approaches were viewed negatively by designers and social scientists at the time, they are now seen as positive and pioneering in creating empathy within the research process.

However, whilst they change the perspective of the designer or researcher by moving them closer to that of the user, recent methods in DT look to empower participants themselves, giving them visibility and a platform for self-expression. It is not enough to simply advocate on behalf of another person, as the ideal is to empower and enable people in order to bring higher level of authenticity and a directness of the 'lived experience' into the ethnographic process.

Methods that involve people are at the core of DT. At the HHCD, 'design ethnography' is used to empower and involve people. This term is being increasingly used to cover the description of a process by which designers conduct research with people using tools and methods drawn from social science [10]. As designers typically have to operate in much shorter timescales than traditional social science, many of the ethnographic methods employed by them have come to be known as 'rapid ethnography' [58]. As a result, designers often evolve existing tools, some of which address the opposing time demands of design and ethnography [37]. The process of design ethnography has been embedded in the HHCD's work over the last 26 years.

Practitioners within research, DT and ethnography, should avoid relationships that promote them as 'so-called' experts, and participants in a project as 'non-experts'. People are often termed 'users' or 'consumers', but this can reduce their contribution to that of a 'research subject' simply responding to questions set by the designer, and following the lines of inquiry as dictated by the process. This removes the human perspective that can so often lead to hidden insight, innovation and design opportunities. Participants should be seen as human beings with a context, lifestyle, attitudes and desires. The terms 'user' or 'consumer' can remove the human perspective which runs the risk of missing deeper connections with an individual. Users just use, consumers only consume, but people live, and the aim of DT is to capture different dimensions of a person's life.

There is a tendency to objectify the people that designers choose to work with, seeing them as 'test subjects' rather than human beings with a context, lifestyle and desires that go beyond their physical representation and as a person with needs, aspirations and ambitions. Design thinkers should shift towards an attitude of peer-to-peer exchange, treating a person as a valued contributor and partner. Only then can contributions go beyond a viewpoint of assumption (Fig. 1).

People are typically seen within the context of their lives, whether at home, at work, in the city, the hospital or whilst travelling. Every person has a context that informs and surrounds them, and a richer ethnographic picture can be formed when people are seen against the background of their connections, their family or even just the architecture of their surroundings. Nothing can replace this type of direct contact [68]. This is especially important when a design thinker aims to empathise with a person who can be radically different from them.

Interaction between a person and their space can also be significantly informative allowing the designer to access contextual issues rather than simply studying the individual, and opening up a wider channel of inquiry. Context also becomes important when dealing with different types of exclusion such as cultural or emotional, or looking at multi-faceted or less tangible issues. The relationship between designers and users of their designs can be historically described as one of producer-consumer, or expert and non-expert. However, co-creative and collaborative viewpoints have begun to engender a more equitable stance in these relationships, moving from a perspective that is empirically objective to one that is

Fig. 1 When researching issues of safety and security with migrant workers in the Middle East, it took days before trust could be built with the project researchers. Dressing in the same clothes was part of a suite of activities to gain trust and build acceptance

more connected. This can result in a heightened level of inventiveness as design thinkers are challenged to seek new, novel or non-conventional solutions. An empathic stance is essential for designers developing their own ethnographic practice in this area [50] (Fig. 2).

Fig. 2 HHCD researchers conduct conversations and interviews with a project participant in their home to gain contextual insights and ensure that they are on an equal footing

2.2 Translate Design Thinking into Design Doing

Some critique of DT exists around actioning frameworks and translating the ‘thinking’ into ‘practice. Although insights have real value, they need to be implemented in order to bring maximum benefit and impact. The traditional processes of design have to be evolved to support the delivery of DT and one prominent approach is called Inclusive Design (ID). This was defined in 2000 by the UK Government as products, services and environments that include the needs of the widest number of consumers [18]. This consumer focus marked it as a business strategy, bringing ideas of people-centred design and a social attitude into the corporate innovation space.

Since 2000, ID has developed in a number of different directions whilst maintaining relevance to the private sector. It has been characterised as a practice, methodology, philosophy and technique, but a key achievement is that it is internationally recognised and used by governments, industry, designers, policymakers, and social and creative organisations. The idea was articulated in a paper delivered at the International Ergonomics Association’s 12th Triennial Congress [12] and is the primary focus of the HHCD at the RCA.

ID is very close in its ideology to two other people-centred design practices, namely Universal Design (UD) and Design for All (DfA). All three began by focusing on issues of accessibility, the needs of older people or those with disabilities, but a variety of different cultural, historical and political factors have affected the exact manner in which these ideals have been interpreted, developed and expressed [67].

ID has progressed to meet a wider range of emerging social challenges that go beyond a focus on age or ability. Today, the people-centred design tools enshrined in ID can be used to address other instances of exclusion such as by gender, socio-economic circumstance, geography, race or locality, amongst many others. Whilst ID has traditionally focused on product design or the built environment, emerging design disciplines such as digital design, service design or experience design are becoming increasingly prominent and more relevant to the innovation landscape today. There is a shift from historical approaches, to defining new theories, practices and experiences within the teaching, research and practice of design. A notion within ID is that design should sit within a multi-disciplinary context, often forming the platform discipline for other departments or sectors to engage with. Within commercial development processes, design has real value when added into the ‘mix’, and DT is key in enabling this.

At the HHCD, designers draw down from research techniques including co-creation approaches, expert consultation, interviews, observation ‘in situ’, testing with prototypes, research kits and more design-specific practices such as cultural probes [27]. If the designer cannot be present to conduct face-to-face research with participants, methods such as diaries or video logging are also used to record material in absentia [60]. Designers are encouraged to derive their own methods and build on existing ones so ensure appropriate engagement throughout the process.

Sometimes design ethnography methods are developed at the HHCD to suit shorter-term, DT-focused applications. *Creative Bridging* has been extensively used in work redesigning the London taxi. This uses a combination of words, user quotes and images associated with modern London to inspire creative designs for the vehicle exterior. *Design Provocation* is another example where props, visuals or sketch ideas are shown to people to stimulate feedback and provoke discussion [23]. This originated in a technology project conducted at the HHCD in 2006 that asked older people to discuss digital technology, something they were unfamiliar with and even fearful of. Creating a set of 'mocked-up' ideas around managing health or money, such as digital medicine bottles and piggy banks, allowed them to step over their apprehension and take ownership of the ideas, even drawing over and modifying them (Fig. 3).

The search is for creative insights rather than a detailed understanding of every aspect of a person's life. The aim is to achieve a diversity of needs, points of view and life positions. It is essential to look at new practice as well as evaluating limitations from existing approaches. Creative speculation within design where the designer as the key protagonist is often at odds with the move to engage users. Facilitating peer-to-peer exchange and community engagement demand deeply different thinking from traditional people-centred methods.

The co-creative stance of ID can raise some questions, such as whether users can be equal partners with designers when they are typically placed in a less powerful position. Or can they be seen as co-inventors with rights to the Intellectual Property within a business context? This all points to the idea that in DT, the practitioner is

Fig. 3 A Design Provocation consisting on an image of a piggy bank with a USB tail that reads out bank information when a credit card was inserted into the slot on its back. This was used to prompt response from older people

ethically bound and responsible for the people they work with, something that needs careful consideration when applied in practice. In summary, ID can be considered as a delivery tool for DT, taking it from the realm of insight, framing and thinking, and into the realm of practice, development and delivery,

2.3 *Create Value and Capture Value*

DT can create value in a number of different ways. It can enable organisations to go beyond the limits of their imagination and overcome innovation barriers [5]. Figure 4 shows a research activity conducted with eight older people as part of a workshop held at the HHCD looking at attitudes to technology and health for a large Japanese company. Initial responses to questionnaires were of limited value, but employing a visual research method allowed researchers to access people's inventiveness. It asked participants to describe what an imaginary fourth hand on a watch would measure about their health and well-being. As well as functional ideas such as steps, heart rate or calories, this inspired aspirational responses such as happiness of family members, proximity of authority figures, or level of stress in the people around them. These ideas enabled a deeper level of communication and insight into what was important and of value to the workshop attendees.

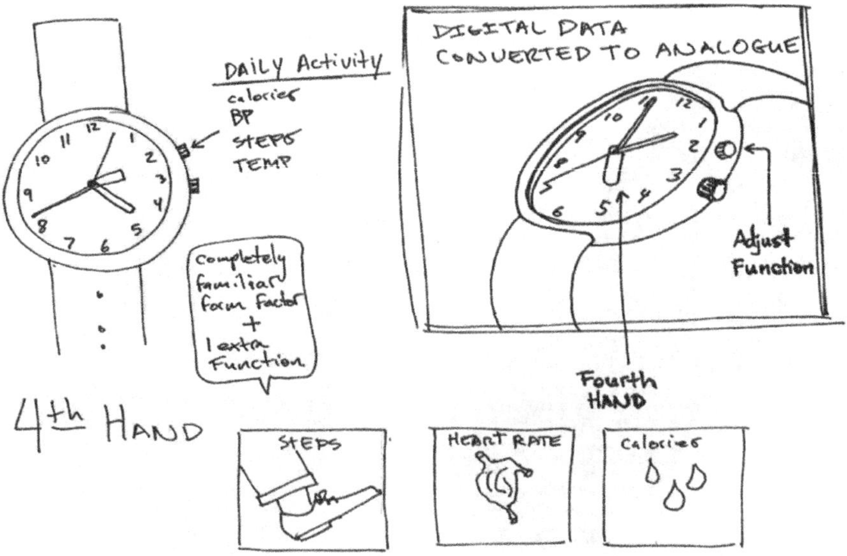

Fig. 4 A DT imagination exercise asking participants to imagine what the fourth hand on a watch would tell them about their health

DT creates value in different instances. It quantifies convergent and divergent ways of approaching problem-solving allowing for a balance of focused methods of assessment with more open-ended exploration [14]. Understanding people's mental models [40] is extremely important and DT can reveal people's actions, as well as uncovering the reasons, motivations and mental perspectives behind them. It is also a tool for business development and social enterprise as discussed earlier in this chapter, contributing to project management and innovation [48]. However, alongside the accepted understanding that DT creates value is the need to look at how it captures value, and can help create new types of evidence-base that support the quantitative, data-driven or numerical information that typically forms the fundament of strategic, organisational decisions.

At the HHCD, a novel way of presenting the people-centred insights that typify DT methods was developed, called an Insight Bank [29]. This was a response to the challenge around how to represent the voice of the user alongside the noise of all the other project considerations [56] and ensure that user experience is valued within decision-making. The Insight Bank is a digital platform that brings user quotes, videos, insights and inspiration together in an easy-access repository that can openly exist online or securely sit behind a firewall on a company intranet. It enables marketers, engineers, designers and technologists within the company to access and hear rich and textured insights from the people who participated in the research, and presents new opportunities for the design ideas. Organisations who have worked with the HHCD on Insight Banks note that they embody a solid, qualitative evidence-base for business decisions, design inspiration as well as research reflection for every employee (Fig. 5).

Another example of a way in which DT can create and capture value is by utilising the visual capability of design. Visualisation can be a powerful tool for expression and communication, particularly when less tangible ideas are discussed or more imaginative directions have to be articulated and researched. One project the HHCD worked on as part of a large European consortium spanning four countries, looked at domestic energy use, bringing aspects of DT into a technologically-driven area of study [3]. Early research indicated that the team needed to understand how people viewed energy, used energy and related to it before any realistic and relevant interventions could be created. Energy's invisibility was a defining characteristic, and one that could help define a person's perspective on their own energy use.

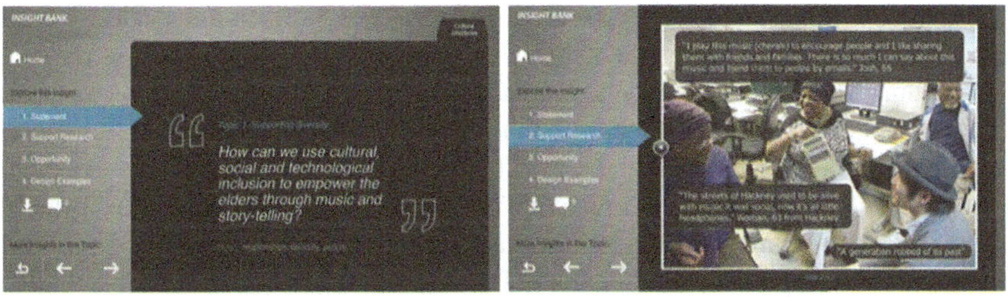

Fig. 5 An example of an Insight Bank

'Keeping energy use visible' was central to the thinking behind home energy monitors [35]—but one that has often been addressed in design through leaping straight to technology-driven interface designs [25] without exploring the issue further in terms of the meanings, social and ecological factors of everyday lived experience [49]. To investigate how people might represent energy's invisibility using their own ideas and sharing their perspectives, a DT method was developed based on the activity of drawing. This mode of visual research is one that social scientists would typically denote as a 'participatory visual method' [34]. People were asked to respond, through drawing or writing on paper, to the question: what does energy look like? As Gray et al. [32] states, '[w]ords become more challenging to visualise as they become less literal', and energy, as a form of dynamism, power, force or activity, might be considered 'an idea that isn't anchored to an object in reality'.

This enabled an exploration of people's mental models and perceptions of energy, and of the infrastructures connected to it. This study was conducted with visitors to the Life Examined exhibition at the Royal College of Art in September 2013 and replicated with students participating in the UK ArtScience Prize at the Silk Mill, Derby in April 2014. A final workshop took place with visitors to the V&A's Digital Design Weekend in September 2014. In each instance, the participants spent the time they needed creating their drawing. Some were informally interviewed about the drawings produced, the subject matter, the materials and the act of drawing. Asking people to draw the invisible offered a very human-centred means of investigating energy and uncovering people's thoughts. Very few drawings contained words or numbers, most representing metaphorical representations of energy drawn from nature, patterns or colour (Fig. 6).

This was in stark contrast to how energy companies communicate to their customers, typically using numerical dialogue, citing Kilowatt Hours as a unit of usage, something very few people could visualise or relate to. This discrepancy highlighted an opportunity to design better ways for energy organisations to understand and engage with their user, potentially affecting more positive and deep-rooted behaviour-change. There is a history of participatory drawing research being used before to explore people's understandings of abstract concepts [55], and building on established design techniques such as this can bring value to DT practice. Visualisation is discussed in detail here but design research contains a plethora of techniques that can be co-opted, from 'low contact' methods such as questionnaires, through 'medium contact' techniques such as interviewing and contextual observation, to 'high contact' ones such as workshops, shadowing and long-term community engagements [23].

Defining the potential value creation that can stem from DT is an ongoing concern and needs further attention. At the HHCD, efforts to visualise a framework have been undertaken (see Fig. 7), lead by Prof. Jeremy Myerson, a noted figure within DT. This is based on conducting and delivering a number of DT projects completed with organisations across a variety of sectors. This helps the framework to be generally applicable to almost any DT endeavour. The diagram splits a circle into quadrants based on two axes. The horizontal axis denotes a shift between 'now'

Fig. 6 A representation of energy resulting from a visual research method

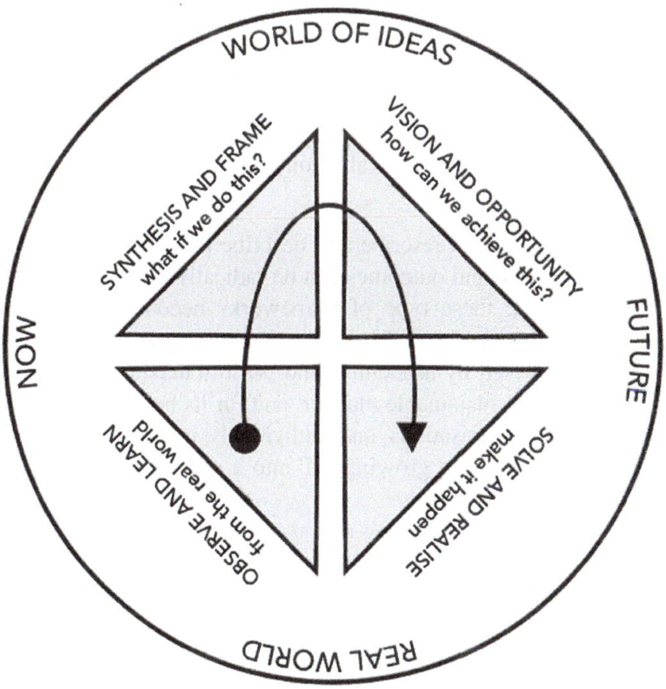

Fig. 7 DT manages a shift of focus between 'now' and the 'future', and the 'world of ideas' and the 'real world'

and the 'future' looking at DT activities that take place today and those that are more speculative and future-facing. The vertical axis attenuates between the 'real world' and the 'world of ideas'. A special characteristic of DT is that it is able to seamlessly move across these axes balancing the real with the speculative and allowing space for creative imagination alongside operational necessities. DT produces a managed approach to enable more intangible but equally important factors alongside established forms of organisational decision-making. Simply put, DT lifts the creative value of any project.

The four quadrants that result from the framework described in Fig. 7. lend themselves to the following DT activities that follow a pathway as shown:

(1) Now-Real World: this is about observing and learning from what is currently happening. Activities are about understanding context and people's behaviours in existing situations and gathering insights that can be evaluated and process. This takes place in the present moment and builds up a real picture of any given situation.
(2) Now-World of Ideas: synthesising and framing the insights that were gathered in the previous stage is the main activity here. Asking questions such as 'what if we do this' talks to the speculative characteristic of this. This is a transformational process that moves a project into more imaginative and unknown spaces.
(3) Future-World of Ideas: here, the focus is on articulating a vision and defining opportunities. Design briefs are typically formulated and answered at this point and a number of creative avenues are delineated and explored. It is important to nourish every idea at this stage as often the most creative ideas come from unfettered ideation.
(4) Future-Real World: The final stage is about solving, realising and delivering ideas back in the real world, making sure that they are relevant to the people who will most benefit from them, and that they are market-appropriate. Activities such as prototyping, evaluation and work-shopping often take place at this point.

There are numerous ways to prescribe and describe the value that DT creates and captures. Definitions vary and outcomes can be radically different. As DT grows in application and interest, these type of frameworks become increasing useful in defining the specific contribution DT makes and allowing the theory and practice to be evaluated and discussed by academics and practitioners alike. This enables DT to mature in a useful and sustainable manner, and for its benefits and affordances to be communicated to organisations and individuals interested in taking up the approach. This is invaluable in growing DT into a mainstream practice.

2.4 *Follow the Arc of Design Thinking*

Co-creation and co-design are regularly presented as signifiers of DT, but design needs to build bridges with other disciplines linking to areas such as social science, anthropology, technology, science, business, policy, the third sector and government. DT can be a powerful way to interact and by embracing other disciplines, the role of the individual designer is impacted, moving from maker to facilitator, from creator to listener, and from instigator to partner. An attitude of DT or 'design enquiry' can ably contribute to other disciplines and demonstrably bring benefits into the co-operative space.

DT needs to take an interdisciplinary approach to design [13], not just looking at different types of design such as product design or graphic design, but across a range of sectors and disciplines. DT is a creative activity that can be appreciated by anyone and done by everyone, but tools, methods, process and structures need to be described that democratise the approach and make it accessible to everyone. The idea of a creative project involving a series of specialist disciplines in a linear way with each sector contributing specific knowledge at a particular point in a process, is something that the world is moving away from. DT presents a more collaborative stance, with diverse individuals working together to equitably contribute ideas in a way that is more involved and less regimented. Although this allows for a heightened sense of creative freedom and more relevant outcomes, frameworks and structures need to be formed that describe the DT process both operationally and empirically.

At the HHCD, researchers have collaborated with Pat Whitney at the Illinois Institute of Technology to develop a way of describing DT that combines the convergent and divergent nature of the UK Design Council's Double Diamond process, typically used by professional creatives, with the innovation stance of DT. The resultant idea, named the Arc of Design Thinking (see Fig. 8), can be co-opted onto any project and has been used within business, government and the third sector by teams of non-creatives including policy makers, civil servants and accountants. Over 40 workshops conducted by the HHCD have used this system to enable rapid

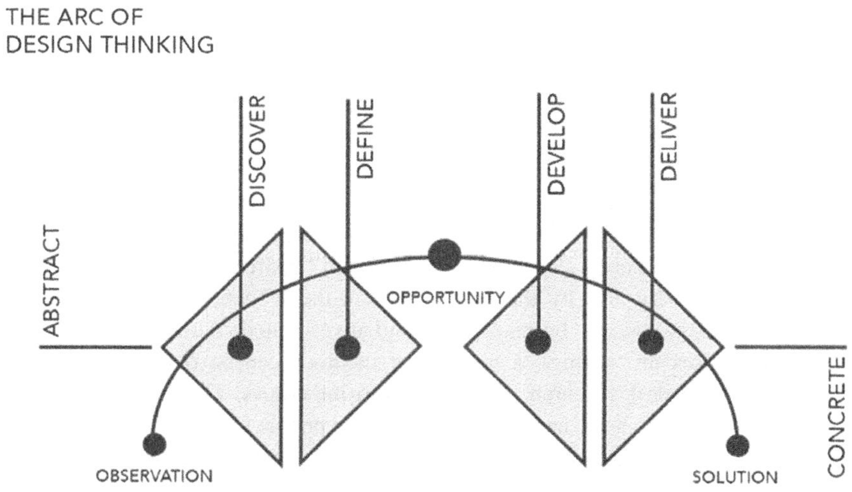

Fig. 8 The Arc of Design Thinking positioned over the UK Design Council's Double Diamond Process

innovation, with participants using DT to address challenging societal issues such as ageing, emergency healthcare or economic division within a city.

An important aspect is that DT is a platform to support creative endeavour and enable teams of disparate individuals to raise the group level of innovation. Of note, is the value that non-experts bring to the process. On a number of occasions, it has been observed that enabling non-experts to engage with an issue and have creative input into the solution, has produced some of the more innovative ideas. Sometimes, expertise has even been seen as a barrier, as nominated experts replay historical perspectives when defining problems, articulating their personal contexts and denying more inventive, non-traditional outputs. The Arc of Design Thinking allows for individual representation across the process, with all levels of expertise valued at different stages. Importantly it brings the voice of the user back into the foreground. As Eikhaug et al. [23] states "people are experts in their own lives" and DT can give users, experts and contributors the tools to not only understand and assess their situation, but also impact and instigate change.

The four stages of the double-diamond process is defined by the Design Council [19, 20] in the UK as follows. Each stage is convergent or divergent depending on whether it is about 'blue-sky thinking' or focusing ideas. Typical activities in each phase are:

(1) Discover phase (divergent): exploration of brief and hypothesis, contextual research and definition of project participants or communities. Looking at the world in a novel way and gathering insights
(2) Define phase (convergent): designing and conducting design ethnography. People-centred briefs defined from the research insights. Curating and implementing the possibilities identified in the Discover phase
(3) Develop phase (divergent): development of a number of ideas through co-creation and design ideation processes. This process of trial and error helps to improve and refine ideas
(4) Deliver phase (convergent): selection of ideas to take forward and delivering outputs in the form of prototypes, services ideas or guidance. The resulting project is finalised, produced and launched.

This process is widely accepted globally as a platform for innovation. The Arc of Design Thinking sits across the four phases with three touchpoints: observation, opportunity and solution. As per Fig. 8 the bottom half and top half of the diagram are spread into two halves denoting 'concrete' and 'abstract' activities.

Observation is an activity that takes place in the 'concrete' world where real challenges and issues can be researched and investigated. Although observation primarily focuses on 'seeing', a number of methods can be deployed including contextual visits, desk research, workshops and interviews. The aim is to create a focused picture of the issue and empathise with the people involved. It is paramount to get this part of the process correctly orientated to ensure that the initial project framing is correct and that the right questions are being answered.

Opportunity is about articulating the scope and focus of the project and recognising the right potential direction to investigate. This is an 'abstract' activity as it involves assimilating the insights gathered in the Discover and Define phases, mapping potential opportunities and then selecting one, often as the project brief or design brief. *Solution* is a touchpoint that brings the project back into a landing phase. It signals the endpoint of the DT process, finishing up in the 'concrete' space with the delivery of a concept, prototype or final design.

The Arc of Design Thinking is a framework that describes how DT can lift a project from remaining in the 'concrete' phases into an 'abstract' space, heightening the possibility for creativity and innovation. Many organisations make the mistake of remaining in an operational space that is more risk-averse, but ultimately less transformational. DT allows for a more imaginative approach within a framework of 'managed risk'. The Arc of Design Thinking describes this pathway presenting the creative possibilities that it affords in a way that designers and non-designers can find approachable. Organisations that have worked with the HHCD state that it takes a level of trust and resilience to stay the course, but the outcomes are typically more valuable, inventive and ultimately applicable.

DT is about achieving balance. Organisations cannot always function if they are in space of challenging themselves all the time as they need to attend to the everyday and balance the 'strategic' activities with operational needs, and incremental discovery with those that occasional periods of disruptive thinking. DT can help employees, colleagues and clients move between two spaces (see Fig. 9.): the first is

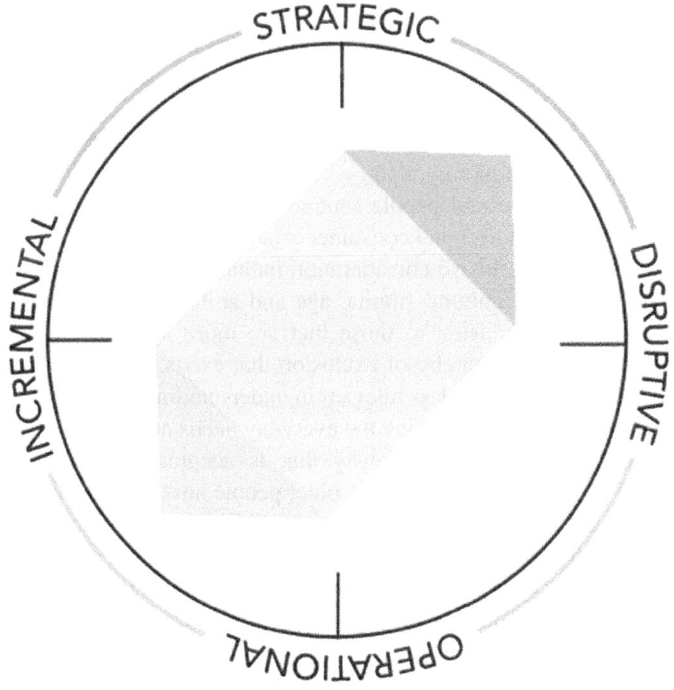

Fig. 9 Organisations need to move seamlessly between being operational-incremental and strategic-disruptive

operationally focused and incremental in terms of innovation, and the second takes a more strategic view that is disruptive in a creative manner. Working in a way that allows both areas to be covered can ensure longevity and create continuity whilst allowing new ideas to surface, be evaluated and then implemented. Innovation can be perceived as risky, however DT offers a secure way to mitigate this.

2.5 Navigate Complexity

DT can help to navigate complex challenges and guide both organisations and individuals through the complicated route towards real innovation. Liedtka et al. [46] state:

> We acknowledge two truths: first, it is often impossible to tell early in the life of an innovation just how big or small it will be; and, second, many small things can add up to something big.

Innovation can be unpredictable, often requiring deviation or even total departure from the 'known' towards the 'unknown'. DT can help to pilot projects, especially in the early stages or when addressing issues that involve systemic change. Too often, the impact of design is limited as it is not implemented at multiple levels whether it is a single project that directly addresses a particular place or community, or whether it is a larger scale endeavour that impacts organisational, cultural or governmental structures. DT is a scalable process that has a role in integrating and influencing these larger agendas whilst also creating positive impact at regional and local level and at smaller scales. DT can step beyond the intricacies that exist and directly enable a project team to get to the root cause of an issue, often creating permanent rather than temporary change.

DT, with its empathic and people-centred stance can be inclusive, especially when looking at different user and consumer groups and there are many groups that can benefit from truly inclusive consideration including diversity across sexuality, race, personal economy, cultural stigma, age and ability. It can help step beyond obvious examples of exclusion to those that are more less recognized and help overcome the ingrained hierarchy of exclusion that exists.

Historic frameworks can be less relevant in understanding current social contexts as they sometimes no longer describe the everyday needs and aspirations of people, institutions and organisations in a way that is accurate or occasionally even recognisable. For most of human history, older people have held the knowledge and wisdom in society, but technology has disrupted this age paradigm with younger age groups typically leading the way in terms of invention and innovation capital. Engaging younger technology entrepreneurs in an empathic and age-inclusive manner is one of many potential benefits that DT can offer. Through a diversity of user engagement, DT can help step beyond preconceived beliefs and traditional forms of deployment, helping to navigate a complex world that can feel like it is in a state of flux.

One of the biggest challenges in DT endeavours is the focus on individuals rather than communities and this presents an increasing layer of complexity and complication. Many projects deal with groups, such as neighbourhoods, families, localities or organisations. DT needs to account for this shift as the majority of methods focus on an expert-user engagement in a way that is primarily person-to-person. Methods need to be derived that account for group behaviour as community-centred design is becoming valued by business, government and the private sector in today's increasingly connected social context.

The rise of digital communities has only supported this. DT needs to evolve methods and hone research intentions to become a trusted part of group conversation. This also assures more sustained engagement allowing members of the community to create, participate and deliver ideas rather than expecting them to simply act as grateful recipients of DT expertise.

In 2013, the author ran a workshop with Sean Donahue from ArtCenter College of Design at the Include 2013 conference in Hong Kong titled "Design and Community". Five specific qualities for working with communities (as opposed to individuals) emerged: Time, Continuity, Ownership, Legacy and Trust. These are discussed below.

'Time' holds a different value for communities than for individuals. Design often works swiftly, but engaging a community for longer periods of time is essential in order to understand the range of relationships that can exist within a group, to see the cycles and rhythms of community interaction, and to identify boundaries and bridges between people. More time is needed to connect with a community than an individual. 'Continuity' looks at building consistent engagements which is important in establishing respectful relationships. Dedicated research teams where personnel remain the same are essential as is giving priority and commitment to existing community events rather than simply inflicting an external research agenda onto a group's activities.

'Ownership' of the process is a significant but often under-considered issue. Even in co-creation and participatory models, designers or researchers can visibly remain as the drivers of the project, with a leadership stance that can prevent true behaviours being revealed from participants. This also reduces the probability of acceptance within a group. Approaches that give collectives the opportunity to lead and direct their own engagements are needed, with designers refraining from being the sole author or facilitators. Working in community-owned spaces can help people feel more comfortable and empowered.

'Legacy' is about developing work that ensures longevity and sustainable outcomes. It is more important to give back to a community than the individual and this needs to be built into a research plan from the start. 'Trust' builds on the previous four values and has to be established at the start of an engagement. Asking a group of individuals to commit and dedicate to working with a designer on issues that can be personal or sensitive means that a strong connection needs to be built and renewed throughout the project with multiple points of contact in the community.

Part of this is about seeing the strengths rather than the deficiencies of a community. Design thinkers should not 'hero' their own research and discoveries, but rather see themselves as another contributor in the conversation. The Asset-Based Community Development Approach [51] supports this move away from traditional needs-orientated frameworks to capacity-focused alternatives. This creates a strong case for DT's potential to redress social inequalities and impact systems at a larger scale, becoming an operational conduit to implement social ideals.

3 Case Studies

3.1 Redesigning the Emergency Ambulance

A key hypothesis within DT is that there needs to be space for other disciplines to engage and even lead design projects as this can add to the socially creative process. A recent HHCD project redesigned the interior treatment spaces of emergency ambulances. This not only closely engaged clinicians and patients, but also co-opted a paramedic onto the design team on a daily basis for over a year. By making the people who deliver the service central to design development, and by giving them a position of authority and visibility in the process, a heightened level of relevance, insight and creativity was achieved.

The paramedic advised the development team on frontline issues and provided immediate feedback, sometimes halting a line of enquiry that was not fit for purpose. An example was the idea of a side-loading ambulance which was highly acclaimed in the design studio. However, the paramedic noted that in most major cities, there is not space next to the vehicle to load a patient due to narrow streets, parked cars and other obstacles. This type of direct engagement was invaluable in accelerating the results towards a realistic treatment space that improved clinical efficiency and enhanced the patient experience.

There were many problems with the design of existing ambulances. The interior was difficult to keep clean given the frequency of use and the resultant lack of opportunity to clean the vehicle. Ambulance crews also suffered from poorly thought-out ergonomics, badly laid out equipment and difficult-to-access storage spaces, all of which affected performance in critical, life-threatening situations.

In the redesigned interior, the stretcher is placed in the centre of the interior, allowing 360 degree access to the patient (see Fig. 10). Equipment and disposables for treating the patient are arranged into treatment packs allowing for easy access and a restocking process that can take place out of the vehicle (see Fig. 11). Patient information is displayed on an overhead touch screen that also integrates with hospital records to facilitate handover and built on the increasing pervasiveness of technology.[26] This multi-disciplinary process of ethnography helped to create and map key insights from ambulance users and then translate them into sketch designs and a full-scale test rig. This all had an impact on safety and security of the vehicle as well as the patients, paramedics and clinical staff who engaged with it.

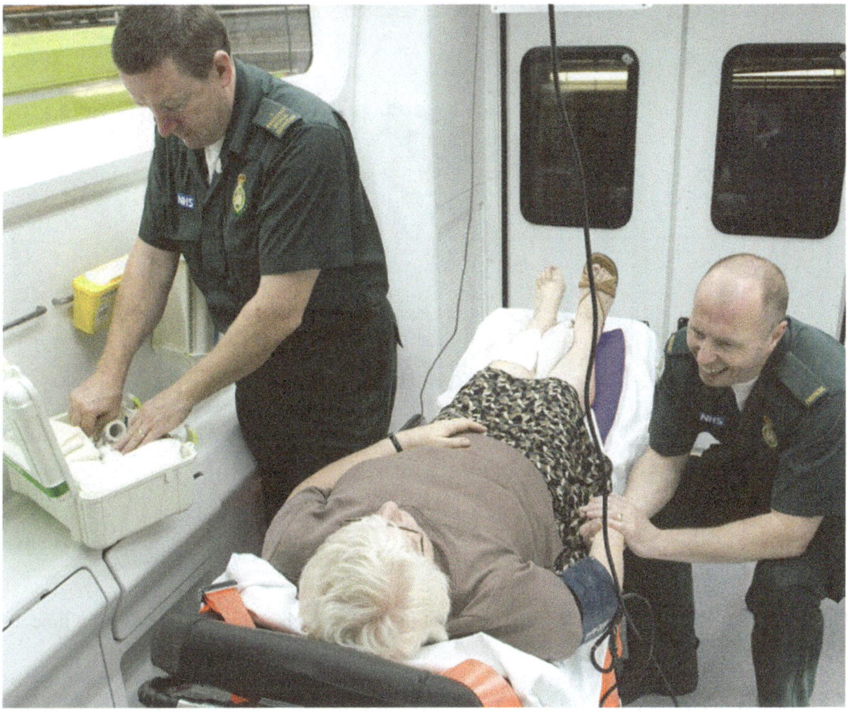

Fig. 10 The new interior allows 360 degree access to the patient. Here a team of paramedics test out the co-designed interior

3.2 Using DT to Understand 'Hopes' and 'Fears' Around Driverless Vehicles

The development of driverless vehicle technology is often met with suspicion. The notion that humans will be made redundant or overly reliant on robotics already reflects a general unease with automated technology that exists. However, the idea that vehicles or machinery no longer need to be supervised by people because they are able to assess and adapt to situations on their own raises more ethical and security concerns. Cautionary sci-fi tales emphasise safety and security concerns.

GATEway (Greenwich Automated Transport Environment), a large research project funded by the UK government, was set up to try and understand how the technical, legal and societal challenges of autonomous vehicles might be tackled. The HHCD and the RCA's Intelligent Mobility Design Centre were tasked by the project consortia with using DT techniques to work with the general public and develop a better understanding of the attitudes towards this emerging future-focused technology. Involving people in conversations surrounding the development of these vehicles meant that user concerns could be taken into account when designing

Fig. 11 Treatment packs that can be replaced at the end of each shift reduce restocking time and complexity

new propositions. This extended to thinking about the kinds of services and infrastructure that will need to be put into place, a number of which were focused on critical areas of personal safety.

The project asked this key question: How does design help to increase the acceptance and adoption of driverless vehicles in urban environments? In order to improve public perception and engagement with the technology, the team led a series of DT workshops to explore the possibilities of what autonomous vehicles could be. As they would no longer need to take a traditional 'car form', ideas such as travelling coffee shops, moving cinemas, enclosed gardens and even mobile showers have been posited as potential vehicles. The workshops helped the team to understand people's hopes and fears about a driverless future, and led to significant insights into how the public might embrace the technology (Fig. 12).

The participant groups included 33 women and 74 men, 5 people between 18 and 24, 39 between 25 and 34, 29 between 35 and 44, 14 between 45 and 54, 10 between 55 and 64 and 7 over 65. Of these:

- 13 had additional needs including visually impairment, reduced hearing or mobility
- 6 classified themselves as cyclists
- 6 as pedestrians
- 9 mainly used buses
- 19 regularly drove a car
- 21 used a mix of various modes apart from a car
- 38 mainly travelled by rail.

Fig. 12 Workshop participants express their hopes and fears around driverless vehicles

Over the eight workshops, people shared over 700 'hopes' and a similar number of 'fears'. Most of the participants said they had a limited or average knowledge of driverless vehicles with only 6% saying that they had expertise. The results of the workshops found that those who are currently excluded from driving, including older people and those with additional needs, might be more attracted to the benefits of autonomous vehicles than other groups. The findings from these initial workshops presented both utopian and dystopian futures and DT was key in eliciting, mapping and understanding responses.

Major hopes centred on the freeing up of time, as cities become less congested and the time during journeys becomes more useful and valuable. Participants felt that driverless vehicles would create healthier and safer city environments leading to less stress, reduced pollution-related illness and more social and convivial streets and journeys. Security issues centred around safer streets, less drink-driving, travel-on-demand, quieter and cleaner vehicles and more accessible and affordable transportation. Some unintended insights focused on the vehicle's potential ability to self-diagnose so that it could take itself for a service before becoming unsafe, noisy or less secure.

Findings around fears and concerns included the impact on employment as 'robots' take over existing jobs and economic activities, fears around data privacy, the ethical judgement of autonomous but supposedly intelligent vehicles, the loss of independence that might come when driverless vehicles control how people use the street, the challenges of cybercrime and the impact that 'on-demand' vehicles will have on people's health as they find it easier and cheaper to get door to door transport rather than walking or cycling around town.

Fig. 13 A driverless vehicle could be a first respondent in an accident or support people with low vision

Quieter vehicles also might make the urban environment less safe for partially sighted or distracted pedestrians and the transition to single surface roadways could make it even harder for people with additional needs to navigate and cross streets. Security concerns included ethical dilemmas around responsibilities in case of an accident, increased potential for antisocial behaviour inside the vehicle, large corporations controlling the transport network based on profit and the increased ability to track travellers. Subtler concerns were also uncovered such as the dangers of being always connected, the dangers of vehicles starting to tout for business and a more disconnected and isolated travel environment where passengers cannot rely on the safety of a 'crowd'.

DT enabled a process to allow workshop participants to express their hopes and fears in a way that was descriptive, imaginative and even unexpected. It further enabled the translation of ideas into opportunities for innovation that challenged the purely functional or technical focus that typically surrounds driverless vehicles and created better social experiences and outcomes that are more relevant to the potential users (Fig. 13).

4 Conclusion

DT enables the practice of design in a number of ways. Firstly, it leads design out of the studio and into the boardroom by delineating the benefits of a creative approach and framing them in a way that business can access and utilise. Secondly, it evolves design practice, adding to the designer's toolbox and ensuring that people-centred, empathy-driven approaches are prominent in mainstream design. Thirdly, it can positively change the perception of design and the role of the designer. Ghajargar et al. [28] found that the role of the designer is still viewed in a primarily aesthetic manner by those from a technical background whilst user experience specialists also found it challenging to accept that a designer has a role in studying user needs. Finally, and most importantly, DT democratises design, delineating inventive methods and activities that can be used by anyone and positively applied to almost any sector, company or corporation.

DT can face internally or externally within an organisation. When internal-facing, it can help with a range of issues from problem-framing to project management, ensuring that employees and colleagues from different backgrounds have equal voice. When applied externally it creates a deeper understanding of context, user behaviour and gives a platform for equality, partnerships and knowledge sharing. It can impact on process, management and strategy and cultivate an environment that encourages exploration and experimentation.

DT also provides an evidence-base that is based on people's attitudes and behaviours rather than simply relying on numbers or statistics. Working with smaller numbers of people to gain richer levels of insights signals a shift in relying on 'big data' to benefitting from 'deep data'. Rather than canvassing hundreds of participants and getting a little information from each of them, DT enables working with fewer people and spending more time to understand the details of their lives, their needs, aspirations and perspectives. Whilst providing real and relevant information to any project, this can also be more time-effective and cost-effective.

Future work and research is needed to map DT's application in a variety of sectors and spaces and there is a need for empirical research [39] to codify and assess its contributions. DT cannot be treated as a "wonder drug" [53] that can solve every issue or foster distinct improvement of all situations. Practitioners need to record challenges and failures as well as successes and discoveries, and the intersection of DT with new areas of application is paramount in evolving current practice.

The application of DT within security further stretches its reach in a way that is timely and appropriate. Security is a critical area of human concern where people's behaviours are often unexpected and variable. This is a complicated field of research and application, requiring a view that spans the holistic through to the specific. DT can help assess risk, design research and articulate concepts that span a range of emerging technologies, global situations and political concerns. It can support endeavours as it is well-placed to understand human needs and perspectives, work creatively with scientific precision and deliver multi-perspective understanding and clarity in one of the most complex and critical fields of study.

Acknowledgements Academic colleagues at the Helen Hamlyn Centre for Design at the Royal College of Art have been invaluable in producing this chapter, primarily Prof. Jeremy Myerson who articulated the Arc of Design Thinking. Dr. Chris McGinley's research and ideas have contributed to the Inclusive Design approach and Dr. Jak Spencer supported efforts in assessing the primary research. External academics and practitioners include Sean Donahue from ArtCenter College of Design (USA) and Onny Eikhaug from the Norwegian Centre for Design and

Architecture, both of whom influenced and supported many of the ideas in this chapter through collaborative work with the author. Prof. Dale Harrow, Dan Phillips and the team at the Intelligent Mobility Design Centre at the RCA were partners on the GATEway project, supporting the creation of the case study.

References

1. Antonelli P (2013) Rotman on design. Rotman Magazine. University of Toronto Press, Canada
2. Beverland M, Wilner S, Micheli P (2015) Reconciling the tension between consistency and relevance: design thinking as a mechanism for brand ambidexterity. J Acad Mark Sci 43 (5):589–609
3. Bowden F, Lockton D, Gheerawo R, Brass C (2015) Drawing energy: exploring perceptions of the invisible. Royal College of Art, UK
4. Brown T (2008) Design thinking. Harvard Bus Rev 86(6):84–92
5. Buchanan R (1998) The study of design: doctoral education and research in a new field of enquiry. In: Doctoral Education in Design: Proceedings of the Ohio Conference, 8–11 Oct
6. Calabretta G, Montaña J, Iglesias O (2008) A cross-cultural assessment of leading values in design-orientated companies. Cross Cult Manage Int J 15(4):379–398
7. Carlgren L, Rauth I, Elmquist M (2016) Framing design thinking: the concept in idea and enactment. Creativity Innov Manage 25(1):38–57
8. Carroll N, Richardson I (2016) Aligning Healthcare innovation and software requirements through design thinking. In: Proceedings of the international workshop on software engineering in healthcare systems, SEHS '16, Texas, USA
9. Chae S (2017) Design thinking in action: changing the public service model. QLA Q 22 (3):15–19
10. Clarke AJ (ed) (2011) Design anthropology: object culture in the 21st century. SpringerWienNewYork, USA, New York
11. Clemmensen T, Ranjan A, Bødker M (2016) How cultural knowledge shapes design thinking—a situation specific analysis of availability, accessibility and applicability of cultural knowledge in inductive, deductive and abductive reasoning in two design debriefing sessions. In: Proceedings of DTRS 11: design thinking research symposium 2016, Copenhagen Business School, Denmark
12. Coleman R (1994) The case for inclusive design—an overview. In: Proceedings of the 12th Triennial Congress, International Ergonomics Association and the Human Factors Association of Canada
13. Cross N (2011) Design thinking. Bloomsbury Academic
14. Curedale R (2015) Design thinking: pocket guide. Design Community College Inc, USA
15. d.school (2017a) https://dschool.stanford.edu/fellows-in-residence/project-fellowship-history-approach. Accessed 19 Nov 2017
16. d.school (2017b) https://dschool.stanford.edu/about/. Accessed 19 Nov 2017
17. Davis J, Docherty CA, Dowling K (2016) Design thinking and innovation: synthesising concepts of knowledge co-creation in spaces of professional development. Des J 19(1):117–139
18. Department of Trade and Industry (2000) Foresight: making the future work for you. Department of Trade and Industry, UK
19. Design Council (2005) The double diamond process model. Design Council, UK. Available at: http://www.designcouncil.org.uk/designprocess
20. Design Council (2005) The double diamond process model, Design Council, London. https://www.designcouncil.org.uk/news-opinion/design-process-what-double-diamond. Accessed Jan 2018

21. Dorst K (2015) Frame innovation: create new thinking by design. MIT Press, Cambridge, Massachusetts
22. Ehn P, Kyng M (1987) The collective resource approach to systems design. Computers and democracy—A Scandinavian challenge. Aldershot, UK, pp 17–58
23. Eikhaug O, Gheerawo R (eds) (2010) Innovating with people: the business of inclusive design. Norwegian Design Council, Norway
24. Fraser H (2012) Design works: how to tackle your toughest innovation challenges through business design. University of Toronto Press, Canada
25. Froehlich JE, Findlater L, Landay JA (2010) The design of eco-feedback technology. In: Proceedings of CHI 2010, USA
26. Fulton-Suri J (2007) Design for people? Design with people? Design by people? Who is designing now? In: Proceedings of include 2007: designing with people conference, London
27. Gaver B, Dunne T, Pacenti E (2009) Cultural probes. In: Interactions, February + March issue
28. Ghajargar M, Mangano G, De Marco A, Giannantionio R (2017) Design thinking applied to data storage innovation: a case study. Des J 20(1):S3776–S3788
29. Gheerawo R, McGinley C, Gorzanelli C (2014) Inventory: 7 years|7 projects. Royal College of Art, UK
30. Gheerawo R, Lee Y (2009) Enabling people—creating inclusive human-computer interactions. In: Proceedings of the 5th international conference on universal access in human-computer interaction—Addressing diversity, UAHCI 2009, held as part of HCI International 2009, USA
31. Gothelf J (2017) Lean vs. Agile vs. Design thinking: what you really need to know to build high-performing digital product teams. Gothelf Corp, USA
32. Gray D, Brown S, Macanfuno J (2010) Gamestorming: a playbook for innovators, rulebreakers, and changemakers. O'Reilly, USA
33. Griffin JH (1962) Black like me. Signet
34. Gubrium A, Harper K (2013) Participatory visual and digital methods. Left Coast Press, USA
35. Hargreaves T, Nye M, Burgess J (2013) Keeping energy visible? Exploring how householders interact with feedback from smart energy monitors in the longer term. Energy Policy 52:126–134
36. Hernández-Leo D, Agostinho S, Beardsley M, Bennett S, Lockyer L (2017) Helping teachers to think about their design problem: a pilot study to stimulate design thinking. Paper presented at: 9th annual international conference on education and new learning technologies. EDULEARN17, Barcelona
37. Hunt J (2011) Prototyping the social: temporality and speculative futures at the intersection of design and culture. In: Clarke AJ (ed) Design anthropology: object culture in the 21st century. SpringerWienNewYork, USA
38. Huq A, Gilbert D (2017) All the world's a stage: transforming entrepreneurship education through design thinking. Educ Training 59(2):155–170
39. Johansson-Sköldberg U, Woodilla J, Centikaya M (2013) Design thinking: past, present and possible futures. Creativity Innov Manage 22(2):121–146
40. Johnson-Laird PN (1983) Mental models. Harvard University Press, USA
41. Kimbell L (2011) Rethinking design thinking. Des Cult 3(3):285–305
42. Kleinsmann M, Valkenburg R, Sluijs J (2017) Capturing the value of design thinking in different innovation practices. Int J Des 11(2):25–40
43. Kotina E, Koria M, Prendeville S (2017) Using design thinking to improve strategic decisions during collaborative sensemaking. In: Design management academy conference, Hong Kong
44. Lawrence G (2001) Hospital beds by design: a socio-historical account of the 'King's Fund Bed', 1960–1975. Ph.D. Thesis, London University
45. Leonard SN, Fitzgerald RN, Riordan G (2015) Using developmental evaluation as a design thinking tool for curriculum innovation in professional higher education. High Educ Res Dev Routledge. https://doi.org/10.1080/07249360.2015.1087386

46. Liedtka J, Salzman R, Azer D (2017) Design thinking for the greater good. Columbia University Press, USA
47. Luchs M (2016) A brief introduction to design thinking. In: Luchs M, Swan K, Griffin A (eds) Design thinking: new product development essentials from PDMA. Wiley, Hoboken
48. Mahmoud-Jouni SB, Midler C, Silberzahn P (2016) Contributions of design thinking to project management in and innovation context. Proj Manage J 47(2):144–156
49. Mazé R, Redström J (2008) Switch! Energy ecologies in everyday life. Int J Des 2(3):55–70
50. McGinley C (2012) Supporting people-centered design through information and empathy. Doctoral thesis, Brunel University, London
51. McKnight JL, Kretzmann J (1990) Mapping community capacity. Center for Urban Affairs and Policy Research, Northwestern University
52. Moore P (1985) Disguised. Word Books, Texas
53. Mootee I (2013) Design thinking for strategic innovation: what they can't teach you at business or design school. Wiley, USA
54. Mubin O, Novoa M, Al Mahmud A (2016) Towards the successful integration of design thinking in industrial design education. In: Proceedings from international conferences ITS, ICEduTech and STE
55. Nemeroff CJ (1995) Magical thinking about illness virulence: conceptions of germs from "safe" versus "dangerous" others. Health Psychol 14(2):147–151
56. Newton K, Riggs MJ (2016) Everybody's talking but who is listening? hearing the user's voice above the noise, with content strategy and design thinking. In: VALA2016: Libraries, technology and the future, VALA, Australia
57. Noel L, Liu TL (2016). Using design thinking to create a new education paradigm for elementary level children for higher student engagement and success. In: Proceedings of DRS 2016, design research society 50th anniversary conference. Brighton, UK
58. Norman DA (1999) Rapid ethnography. In: Aldersey-Williams H, Bound J, Coleman R (eds) The methods lab. Royal College of Art, UK
59. Peterson M, Hempler NF (2017) Development and testing of a mobile application to support diabetes self-management for people with newly diagnosed type 2 diabetes: a design thinking case study. BMC Med Inform Decis Mak 17:91
60. Pink S (2005) Dirty laundry. Everyday practice: sensory engagement and the constitution of identity. Soc Anthropol 13(3):275–290
61. Porritt J (2006) Capitalism as if the world mattered. Earthscan, London
62. Roberts JP, Fisher TR, Trowbridge MJ, Bent C (2016) A design thinking framework for healthcare management and innovation. Healthcare 4:11–14. https://www.sciencedirect.com/science/article/pii/S221307641500113X.Accessed Jan 2018
63. Sanders EBN, Dandavate U (1999) Design for experience: new tools. In: Overbeeke CJ, Hekkert P (eds) Proceedings of the first international conference on design and emotion, The Netherlands: TU Delft
64. Sanders EBN, Stappers PJ (2014) Editorial. CoDesign Int J CoCreation Des Arts 10(1)
65. Sewchurran K, Philipp F, Baets W, McDonogh J (2016) Could more thoughtful practice of complexity, design thinking and values-based organising address some of the limitations of current management and organising paradigms? Complex Leadersh Manage 3(3):169–181
66. Szczepanska J (2017) Design thinking origin story plus some of the people who made it all happen. https://medium.com/@szczpanks/design-thinking-where-it-came-from-and-the-type-of-people-who-made-it-all-happen-dc3a05411e53. Accessed Jan 2018
67. Vavik T, Gheerawo R (2009) The challenges in universal design. In: Vavik T (ed) Inclusive buildings, products & services: challenges in universal design. Tapir Academic Press, Norway
68. Warbuton N (2003) Everyday inclusive design. In: Clarkson J, Coleman R, Keates S, Lebbon C (eds) Inclusive design—design for the whole population. Springer-Verlag, UK

An Integral Futures Lens on Future Security Issues

Craig Perry and Andy Hines

Abstract Conventional analysis of security issues tends to take a disjointed approach through tightly focused tactical lenses. Increasingly, however, security issues are embedded in a wider and diverse range of factors that may escape conventional analysis. An integral futures perspective to exploring security issues is proposed that provides a wider lens by taking an integrated holistic approach that explores individual and collective aspects of emerging issues from both an interior (e.g., motivation, culture) and exterior (e.g., behavior, systems, and infrastructures) perspective. We believe this approach provides a fresh way to identify issues as well as providing a more holistic explanatory framework.

Keywords Integral futures · National security

1 Literature Review

The most recent U.S. National Security Strategy outlined a remarkably comprehensive agenda, playing up America's perceived advantages in diplomacy, development, intelligence, science and technology, and more. Beyond simply advocating for a strong national defense and homeland protection, its conception of security included such nonmilitary threats as natural disasters, pandemic diseases, and climate change. It also emphasized the importance of "soft power" derived from America's economic prosperity and respect for universal values, within a cooperative international order. Acknowledging that long-term transitions are underway and unfolding over decades, the 2015 NSS attempted to position America to

C. Perry · A. Hines (✉)
University of Houston Foresight Program, 4235 Cullen Boulevard
Room 110, Houston, TX 77204, USA
e-mail: ahines@uh.edu

A. J. Masys (ed.), *Security by Design*, Advanced Sciences and Technologies for Security Applications, https://doi.org/10.1007/978-3-319-78021-4_3

influence their trajectories, seize the opportunities they create, and manage the risks they present in the future.[1]

By contrast, the U.S. National Military Strategy published that same year focused on just two principal threats—revisionist states challenging international norms and violent extremist organizations undermining international security—arrayed across a "continuum of conflict" from non-state conflicts to interstate wars.[2] This is a very different "continuum of conflict" than that described in U.S. military doctrine, which encompasses everything from peacetime military engagement, security cooperation, and deterrence to large-scale combat operations. Yet this seemingly smooth spectrum hits a snag at the threshold of armed conflict, when "normal cooperative and competitive environments" give way to combat. Moreover, U.S. doctrine defines "conflict" as an armed struggle or clash,[3] which begs the question of where on this conflict continuum peace supposedly resides. More jarring is the realization that the U.S. military officially recognizes only two basic forms of warfare—traditional and irregular—and conceives of both exclusively in terms of violent struggle.[4]

Western military thinking has a tendency to draw sharp lines between war and peace, as if these two states of human affairs were inherently distinct. Advocates of this position often cite Carl von Clausewitz, the 19th-century Prussian military strategist, in arguing that war is the continuation of politics by other means—suggesting that once war begins, politics end. However, Clausewitz actually believed there is an indissoluble connection between politics and war, which ought to be treated holistically [1]. This mistaken conception of discontinuity is likely based on a bad translation of Clausewitz, who actually wrote that war is the continuation of policy "with"—not "by"—other means, with the transition from peace to war involving the addition of the military instrument of national power to ongoing diplomatic, informational, and economic efforts.[5] Compartmentalizing politics and war in a linear sequence—first comes politics, then war, then a return to politics with peace—fails to recognize that politics and war interact in a complex, continual feedback process [2].

As insightful as Clausewitz may have been, however, he did not anticipate the full complexity of modern warfare. By the end of the Cold War, critics had begun questioning whether his theories were still relevant with the advent of nuclear weapons, transnational constabulary warfare, and modern statecraft [3], and such critiques only intensified in the era of terrorism, insurgencies, and civil wars that

[1]The White House. *National Security Strategy*. February 2015.

[2]Chairman of the Joint Chiefs of Staff. *The National Military Strategy of the United States of America 2015: The United States Military's Contribution To National Security*. June 2015.

[3]Joint Staff. *Joint Publication 3-0, Joint Operations*. January 17, 2017.

[4]Joint Staff. *Joint Publication 1, Doctrine for the Armed Forces of the United States*. March 25, 2013.

[5]Holmes, James R. "Everything You Know About Clausewitz Is Wrong." *The Diplomat*, November 12, 2014, thediplomat.com/2014/11/everything-you-know-about-clausewitz-is-wrong. Accessed June 10, 2017.

followed. Yet other scholars have countered that the timeless principles of war Clausewitz outlined nearly two centuries ago—a trinity of violence, chance, and rational purpose, coupled with the interplay of civilian populations, armed forces, and their political leaders—are still applicable to war in the modern era. After all, even terrorist networks pursue rational or instrumental motives, and they depend on a wider populace for legitimacy, recruits, financing, sanctuary, intelligence, and other material support. Although defeating such adversaries has less to do with inflicting destruction on opposing armies than winning "hearts and minds" of noncombatants, it is not fundamentally different from the sort of conflicts Clausewitz described [4].

Unfortunately, the U.S. military has struggled to make sense of these new types of conflict. The Department of Defense introduced the doctrinal concept of "military operations other than war" in the 1990s to encompass everything from humanitarian assistance and noncombatant evacuation to counterterrorism (CT) and counterinsurgency (COIN). Whereas "war" was meant to achieve national objectives or protect national interests, "MOOTW" focused on deterring war and promoting peace, and were supposedly more sensitive to political considerations[6]—as if national interests and objectives could ever be completely divorced from politics. By 2011, U.S. joint doctrine was updated to acknowledge that the Department of Defense had "learned through the difficult experiences of both Iraq and Afghanistan that success is not only defined in military terms," but also involves activities to "protect and defend the population, facilitating the personal security of the people and, thus, creating a platform for political, economic, and human security."[7] Such "irregular" warfare experiences also reinforced the value of restraint, perseverance, and legitimacy, which have since been incorporated into joint doctrine alongside the classic principles of war.[8]

The U.S. military does field some niche capabilities designed to prosecute irregular warfare. Special operations forces, for example, consider "the totality of the cognitive, informational, physical, cultural, and social aspects of the operational environment to influence the local population's behavior," bringing to bear a different mindset and capabilities than traditional warfare methods—including less reliance on violence to accomplish their objectives.[9] Similarly, cyberspace operations involve the projection of power by the application of force in and through cyberspace, with nary a mention of violence in the applicable doctrine.[10] The same goes for information operations, which are designed to influence, disrupt, corrupt, or usurp adversary decision-making.[11] Although these offensive military activities

---

[6]Joint Staff. *Joint Pub 3-07, Joint Doctrine for Military Operations Other Than War*. June 16, 1995.

[7]Joint Staff. *Joint Publication 3-07, Stability Operations*. September 29, 2011.

[8]Joint Staff. *Joint Publication 3-0, Joint Operations*. January 17, 2017.

[9]Joint Staff. *Joint Publication 3-05, Special Operations*. July 16, 2014.

[10]Joint Staff. *Joint Publication 3-12 (R), Cyberspace Operations*. February 5, 2013.

[11]Joint Staff. *Joint Publication 3-13, Information Operations*. November 20, 2014.

do not involve violence, they must be accounted for by any reasonable conception of conflict and war.

In 2016, the Heritage Foundation offered an alternative "contemporary conflict spectrum" that more fully captures the range of challenges we face in the modern security environment. The author, Frank G. Hoffman, cautions that the U.S. national security community should avoid narrow categorizations when it comes to war. "The black-and-white distinction between war and peace, or traditional war and irregular war, makes for nice, simple boxes, but the real world is not so easily categorized. In fact, some adversaries seek to exploit U.S. paradigms and the gaping institutional seams that they create." For example, so-called "gray zone" conflicts involve "deliberate multidimensional activities by a state actor just below the threshold of aggressive use of military forces," where "adversaries employ an integrated suite of national and subnational instruments of power in an ambiguous war to gain specified strategic objectives without crossing the threshold of overt conflict [5]." However, this more nuanced perspective on conflict has yet to make serious inroads into the U.S. national security community.

2 A Brief Overview of Integral Futures Analysis

As modern societies grow increasingly complex and interdependent, risks to security tend to multiply accordingly. With each innovation come new vulnerabilities to be targeted by adversaries, exploited by criminals, or simply triggered by chance. Furthermore, the proliferation of interconnected networks—whether physical or virtual—can expose entire nations to threats as disparate as diseases, cyber attacks, and "fake" news. While a strong military may be a prerequisite for engaging in—or better yet, deterring—armed conflict among states, it is typically not enough on its own to defeat terrorist organizations intent on sowing fear and confusion, and it would likely be of little use if an adversary managed to subvert the national will through other, more nefarious, means. To address such a bewildering array of actual and potential threats, we must take a more holistic approach to security.

The "integral" perspective pioneered by Ken Wilber offers just such an approach. According to this universal theory, virtually any phenomenon or issue can be understood as possessing both interior and exterior qualities, as well as having a singular and plural nature.

- The *interior* of a phenomenon has no spatial location (shared feelings, thoughts); requires interpretation
- The *exterior* of a phenomenon does have spatial location and it can be seen by our senses (behavior, physical objects, brain wave patterns); no interpretation required—dealing with measurable, objective facts.

- The *individual* aspect refers to all phenomena taken in *singular* form (a feeling, a tree)
- The *collective* aspect refers to phenomena taken in *plural* form (shared feelings, a forest)

Put simply, the quadrants suggest that everything has an inside and an outside, and is singular and plural. For example, each person possesses a subjective consciousness, while also displaying objectively observable behavior and physical features. And he or she in turn forms part of a larger collective which also has internal (cultural) and external (social) aspects. Taken together, these four dimensions—or quadrants, formed by the intersection of the interior/exterior and individual/collective axes—provide a comprehensive perspective on reality [6] (See Fig. 1).

Integral theory is developmental; that is, it posits a consistent direction of change over time, with each of the quadrants containing progressively more complex levels or stages of development. Thus, establishing the current level of development of an issue within a quadrant suggests its potential future direction of development, i.e., to the next stage. For example, Integral theory incorporates the "Spiral Dynamics" model of Clare Graves and Don Beck, which extrapolates hierarchical levels of personal development to organizations and societies as they unfold into ever-greater complexity.[12]

Richard Slaughter first applied Integral theory to the field of foresight nearly two decades ago, encouraging a holistic approach that incorporates multiple points of view and offers a source of methodological innovation for practitioners [7]. Integral Futures, then, applies an "all quadrants, all levels" methodology to better anticipate the variety and complexity of plausible futures.

Applying Integral Futures theory to security issues can take many forms. At the most basic level, we could examine a person's sense of personal security as a combination of several factors, including:

- Individual fears of perceived threats, such as physical violence or cyber crime;
- Cultural norms and trends that reinforce or allay such perceptions;
- Personal behavior meant to mitigate these threats; and
- Societal factors, like law enforcement and immigration policy, which contribute to public safety.
- We could then factor in the relative developmental levels of the various individuals and groups involved to provide a multidimensional model of their interaction.

Zooming out, we could use this same Integral Futures approach to better understand how an individual state functions within the global security environment.

---

[12]Beck, Don. "What Is Spiral Dynamics Integral?" Integral Naked, 2003, www.spiraldynamics.net. Accessed.

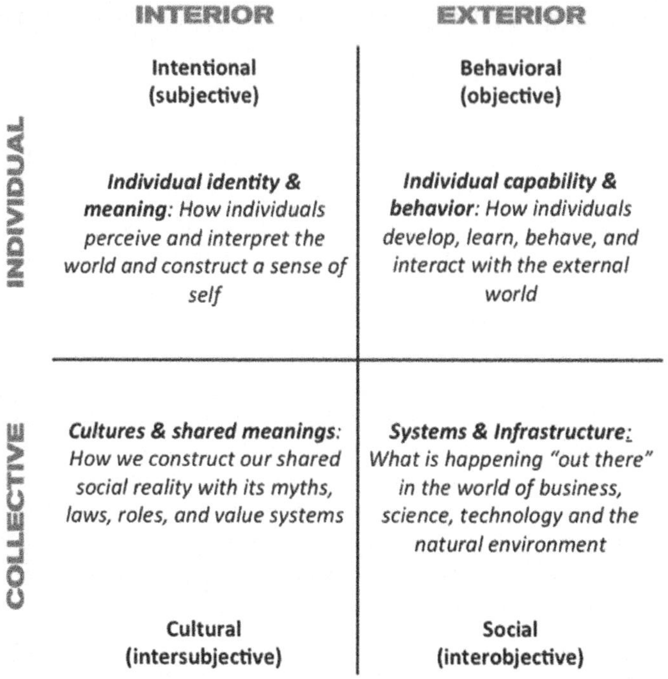

Fig. 1 Quadrants in integral theory

How might we employ Integral Futures theory in the realm of national security? Here we must first carefully consider our terms of reference by analyzing how a state provides security for its citizens through an "all quadrants, all levels" lens.

- Whatever form of governance prevails in a particular territory, a head of state or government and other senior leaders will be entrusted with maintaining the sovereignty of the state, and the internal decision-making processes of this "national command authority" are analogous to those of an individual in the Integral model.
- Next, various organs of state, including government ministries and the military, execute these national security policies, forming a second quadrant of external, observable behavior.
- The direct and indirect effects of these policies throughout society constitute a third quadrant, encompassing everything from border control to economic activity to education levels.
- Finally, public perceptions and cultural norms shape not only the formulation of security policy, but also determine who gets to serve in national leadership within most political systems (See Fig. 2).

We could expand this analysis even further by considering how well developed various elements of the state are in terms of their security perspective. For example,

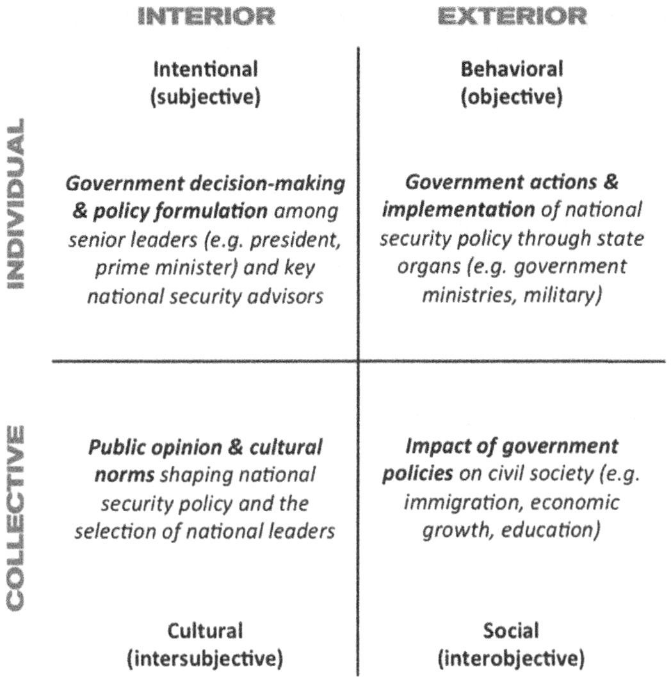

Fig. 2 Integral futures model of national security

political leaders who adopt an aggressive, nationalist foreign policy are less likely to cultivate enduring alliances or harness the full potential of international organizations, and so increase the exposure of their states to certain types of threats. Similarly, cultures that are politically polarized into adversarial "tribes" may become susceptible to foreign propaganda intended to sway public opinion. On the other hand, states that benefit from global trade while providing their citizens robust social safety nets to mitigate economic inequality may face a different set of risks.

When viewed from this holistic perspective, national security becomes a complex whole-of-government endeavor, with potential threats looming from multiple, often unconventional, directions. Military doctrine has traditionally focused on only certain external aspects of national security, such as the armed forces, economic resources, and demographics, while neglecting other potential sources of national power. More critically, almost no attention has been paid to the internal dimensions of the state—from public confidence to policy formulation—that undergird the military-industrial complex. Ironically, the U.S. military, which is renowned for its proven ability to directly target an adversary's most critical center of gravity—its leadership—and produce "shock and awe" among the population, too often fails to recognize the risk that others might do the same to us through more indirect means.

3 How Integral Futures Might Be Applied to Security Issues

Security challenges are inherently interconnected; what war achieves through direct application of violence may sometimes be accomplished by indirect actions aimed at the same ultimate political objectives. As nations bolster their defenses against certain kinds of threats, adversaries invariably seek out new ways to defeat, bypass, or otherwise nullify these protective measures. A security regime that fails to anticipate and mitigate this multiplicity of risks is doomed to fail when confronted by determined and resourceful opponents. An Integral Futures approach to security, on the other hand, treats the state as an organic and indivisible whole potentially susceptible to threats from all quadrants and all levels. An integral approach insists that one not focus on just one quadrant. If an analysis begins with culture, for example, it must also consider the social or institutional aspects, as well as the influence of individuals, i.e., how they are motivated and what are they actually doing.

One state that has seemingly adopted a holistic conception of security is the Russian Federation. In 2013, Gen. Valeriy Gerasimov, chief of the Russian general staff, declared that the very "rules of war" have changed in the 21st century, given the growing role of asymmetric and indirect operations in modern warfare—including "the broad use of political, economic, informational, humanitarian, and other nonmilitary measures." Gerasimov concluded that, "in many cases, they have exceeded the power of force of weapons in their effectiveness" [8]. Ironically, Gerasimov framed this doctrine as a necessary response to Western-backed "hybrid" warfare, which he blamed for so-called "color revolutions" during the Arab Spring and throughout the former Soviet Union. "If you call what's going on now a hybrid war, let it be hybrid war," a Kremlin spokesman recently conceded. "It doesn't matter: It's war."[13]

Russia subsequently employed just such tactics in Ukraine and Syria, and Moscow and its proxies routinely engage in "information confrontation"—including both cyberspace and psychological operations—to shape perceptions and manipulate the behavior of target audiences around the world. Russian propaganda, for example, is intended to influence, confuse, or demoralize domestic and foreign audiences through the work of pro-Kremlin "news" outlets, social media bots and trolls, and paid journalists in Western and other foreign media.[14] These efforts came to light during the 2016 U.S. presidential election, when a Russian influence

[13]Rutenberg, Jim. "RT, Sputnik and Russia's New Theory of War: How the Kremlin built one of the most powerful information weapons of the 21st century—and why it may be impossible to stop." *The New York Times*, September 13, 2017, https://www.nytimes.com/2017/09/13/magazine/rt-sputnik-and-russias-new-theory-of-war.html. Accessed September 23, 2017.

[14]Defense Intelligence Agency. *Russia Military Power: Building a Military to Support Great Power Aspirations*. 2017.

campaign aimed to undermine public faith in the democratic process and tip the scales in favor of a particular candidate.[15]

The People's Republic of China has also adapted its military doctrine to incorporate nonmilitary means of influence—embodying the indirect approach to warfare first articulated by the classic Chinese military theorist Sun Tzu over two millennia ago. The People's Liberation Army unveiled its "three warfares" strategy—encompassing public opinion warfare, psychological warfare, and legal warfare—in 2003, intended to control the public narrative and influence perceptions to advance China's interests, while compromising the capability of opponents to respond [9]. This approach offers China a new form of "non-kinetic" weaponry that can be combined in a highly synergistic way; for example, to advance its territorial claims in the East and South China Seas, Beijing is advancing false legal arguments, deploying civilian flotillas, and broadcasting propaganda portraying itself as a victim of foreign powers.[16]

State actors are far from the only adversaries that can undermine national security through asymmetric, indirect means. Terrorists, for example, employ violence—or the threat of violence—to instill fear and coerce governments or societies.[17] This violence need not be particularly destructive or target critical capabilities to effectively terrorize a population or undermine its confidence in government forces. Similarly, insurgents wage protracted politico-military struggles designed to weaken the control and legitimacy of established governments, supplementing political, economic, and psychological actions with subversion, sabotage, and armed conflict to achieve its political aims.[18] Terrorists and insurgents typically avoid engaging the state's security apparatus head-on, and may even deliberately provoke a heavy-handed response so as to further discredit the government and spur recruiting among the population. In other words, they, too, are pursuing a holistic approach to warfare.

An Integral Futures approach to security would also reveal new opportunities to advance national interests through military—and nonmilitary—operations other than war. The United States and its democratic allies wield considerable "soft power"—a term coined by Joseph Nye to describe a state's ability to attract, co-opt, and shape the preferences of others without resorting to coercion or other inducements [10]—that has proven appealing to countless citizens of authoritarian regimes around the world. Broadcasters like Voice of America and BBC, as well as the proliferation of online and satellite media, offer engaging alternatives to state-run propaganda outlets and can subtly influence how these audiences think and feel about the West and their own governments. Encouraging such information flow into

---

[15]Office of the Director of National Intelligence. *Assessing Russian Activities and Intentions in Recent US Elections*. January 6, 2017.

[16]Navarro, Peter. "China's Non-Kinetic 'Three Warfares' Against America." *The National Interest*, January 5, 2016, http://nationalinterest.org/blog/the-buzz/chinas-non-kinetic-three-warfares-against-america-14808. Accessed September 9, 2017.

[17]Joint Staff. *Joint Publication 3-07.2, Antiterrorism*. November 24, 2010.

[18]Joint Staff. *Joint Publication 3-24, Counterinsurgency*. November 22, 2013.

a dictatorship like North Korea, for example, could potentially hasten the end of this rogue regime without firing a shot [11].

Information operations have always been a part of warfare. However, the advent of the Internet and social media has fundamentally transformed the information environment, facilitating the dissemination of disinformation with unprecedented speed and reach, aimed at exploiting the cognitive vulnerabilities of target audiences. The ease, efficiency, and low cost of such efforts have allowed states and non-state actors alike to effectively "weaponize" information, requiring a coordinated "whole-of-nation" approach—which Rand Waltzman of RAND Corporation has dubbed "cognitive security"—to counter this threat.[19]

Integral Futures offer just such an approach, accounting for a full array of instruments of national power to address even the most insidious threats. It explicitly acknowledges the internal dimensions of the state, including public opinion and leadership decision-making, and treats risks in these domains as seriously as traditional military threats. This type of holistic security perspective appears to have paid off during the recent French presidential and German parliamentary elections, when national governments effectively warned their voters about the potential for Russian meddling.[20,21] A 2016 RAND Corporation study offered additional recommendations on how to counter the "firehose of falsehood" of Russian propaganda,[22] which are entirely consistent with an Integral Futures approach.

4 Conclusion

Psychologist Abraham Maslow is often credited with coining the phrase, "if all you have is a hammer, everything looks like a nail." This is a shortcoming of most Western military doctrine, which offers a relatively narrow and fragmented perspective on threats to national security.

An Integral perspective can help policymakers develop a more holistic approach to assessing threats as well as developing more comprehensive responses. An Integral approach could be especially useful in what we used to call "military operations other than war" such as COIN and CT operations. COIN is all about "winning hearts and

[19]Waltzman, Rand. "The Weaponization of Information: The Need for Cognitive Security." RAND Corporation, April 27, 2017.

[20]McKew, Molly K. "The Gerasimov Doctrine: It's Russia's new chaos theory of political warfare. And it's probably being used on you." *Politico*, September 5, 2017, http://www.politico.eu/article/new-battles-cyberwarfare-russia/. Accessed September 23, 2017.

[21]Reinbold, Fabian. "Germany Prepares for Possible Russian Election Meddling." *Spiegel Online*, September 7, 2017, http://www.spiegel.de/international/germany/how-germany-is-preparing-for-russian-election-meddling-a-1166461.html. Accessed September 23, 2017.

[22]Paul, Christopher and Miriam Matthews. "The Russian 'Firehose of Falsehood' Propaganda Model: Why It Might Work and Options to Counter It." *RAND Corporation*, 2016.

minds" and building partner capacity to provide its own security—which has everything to do with the upper-left "intentional" and lower-left "cultural" quadrants of the Integral model. CT is typically more kinetic, but it also requires a sensitivity to local culture to ensure we don't inadvertently inspire more recruitment and facilitation.

Integral Futures offers a more compelling perspective, providing a flexible and comprehensive approach to emerging security issues that accounts for both their internal and external dimensions as well as their individual and collective aspects. Such an approach is much more suitable for confronting asymmetric nonmilitary threats, such as propaganda and cyberspace operations, that target the state from unexpected directions. We highly encourage the U.S. Department of Defense, Allied militaries, and their interagency partners adopt such an Integral perspective into their national security processes.

We hope that this thought piece will stimulate interest in an Integral approach to security, and look forward to reader feedback.

References

1. Waldman, T (2010) Politics and war: Clausewitz's paradoxical equation. Parameters 40(3):5–8. Autumn 2010
2. Beyerchen AD (1997) Clausewitz, nonlinearity, and the importance of imagery. In: Alberts DS, Czerwinski TJ (eds) Complexity, global politics, and national security. National Defense University, p 64
3. Shephard JE Jr (1990) On war: is Clausewitz still relevant? Parameters 20(3). Sep 1990
4. Schuurman B (2010) Clausewitz and the 'new wars' scholars. Parameters 40(1):97–99. Spring 2010
5. Hoffman FG (2016) The contemporary spectrum of conflict: protracted, gray zone, ambiguous, and hybrid modes of war. In: The heritage foundation. www.index.heritage.org/military/2016/essays/contemporary-spectrum-of-conflict. Accessed 10 June 2017
6. Esbjörn-Hargens S (2009) An overview of integral theory: an all-inclusive framework for the 21st century. Integral Institute, Resource Paper No. 1, March 2009, pp. 1–24
7. Collins T, Hines A (2010) The evolution of integral futures: a status update. World Future Review 5–16. June–July 2010
8. Gerasimov V (2016) The value of science is in the foresight: new challenges demand rethinking the forms and methods of carrying out combat operations. Military Review (trans: Coalson R), Jan–Feb 2016, pp 23–29
9. Kania E (2016) The PLA's latest strategic thinking on the three warfares. China Brief 16(12). 22 Aug 2016
10. Nye J (1990) Bound to lead: the changing nature of American power. Basic Books, New York
11. Malinowski T (2017) How to take down Kim Jong Un. Politico Magazine. 24 July 2017. http://www.politico.com/magazine/story/2017/07/24/how-to-take-down-kim-jong-un-215411. Accessed 10 Sep 2017

Craig Perry served in the U.S. Air Force for over two decades, culminating his military career as a strategic foresight analyst at the Joint Information Operations Warfare Center (JIOWC). He was first introduced to Integral theory as a graduate student in the University of Houston's foresight program, and immediately recognized its potential utility for understanding national security and international relations. He currently lives in the United Kingdom.

Andy Hines is Assistant Professor and Program Coordinator for the University of Houston's Graduate Program in Foresight.

Innovations in Active Shooting Emergency Medical Response and Triage

Bradley Keating and Anthony J. Masys

Abstract The high impact and low frequency 'black swan' events such as terrorist attacks and active shooters are becoming the new normal. The terrorist events in Kenya (2013), Paris (2015, 2016), Belgium (2016), USA (2015 and 2016), UK (2017) highlight how 'unexpected events often audit our resilience; everything that was left unprepared becomes a complex problem, and every weakness comes rushing to the forefront' [20]. The recent mass casualty shooting events in the USA and around the world [6] have certainly highlighted national security issues globally and have had a profound effect on communities. The response to such an event crosses organizational and disciplinary boundaries such as law enforcement, fire/rescue, EMS and medical centers and as such contributes to potential erosion of efficiency and effectiveness in response stemming from matters pertaining to coordination, collaboration and cooperation. With the rise in prevalence of mass shootings, Emergency Medical providers may now find themselves responding to an active shooter event. Traditional triage and victim management techniques leveraged from military experiences are being re-examined. As noted by Smith et al. [14], 'A recent analysis of mass shooting events noted a wounding pattern different from military experience with very few cases of life-threatening extremity hemorrhage; therefore, extrication and transfer to definitive care needs to be a priority in addition to any in-place care provided'. As noted in van Rein et al. [19] 'prehospital trauma triage is essential in providing appropriate care for patients at risk from severe injury to improve their chance of survival'. This chapter will discuss the requirements for a more 'tactically oriented' medical response in an active shooter incident, and presents novel triage and scene management techniques to facilitate a more rapid and effective response to maximize victim survivability.

Keywords Mass shooting · Triage · Terrorism

B. Keating
College of Public Health, University of South Florida, Tampa, FL, USA

A. J. Masys (✉)
University of South Florida, Tampa, FL, USA
e-mail: tmasys@health.usf.edu

A. J. Masys (ed.), *Security by Design*, Advanced Sciences and Technologies for Security Applications, https://doi.org/10.1007/978-3-319-78021-4_4

1 Introduction

'Things that have never happened before, happen all the time'.

(Sagan [12]:12)

The quotation from Sagan [12] certainly resonates with the recent occurrences of active shooters and suicide bombings. For example:

- On 22 May 2017, a suicide bomber detonated a IED at the Manchester Arena during an Arianna Grande concert. The attack killed 22 people and injured upwards of 250 people.
- On 12 June 2016, an active shooter killed 49 people and wounded 58 others in a terrorist attack inside a nightclub in Orlando Florida.
- On 2 December 2015, a terrorist attack (mass shooting) in San Bernardino, California killed 14 people and injured 22 others.
- On 13 November 2015 a series of coordinated attacks (mass shooting and suicide bombing took place in Paris, France. The attackers killed 130 people and injured 368 people. This follows a previous attack in January 2015 at the Charlie Hebdo offices in France where an active shooter killed 17 people and wounded 22 others.
- On 22 March 2016, a coordinated attack of suicide bombings took place in Belgium resulting in 32 persons killed and more than 300 persons injured.

Such mass casualty events are a constant reminder of the vulnerability that exists in our communities and the complexity regarding the management of such events from a law enforcement and EMS perspective.

The threat landscape associated with terrorism is continually evolving. ISIS and Al Qaeda inspired attacks are borderless and global. Similarly lone wolf attacks (inspired by a terrorist ideology) can have devastating effects within the community and resonate regionally and globally. As described by Straub et al. [16]:1 '…during the last decade, individuals motivated by a range of ideological beliefs and individual factors have engaged in horrific acts of mass violence targeting innocent civilians in communities across the United States. These attacks, according to a report by the Congressional Research Service in 2015, have **increased in frequency** as well as lethality'.

These 'Black swan' events represent the unpredictable and thereby necessitates the requirement to readjust to their existence and the need to consider the extremes [17]:xx. Weick and Sutcliffe [20] argue that 'unexpected events often audit our resilience; everything that was left unprepared becomes a complex problem, and every weakness comes rushing to the forefront'. What emerges from the analysis of such 'black swan' events is that they challenge traditional casualty management processes. The complexity of such casualty management scenarios associated with active shooters and terrorist bombings stems from the hostile environment in which they evolve in. As noted in van Rein et al. [19] 'prehospital trauma triage is essential in providing appropriate care for patients at risk from severe injury to improve their chance of survival'. The mortality associated with penetrating trauma and blast injuries necessitates the requirement to access patients quickly in order to render external hemorrhage control (e.g., tourniquets) and rapid evacuation to an appropriate trauma center. Van Rein et al. [19] argue that 'incorrect triage results in undertriage and overtriage'. In fact the study by van Rein et al. [19] noted that nearly all the prehospital trauma triage protocols had significant deficiencies. With regards to triage during an active shooter scenario, additional complexity stems from the threats to the responders themselves operating in a hostile environment.

2 Triage in Context

As described in Turner et al. [18], the FBI defines "active shooters" as 'an individual actively engaged in killing or attempting to kill people in a confined and populated area. The US congressional research service report on public mass shootings describes it as an incident in a public place, killing four or more people in an indiscriminate manner'.

In a mass casualty event associated with an active shooter, triage becomes an essential response process given a dynamic complex threat environment characterized by insufficient resources to support the needs of the victims. What differentiates triage in an active shooter scenario from other disaster contexts is the mechanism of injury and the threat vector that intentionally seeks to '…maximize the number of casualties and the lethality of injuries compared to conventional blunt trauma' [8]. Recent analysis of mass shooting events described by Hick et al. [8] noted '…a wounding pattern different from military experience with very few cases of life-threatening extremity hemorrhage; therefore, extrication and transfer to definitive care needs to be a priority in addition to any in-place care provided'.

This chapter focuses on the 'design' of emergency medical services (triage) in support of terrorist and active shooter event response scenarios.

3 Case Study Vignettes

In today's complex threat landscape, the number of active shooter events is increasing in quantity and lethality. A delay in emergency medical services to the victims of such attacks can be fatal. As described in Hick et al. [8], evaluation of past terrorist and active shooter events shows that 'EMS and first responders, when confronted with a shooting or other violent situation, historically staged a safe distance away and waited until law enforcement had secured the scene. At that point they entered the scene and began treating the injured. In a mass casualty event, the classic triage and treatment areas were established with patients eventually being transported to facilities after proper sorting and on-scene treatment'. Examples of mass shooting incidents include:

- In 2013 an event took place at LAX in which a TSA officer was shot. The victim was visible from the doors of the terminal, but response protocol did not allow for EMS to enter until the scene was safe. By the time EMS was escorted in, the officer had exsanguinated [8].
- The Pulse nightclub shooting of 2016 saw nearly a dozen people take shelter in the bathroom. Initial responders made contact with the shooter but then held a perimeter within the nightclub. By the time PD removed the threat many victims perished.
- At the Utoya Island shooting in 2011, limited interagency communication and a persisting threat by the shooter resulted in EMS staff unable to get onto the island and access casualties for two hours and seven minutes [18].
- Similarly in the report on the Columbine Shootings of 1999, injured victims were still stranded in the school over two hours after the shooters had committed suicide [18].
- In the UK after the Cumbria shootings in 2010, Chesterman's report identified interoperability of the police and ambulance as a 'highly significant issue leading to extensive delays of ambulances reaching patients and police being left to transport the injured' [18].

Finding the right balance in triage between safety and security is a challenge. Poor patient outcomes are not only shaped by the inability to conduct timely triage but also emerge from over triage. For example, the Virginia Tech shooting in 2007, the rate of over-triage was 69% [18]. High levels of over-triage may deprive severely ill patients of the resources needed to treat them, and also put pressure on the surge capacity of the trauma hospital. Similarly, high rates of over triage were also evident at the Fair Child Massacre. The Fort Hood report concludes that a lack of scene safety, chaotic triage organization and communication failures led to fear amongst the emergency staff, which compromised triage and led to the inappropriate or under-triage of several patients [18].

These vignettes highlight the competing agendas (safety and security) and complexity of active shooter scenarios thereby revealing the critical role effective triage has in patient outcomes. The following discussion will highlight some triage approaches and present an alternative innovative approach termed RAMP that addresses the challenges associated with delays in triage and over triage.

4 Discussion

Rejecting the old paradigm of ***'Stage and wait'***, Smith and Callaway [15] argue that 'in trauma, time counts. Essentially, every minute with uncontrolled life-threatening traumatic injury decreases the casualty's chance of survival'. With an increase in the frequency of mass shootings in recent years, EM providers may find themselves responding to an active shooter scenario. Lessons learned from the incidents in the Paris theatre shooting, Columbine school shooting, and Orlando nightclub shooting provide opportunities for reflection upon current active shooter and triage protocols. The disconnect between law enforcement and prehospital EMS response objectives often leads to more negative outcomes for those injured in the shootings. Standardized and agreed upon policies need to be created that address the needs of all agencies involved while still providing mission specific tasks for threat containment and expedited patient triage and treatment. A number of initiatives have examined the problem space associated with active shooter EM response. The case study vignettes described help to situate and contextualize the problem space and establishes the impetus to support the design of an innovative emergency medical service triage solution for such complex threat and security environments associated with active shooters.

5 Innovations in Mass Casualty Triage

The Hartford Consensus [9] represents the outputs from a group of subject matter experts to Enhance Survivability from Mass Casualty Shooting Events. The committee was initiated by the American College of Surgeons in an effort to ensure that victims receive expeditious treatment of their injuries.

Within the Hartford Consensus the acronym **THREAT** emerged to describe the needed response to active shooter and intentional mass casualty events.

T—threat suppression.
H—hemorrhage control.
RE—rapid extrication to safety.
A—assessment by medical providers.
T—transport to definitive care.

As described by Jacobs [9] '...within the framework of THREAT, there exists the opportunity to improve survival outcomes for the victims of active shooter and intentional mass casualty events through mutual collaboration and reinforcing responses'. The framework is predicated on collaboration, cooperation and coordination across the EMS and law enforcement domains. Central to the findings of the Hartford Consensus, was the requirement for timely effective hemorrhage control.

6 "Battlefield Fallacy": Wounding Patterns Differ from Battlefield

The analysis of active shooter events reveals the 'Battlefield fallacy' regarding the wounding patterns. Civilian wounding patterns were predominantly to the head/chest (72%), compared to combat wounding patterns of 48%. As reported by Smith et al. [14] of those who were wounded in civilian active shooter events, the case fatality rate was 46.5%. Despite the assumption being taught that extremity hemorrhage is the most preventable cause of death, the study found that of the 12 active shooter events studied there were zero deaths from extremity wounds. Most common preventable cause of death was pneumo/hemothorax. This treatable cause of death was found to be responsible for nearly 90% of the preventable deaths.

7 Scene Management

No matter how well trained the responding crews may be, a poorly managed scene logistically creates challenges that may prove difficult to overcome. If initial units and command do not establish travel corridors and locations for secondary triage and transport quickly, a scene may become unmanageable in regards to the shear amount of units on scene. This creates a log jam effect in which secondary units cannot gain access and EMS transport units may be required to walk some distance with stretchers before being able to remove victims for further care at a hospital.

Those units who start in a staging location who then attempt to move into an already overcrowded scene often face challenges positioning their apparatus in a manageable location. Additionally, all units coming in after the initial response will have difficulty in gaining ingress and egress from where the casualty collection point will be. To prevent this from occurring initial units on scene need to assign team leaders for triage, transport, and staging as soon as possible to handle the influx of units accordingly.

As mentioned previously, a bottleneck of units can cause havoc on scene management. Transport units need to be able to enter and exit a scene as quickly as possible. Failure to establish corridors for this to occur delays the ultimate extraction of victims from the scene as crews may have to walk into a scene from some distance away to get a patient.

A number of factors pertaining to scene management emerged from the analysis of case studies regarding active shooters. At the macro level, the problem space associated with active shooter and EMS must be considered within the concept of a security ecosystem. This systems view recognizes the actors that must work in concert to ensure safe and effective response. This 'tactical triage' within the security ecosystem must consider elements of interoperability, command and control, communications, safety and security, individual and team training across the disciplines of emergency medical services and law enforcement, shared situational awareness, adaptive practices and the necessity for reflective practices both within and across disciplines. In addition, scene management includes the effective management of social media in supporting information sharing, situation awareness and public alerting. Misinformation can have significant impact on operational effectiveness. Numerous cases have shown weaknesses in these areas.

Some best practices that resonate with the 'security ecosystem' elements above were noted in the EM response to the 2013 Boston Marathon Bombing [1]. These include:

- Integrated Public Safety Planning Activities In Support of Marathon Day
- Implementation of a Comprehensive Medical System Supporting the Marathon
- Activation of a Multi-Agency Coordination Center in Support of Marathon Activities
- Participation in Multi-Jurisdictional Exercises
- Hospital Participation in Emergency Preparedness Planning and Exercises
- Pre-Race Assessments of Route by Joint Hazard Assessment Teams (JHATs).

A mass shooting event will undoubtedly involve resources from numerous agencies, including law enforcement, fire department, and EMS. Where confusion lies is who exactly is in charge of the scene? Traditionally law enforcement was in control until the scene was deemed safe. This process could take precious minutes away from the victims receiving the care they need. As noted in Columbine and Orlando this process took over 3 h. Operations are not integrated, law enforcement operates with separate goals than EMS, and vice versa. In order to improve our responses these goals need to be integrated. The first step towards accomplishing this is to have a unified command structure.

8 Ineffective Triage

The deficiencies associated with many of the current accepted triage methods (e.g., SALT: Sort, assess, life-saving interventions, treatment; and START: Simple triage and rapid treatment) demonstrate the need for a fundamental shift in how active shooter incidents are approached. Current triage methods show only a 55–65% accuracy rate in their abilities to accurately categorize a patient, and are over-complicated, which leads to delays in triage, treatment and evacuation of the casualty [3]. The need for greater training of all members involved in a response is described in a recent study that showed only 37% of first responders felt adequately trained for a mass casualty event [10]. Even with training, studies of both SALT and START triage methods showed that both methods were neither sensitive nor specific in predicting patient outcome.

For example, START triage uses capillary refill as one of its criteria for category selection for the victim. When studied, capillary refill in adults was shown to only be 64% sensitive in identifying hypovolemic patients from blood loss [13]. It was also found that lighting conditions can affect the accuracy of capillary refill. In settings with poor lighting it was found that the capillary refill time was incorrect in over 66% of cases [4].

Currently marketed triage tags are overly complicated both in their application and in required information. Many now also include a bar code system in an effort to track to patient destinations. In reality this system only serves to slow the removal of victims from a scene. In a mass casualty event undoubtedly there will be law enforcement personnel at each hospital that has patients transported to it that can work towards identification once the patients have received care. Additionally, many triage tags require the tearing of finely perforated edges to identify which triage category the patient is. During a high stress event causing a sympathetic response the use of fine motor skills become diminished. This can make tearing these tags difficult for responders.

Several of the commercially available triage tags currently on the market also have multiple areas to be filled out by responders. These include sections for vitals, marking of injuries, contamination status, possible exposures, etc. While this info is important to have, in a mass casualty event the time taken to document all this on a tag delays the removal of victims from the scene and definitive care.

9 Triage Solution Space

Many of the potentially survivable injuries suffered in a mass shooting event require interventions within minutes of occurring in order for the patient to live. Traditionally, the approach has always been to wait for the scene to be cleared before care can be rendered. This mentality has led to unnecessary deaths from victims who could have easily survived their injuries if attended to sooner. In a mass shooting event, the delaying of entry by EMS providers could lead to a higher rate of mortality for victims. A wound to the femoral artery can cause exsanguination within 5 min, a brachial artery can bleed out in 8–10 min, a simple and easily correctable airway obstruction such as the tongue can cause brain damage in 6–10 min [11]. All of these injuries or impairments can be addressed by even the most novice of providers in only seconds. So, the most obvious question raised is "how do we get to the victims faster?"

To gain access to victims faster there needs to be an integrated response from both law enforcement and EMS. There needs to be a paradigm shift in how these events are handled, with the old system of waiting for scene clearance being removed and replaced by integrated entry teams and use of the zone approach to active shooter scenes.

The zone approach to mass shooting scenes is not new, and has been taught for several years now within the United States. By breaking the scene down into three distinct but fluid zones it makes scene management for command far easier as well as having the added benefit of knowing where various unit assignments and tasks will occur.

10 Zones of Care and Care Under Fire

The areas of operation for a tactical mission are usually based on the threat level (Fig. 1). Traditionally, terms such as inner and outer perimeter have been used. The inner perimeter is generally the tactical area of operations controlled by the tactical team. Entry or exit in this area is tightly controlled, and the personnel functioning within this location are at the highest risk of injury. The outer perimeter is the larger area of law enforcement operations, and encompasses the inner perimeter. While the concept of static inner and outer perimeters is useful in planning tactical operations, its application to tactical medical support is limited. As the incident evolves, areas of safe refuge or egress may rapidly change. Stratifying the medical areas of operations based on risk better reflects the dynamic process of treating the injured in the tactical arena. The area with the greatest risk is the hot zone, where an immediate threat of injury is present. This risk may be secondary to a known threat in the area or due to hazardous materials. Patient assessment and treatment in the hot zone is inherently dangerous.

Usually, patient extraction, opening an airway, or applying a tourniquet for life-threatening hemorrhage are the only acceptable interventions in this situation, and even these procedures may be delayed because of the potential risk for further injury to both the patient and the rescuer.

The cold zone is the area with neither significant danger nor threat. Medical care in this area parallels that in the routine EMS world; patients may be assessed and treated without risk to either the patient or the EMS personnel.

The warm zone is the area of a potential, but not an immediate or direct, threat. Medical care provided in this zone is dictated by assessing the risk/benefit ratio. Decisions regarding spinal immobilization, intubation, or intravenous therapy rather than immediate extraction depend on the perceived level of threat. Thinking in terms of graduated zones provides a basis to critically analyze medical treatment options in a highly dynamic environment.

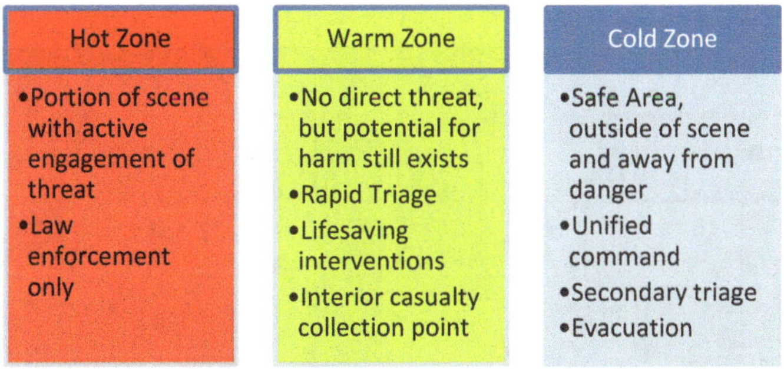

Fig. 1 Zone approaches in active shooter situations (derived from Hartford Consensus Compendium [7])

In the context of a mass casualty, or truly any significant trauma patient treatment on scene should be limited to life saving interventions only. These are defined as interventions that address immediate life threats. These threats include uncontrolled hemorrhage, airway obstruction, tension pneumothorax, and sucking chest wounds. With the exception of a tension pneumothorax all of these injuries can be treated by someone with basic trauma training.

Figure 2 provides an example of an active/mass shooting scenario at a theatre. The Black dot represents the shooter moving through the building. The (X)s on the ground represent victims who have been shot. The colors denote the Hot Zone (Red), Warm Zone (Yellow), and the Cold Zone (Blue) for how the scene should be divided up and approached by responders. Additionally, the formation of the integrated rescue team should be observed, with the law enforcement officers leading the movement and a third officer watching the rear guard.

The Hot zone is defined as the area of immediate threat; this is the location that the shooter is actually located. This zone is reserved for law enforcement only, with the first several law enforcement units arriving on scene making a rapid entry and engaging the threat directly. Their primary task is to either eliminate or contain the active shooter. During the active engagement of the threat there is to be no EMS activity within this zone, with the possible exception being a Tactical or SWAT medic depending on local protocols.

The warm zone is the area of operation in which currently EMS has the least amount of training and experience operating within. This zone is located within the building or location of the active shooter, with the distinction that it is not in the immediate area of the threat. This area historically was off limits to EMS as it is still in an area of potential danger for crews. The warm zone utilizes Integrated Entry Teams to make rapid scene entry, perform lifesaving interventions, as well as primary triage and extraction to a casualty collection point.

An Integrated Entry Team also sometimes referred to as a Rescue Task Force incorporates both law enforcement and EMS into small, mobile teams. The team ideally consists of three officers, with two taking the lead and one watching rear guard with two EMS providers located in the middle. All movements are led by the two lead officers with the EMS following closely behind. The team moves from

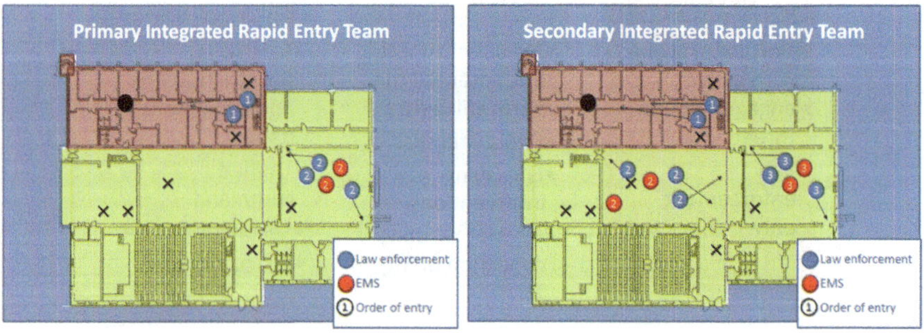

Fig. 2 Active shooter scenario in a theatre

victim to victim triaging and performing life saving interventions, spending less than 60 s with each patient. They move quickly and carry minimal equipment. It is highly suggested that EMS have access to, and wear a ballistic vest and helmet when operating within this zone. This zone will also be where the interior casualty collection point is located. This point, usually near the exit if possible is where the triaged patients will be brought for removal from the scene by secondary entry teams.

The primary Integrated Entry Team's sole function is the rapid assessment, triage, and life saving intervention of the victims and establishment of where the casualty collection point will be. It is incumbent of secondary entry teams consisting of the same make up of officers and EMS to then remove the victims in order of triage category (reds first, then greens) to the casualty collection point for extraction.

The Cold Zone is the area of the scene that is away from danger and in no immediate threat. This is usually outside the building but could be located up to several blocks away. Within this area unified command will set up, evacuation will take place, unit staging will occur, and if necessary a secondary triage or treatment sector will be established.

11 Active Shooter Contextual Treatment Protocols

As discussed, the wounding patterns associated with active shooters scenarios differs from battlefield injury and thereby requires different mental models and patient protocols. It was reported by Smith et al. [14] that civilian wounds were predominantly to the head/chest (72%), compared to only 48% for those wounded in combat.

11.1 Hemorrhage Control

Patients suffering from uncontrolled hemorrhage need to have care rendered within minutes if they hope to survive. This hemorrhage can occur both internally or externally, or both. Unfortunately in the prehospital environment only external hemorrhage can be addressed adequately. There are two primary methods that can be utilized to control hemorrhage, tourniquets and wound packing with hemostatic agents.

11.2 Tourniquet

Traditionally up until the last decade, the use of tourniquets on limbs was to be avoided at all cost as it would mean the loss of that limb if applied. This mindset led to the needless deaths of potentially survivable wounds both in the military and civilian setting. Thankfully this mentality has changed and the acceptance of early and aggressive tourniquets is being taught globally. The use of tourniquets has been shown to be both safe and effective when left on for four hours without removal, with some studies showing longer time periods without negative consequences.

A tourniquet should be applied to any patient who is suffering from a significant limb hemorrhage, arterial or venous. There are many versions of tourniquets currently on the market, each with their own advantages and disadvantages. When choosing one makes sure it can be easily applied and secured by providers and will not come free when moving the patient. Regardless of the tourniquet type they should be applied as high up on the limb as possible. Traditionally it was taught to only place it a few inches above the wound but this can lead to potentially fatal consequences as a severed artery may retract behind the tourniquet and continue to bleed internally. A tourniquet is applied appropriately when distal pulses are lost, or more importantly the bleeding stops. In large vessels such as a femoral artery it may take two tourniquets to get enough pressure to stop the bleeding.

Once applied this intervention should be reassessed frequently to ensure it has not loosened or that the bleeding has continued. The most important piece of advice for this is that "it's not the massive bleeds that will kill people with a tourniquet; it's the slow oozing that will." What this means is that it is easy to recognize the massive bleeding and address it, but it is more difficult to notice the slow oozing bleeding that could ultimately cause the patient to exsanguinate.

11.3 Wound Packing

In the junctional areas of the body that cannot be addressed by a tourniquet (groin, armpit, angle of neck) the use of wound packing can be used to control hemorrhage. Many options exist, with the most efficient ones containing a hemostatic agent that expedites the clotting within the wound, though it should be noted that regular gauze can be used, to a lesser effect. The process for packing a wound is relatively the same regardless of the product and involves physically packing the gauze tightly into the cavity created by the wound. EMS personnel must insert their fingers into the wound to ensure that the gauze is packed directly onto the bleeding vessel. Once it has been filled tightly pressure must be kept on it for a minimum of 3 min and then secured with a dressing to keep it in place.

11.4 Tension Pneumothorax

A tension pneumothorax is an immediate life threat that must be recognized and addressed by EMS if the patient is to have a chance at a positive outcome. Within a tactical environment of a mass shooting using traditional methods such as lung sounds or percussion may not be possible. Within this context, EMS should recognize the potential for the injury if there is mechanism as well as lack of a radial pulse on affected side, possible paradoxical chest wall movement, difficulty breathing, and tachycardia. Anybody with penetrating chest trauma should be decompressed on site if there are any potential signs of a tension.

11.5 Basic Airway Control

Basic airway control can be accomplished in unconscious patients with relative ease by the provider. Any patient who is unresponsive should be placed on their side as to prevent their head from tilting down and causing an obstruction and allowing any vomitus or fluid to drain out of the mouth and not into the airway. This is called the recovery position. The addition of opening their airway using a jaw thrust maneuver in combination with a nasopharyngeal airway (NPA) can provide a decent amount of airway control. NPA sizes vary but having numerous 26 French sized catheters in your kit is fairly standard and will fit most adult patients. The use of oropharyngeal airways (OPA) should be avoided as patient's who lack a gag reflex have a higher mortality rate and will most likely be categorized as expectant/black.

12 Scene Management Techniques

Mass shootings are in themselves chaotic in nature and as such must have a fluid approach to how they are addressed by EMS and law enforcement. Traditional scene management methods utilized within the fire service do not typically function effectively in these environments. Several key points to be addressed in scene management are discussed below.

12.1 Minimal Equipment

A scene in which potentially dozens of units involved can quickly lead to a bottleneck with equipment. Crews arriving need not carry all of their medical equipment into the scene, items such as defibrillators, advanced airway and medication kits are not only a burden to carry but also can lead to over-treatment by providers on scene. First arriving crews should only bring into the scene a "life saving intervention kit." This kit should be in a bag or backpack that can be utilized swiftly and with ease of access and mobility. Item within this kit should include the following: tourniquets, hemostatic gauze, triage tape, 10 g 3-in. needles, and 26 French NPAs. This equipment is used only to address the immediate life threats addressed previously. Later arriving crews can bring in further gear as necessary. Transport capable units coming into the scene should only bring their stretcher or a litter to move the patients back to their rigs for evacuation to a hospital. Time on scene for these units should be limited to only as long as it takes to get patients into their ambulances.

13 Innovation in Triage: RAMP Triage Technique

Once EMS personnel make entry and contact victims, they will need to implement a rapid and effective triage system. The Glasgow Coma Score (GCS) has been shown to be an adequate indicator of patient survival to hospital discharge, with a specificity and sensitivity of around 0.68. A GCS < 6 has a poor prognosis, while a GCS ≥ 6 has a more favorable one [2]. This is supported by data on battlefield casualties collected by the Prehospital Trauma Life Support Committee (2016). In those with an inability to follow basic commands, in addition to a weak or absent radial pulse, the mortality rate approaches 92%. In a mass shooting event, a novel "Rapid Assessment of Mentation and Pulse, or RAMP" triage technique (Fig. 3) is proposed, in which EMS determines if the casualty can follow a basic command, such as "squeeze my fingers", combined with the assessment of a radial pulse to establish a triage category. The benefit of this simplified process is speed and ease of use. When there are dozens of patients, spending 60 s with each patient is not only unfeasible; it may cost the lives of those awaiting triage. To assist in simplification, the use of a multi-colored system that included 4 levels of treatment priority is removed. The remaining categories are expectant/deceased, urgent, or delayed. With this system, those labeled urgent will receive the first care. Then, once the urgent patients have been treated, the delayed category will be tended to. After all urgent and delayed patients are transported, the expectant patients can be reassessed to determine if signs of life still remain, if so, then they will be treated and transported.

RAMP triage begins with approaching the nearest victim who does not have obvious signs of death (i.e. decapitation, incineration, dismemberment, etc.). As you approach the patient observe them for any obvious immediate life threats such as excessive hemorrhage, airway compromise, or open chest wound. Any immediate life threat found should be immediately addressed via tourniquet, wound packing, chest seal, nasopharyngeal airway, or needle decompression. As you address life threats or if none exist you can begin the triage by asking the patient to follow a very basic command such as "squeeze my fingers" or "look in my eyes" to assess mentation. From there you would assess the presence or absence of a radial

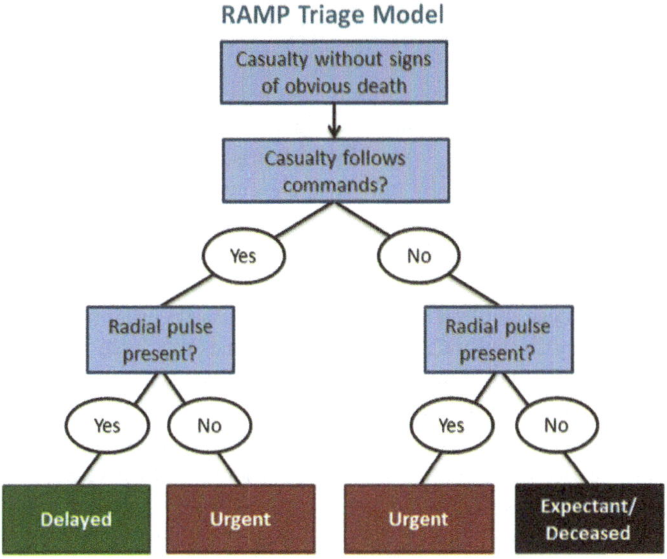

Fig. 3 RAMP triage model

pulse. If both of these items are intact the patient is labeled as delayed or green and they have a two foot long piece of green streamer tape tied to their upper extremity. These patients may not be completely stable or may have the potential for decompensation at a later time; however, in the mass casualty context they can be delayed until the most critical are removed. Those with only either the ability to follow commands or a radial pulse, but not both, are deemed critical or red. These will be the patients who are removed first from the scene and taken to the casualty collection point and then further sorted for transport. If a patient lacks both the ability to follow commands and a radial pulse they are labeled as expectant/deceased or black. As noted above the absence of both of these items has a 92% mortality rate associated with it. In the context of a mass casualty they are presumed to be un-savable. If time or resources were available they may have had a different outcome, but in a situation where the greatest good must be done for the most people their care would cause a greater overall loss of life. Once all the surviving patients have been removed from the scene it is appropriate for these patients to be reassessed and if still alive then transported.

The benefits of this technique are clear. In a world where large scale incidents necessitate that EMS providers triage dozens if not hundreds of victims quickly time is of the utmost importance. Traditional techniques can be time consuming; using RAMP triage a crew of two can triage numerous patients with minimal delays. Another benefit is that it is simple, once taught to even the most inexperienced provider it can be easily recalled when an event occurs. Current methods require charts or cards to refer to. Assuming the crew remembers to use the cheat cards it still delays the triage of all patients.

14 Triage Tape

The move to triage tape and not cards is also necessary to increase speed. It also offers the benefit of being a cost saver for agencies as color coded streamer tape is cheaper than traditional triage tags. Enough tape can be carried for incidents involving hundreds of victims by only a single crew, without worry of needing to get more as they would if using cards. The process is simple and after one member triages the patient the second tears off a two foot piece of tape and ties it to the patient's upper extremity.

The RAMP process and treatment protocols discussed certainly resonate with the Tactical Emergency Casualty Care (TECC) guidelines that have articulated a set of principles for trauma management in high-threat prehospital environments [5]:113. As well, the recommended 'security ecosystem' elements for effective casualty management in high-risk environments such as active shooters draw upon lessons learned from the Boston Marathon Bombing response and active shooter incidents lessons learned described.

The RAMP protocols have undergone initial feasibility test and evaluation showing promising results. Further validation and verification (V&V) and developmental test and evaluation is recommended for RAMP. Additionally for a comprehensive solution, it is recommended to conduct table top exercises (TTX) to reflect upon establishing guidelines and procedures for tactical employment of EMS in active shooter scenarios and explore optimization of logistical staging and command and control.

15 Conclusion

In the context of responses, a mass shooting is one of the most feared for EMS and law enforcement. These low probability, high consequence events need to be prepared for by unifying the response among all agencies that will be on scene. Frequent training is imperative to ensure skills are up to date and equipment and tactics are functional. By implementing a simplified triage system and integrating EMS response with law enforcement, the capability to perform a rapid triage, treatment, and evacuation of victims may decrease mortality secondary to exsanguination or other preventable causes. Being reactive in the current geopolitical environment to mass shooter incidents is no longer a viable option; rather, a proactive approach to triage and integrated response training is required. The lack of integrated planning and preparation before a mass shooting event occurs has the potential to cost the lives of victims and responders alike. This chapter proposes the application of the RAMP model and an integrated scene management approach to better manage mass casualty active shooter incidents.

References

1. After Action Report for the Response to the 2013 Boston Marathon Bombings (2013) http://www.mass.gov/eopss/docs/mema/after-action-report-for-the-response-to-the-2013-boston-marathon-bombings.pdf
2. Al-Salamah M, McDowell I, Steill I, Wells G, Perry J, Nesbitt L, Al-Sultan M (2014) Initial emergency department trauma scores from the OPALS study. Acad Emerg Med 11 (8):834–843. https://doi.org/10.1197/j.aem.2004.03.008
3. Bhalla M, Frey J, Rider C, Nord M, Hegerhorst M (2015) Simple triage algorithm and rapid treatment and sort, assess, lifesaving interventions, treatment, and transportation mass casualty triage methods for sensitivity, specificity, and predictive values. Am J Emerg Med 33 (11):1687–1691. https://doi.org/10.1016/j.ajem.2015.08.021
4. Brown L, Prasad N, Whitely TW (1994) Adverse lighting condition effects on the assessment of capillary refill. Am J Emerg Med 12(1):46–47. Retrieved from https://www.ncbi.nlm.nih.gov/pubmed/?term=brown+prasad+whitley%2C+capillary
5. Callaway DW, Smith ER, Cain J, Shapiro G, Burnett WT, McKay SD, Mabry R (2011) Tactical emergency casualty care (TECC): guidelines for the provision of prehospital trauma care in high threat environments. J Spec Oper Med 11. Edition 3/Summer/Fall 11
6. Crime Prevention Research Center (2016) Comparing death rates from public mass shootings and mass public violence in the US and Europe. Retrieved from http://crimeresearch.org/2015/06/comparing-death-rates-from-mass-public-shootings-in-the-us-and-europe/
7. Hartford Consensus Compendium (2015) **BULLETIN** supplement strategies to enhance survival in active shooter and intentional mass casualty events: a compendium. Bull Am Coll Surg 100(1S). https://www.facs.org/~/media/files/publications/bulletin/hartford%20consensus%20compendium.ashx
8. Hick JL, Hanfling D, Evans B, Green berg S, Alson R, McKinney S, Minson M (2016) Health and medical response to active shooter and bombing events. https://nam.edu/wp-content/uploads/2016/06/Health-and-Medical-Response-to-Active-Shooter-and-Bombing-Events.pdf
9. Jacobs L (2014) Joint committee to create a national policy to enhance survivability from intentional mass casualty shooting events. The Hartford Consensus II. Am Coll Surg 218 (3):476–478. http://www.naemt.org/Files/LEFRTCC/Hartford%20Consensus%20Call%20to%20Action.pdf
10. Jones J, Kue R, Mitchell P, Eblan G, Dyer K (2014) Emergency medical services response to active shooter incidents: provider comfort level and attitudes before and after participation in a focused response training program. J Prehospital Disaster Med 29(4):350–357. https://doi.org/10.1017/s1049023x14000648
11. Porayko D (2017) Exsanguination time from damage to major arteries. Retrieved from www.biology.stackexchange.com
12. Sagan S (1993) The limitation of safety—organizations, accidents and nuclear weapons. Princeton University Press, Princeton, New Jersey
13. Schriger D, Baraff L (1991) Capillary refill: is it a useful predictor for hypovolemic states. Ann Emerg Med 20(6):601–605. https://www.ncbi.nlm.nih.gov/pubmed/2039096
14. Smith ER, Shapiro G, Sarani B (2016) The profile of wounding in civilian public mass shooting fatalities. J Trauma Acute Care Surg 81(1):86–91
15. Smith ER, Callaway DW (2014) Tactical emergency casualty care: the need for & evolution of civilian high threat medical guidelines. When time matters most. An exclusive editorial supplement to JEMS, Law Officer, Fire Rescue and Fire Engineering, pp 10–15. October 2014
16. Straub F, Zeunik J, Gorban B (2017) Lessons learned from the police response to the San Bernardino and Orlando terrorist attacks. CTC Sentinel 10(2017):1–7
17. Taleb NN (2007) The black swan: the impact of the highly improbable. Penguin Books Ltd., London, UK

18. Turner CC, Lockey DJ, Rehn M (2016) Pre-hospital management of mass casualty civilian shootings: a systematic literature review. Crit Care 20:362. https://www.ncbi.nlm.nih.gov/pmc/articles/PMC5101656/pdf/13054_2016_Article_1543.pdf
19. Van Rein EAJ, Houwert RM, Gunning AC, Lichtveld RA, Leenen LPH, van Heijl M (2017) Accuracy of prehospital triage protocols in selecting severely injured patients: a systematic view. J Trauma Acute Care Surg 83(2):329–339
20. Weick KE, Sutcliffe KM (2007) Managing the unexpected: resilient performance in an age of uncertainty, 2nd edn. Wiley, California, USA

Mass Migration, Humanitarian Assistance and Crisis Management: Embracing Social Innovation and Organizational Learning

Miriam Kaundert and Anthony J. Masys

Abstract As described by Moore and Westley [47] 'complex challenges demand complex solutions. By their very nature, these problems are difficult to define'. During the past 10 years the number of international migrants has been increasing. By 2015 there were 244 million international migrants and in 2015 alone upwards of 28 million people were displaced in 127 countries stemming from issues pertaining to conflict, violence and disasters [3]. In Asia/Pacific region, extreme weather disasters and climate change have been cited as influencing factors in mass migration of 'climate refugees'. In fact it has been projected that over the next century there will be tens of millions of such "climate refugees". The current surge in migration to the European Union (EU) is rapidly becoming the largest and most complex facing Europe since the Second World War. Between January 2015 and February 2016, over 1.1 million people made their way to the EU, escaping conflict and poverty in their countries. Moore and Westley [47] argue that 'urgent calls for "innovation" have become increasingly frequent as people begin to recognize both the need for fundamental change in human beings' relationships with each other and their environment, and the catastrophic ramifications of inaction or ill-informed decisions as to how such change might be effected'. This chapter examines the complex causality regarding mass migration of refugees and leverages social innovation and organizational learning as theoretical frameworks to support the design of humanitarian assistance and crisis management.

Keywords Mass migration · Refugees · Climate change · Conflict Social innovation · Organizational learning

M. Kaundert (✉) · A. J. Masys
University of South Florida, Tampa, FL, USA
e-mail: tmasys@health.usf.edu

A. J. Masys (ed.), *Security by Design*, Advanced Sciences and Technologies for Security Applications, https://doi.org/10.1007/978-3-319-78021-4_5

1 Introduction

According to World Humanitarian Data and Trends, the overall number of people forcibly displaced or affected by either violent conflicts or natural disasters has exceeded 200 million. An average of 30,000 new displacements occurs daily. The UNHCR Annual Global Trends Report, *World at War*, noted that worldwide displacement was at the "highest level ever recorded". Global trends point out that the number of displaced people has been rising. A human security condition has emerged. The recent trends pertaining to the inflow of the migrants and refugees into Europe stem from a combination of several factors such as political instability, social unrest, violence, socio-economic conditions. In particular, the Syrian crisis has become 'the world's single-largest driver of displacement' since early 2011. At the same time, persisting instability and conflict in Afghanistan, Somalia and other parts of the world are keeping millions of people displaced. The UNDP report, 'Syria Alienation and Violence: Impact of Syrian Crisis Report,' notes that armed-conflict, economic disintegration and social fragmentation have transformed the human geography of Syria making it into a most significant human security issue.

In addition to the Syrian crisis, mass migration stemming from North Africa is being fuelled by conflict, socio-economic hardship, famine and persecution. In Afghanistan, Somalia, South Sudan, the Central African Republic, Mali, Nigeria and Eritrea, protracted conflicts and crises threaten the lives of millions of civilians, forcing many to leave their home country in search of safety and security in Europe and elsewhere [45: 2].

As described in Masys [38–41], the complexity lens is a key approach to examine the security ecosystem within the context of human security. This presents an opportunity for new strategic possibilities regarding human security management that leverages new ways of thinking (social innovation). The complexity landscape that characterizes the humanitarian crisis aetiology emphasizes that there exists a '…profound disproportionality of 'causes and effects'' [71]. Such systems possess a history that irreversibly evolves and in which past events are never 'forgotten'. Failure to recognize or understand the complex interdependencies associated with the security ecosystem and human security can result in making incorrect assumptions regarding attribution and contribution of events and decisions. This has been demonstrated in Taylor and Masys [63]. Ramalingam and Jones [53: vii] describe how '…much development and humanitarian thinking and practice is still trapped in a paradigm of predictable, linear causality and maintained by mindsets that seek accountability through top-down command and control. Recent years have seen more emphasis on the mechanistic approaches of this paradigm and the kinds of procedures which are increasingly questioned by successful private sector organization's. Local actions and decisions can have regional and global impacts. As noted in Goldin and Mariathasan [17] 'systemic analysis must examine nodes, pathways, and the relationships between them, because catastrophic changes in the overall state of the system can ultimately derive from how it is organized—from feedback mechanisms within it and from linkages that are latent and often unrecognized'. Here we argue for a systemic view of the complexity landscape associated with human security and humanitarian crisis leveraging the insights from social innovation.

2 Overview of Migration and the Humanitarian Crisis

"We can do that!" ("*Wir schaffen das!*") [5] is the infamous quote of German Chancellor Angela Merkel regarding Germany's involvement in the 2015 European refugee crisis, currently the largest movement of migrants and refugees since World War II [67]. She modified her statement later [57]—but the initial declaration remains, to this day, tightly connected to the current challenges of migration in Europe. What is it, that "we can do"—and how can it be "done"?

Human migration, most broadly defined as the "movement of a person or group of persons from one geographical unit to another across an administrative or political border, who wish to settle definitely or temporarily in a place other than their place of origin" [28], has been happening for thousands of years. Spanning from pre-modern times to the imperialist era and beyond to today's industrialized age, it has always been accompanied by various regulatory means and socio-cultural arrangements, such as indentured labor in British North America [75]. Up to WWII and shortly after, it was mostly bilateral understandings which regulated migration and (re)admission of displaced persons [7] or even unilateral expulsion [32, cited in Coleman 2009]. Due to the enormous number of displaced persons, migrants and refugees after the end of the war [24], multilateral agreements became necessary, sparking the creation of the International Refugee Organization (IRO) which would later grow into the United Nations High Commissioner for Refugees (UNHCR). The necessity of international cooperation for managing migration appears acknowledged today. Contrary to refugees, whose definition and rights are stated in the 1951 Refugee Convention Relating to the Status of Refugees and the 1967 Protocol Relating to the Status of Refugees, there is no recognized international legal definition of migrants. They are usually described as persons who leave their home for, mostly, other countries for a variety of reasons [52], where they are often perceived as a threat to the sovereignty, the economic and welfare system, as well as to culture and security of the receiving state [9, 20]. Due to the ambiguous definition of status, rights and duties it remains important to differentiate between migrants and refugees. Legally, asylum applicants, refugees and migrants are part of the group 'foreigners', a naming they share with guest workers or lawful residents of other citizenship. The lack of a clear definition of migrants therefore includes potential to lead to a variable number of asylum seekers, refugees and migrants, depending on a particular national definition [52]. Migratory flows are regarded as a network phenomenon by Tranos et al. [65], mirroring the global socio-economical and political environment, catalyzing change by increasing socio-cultural diversity, an aspect which can be regarded as an advantage as well as problematic [28].

Migration may be based on other challenges, such as the search for labour (be it skilled, unskilled) or business and research, short-term relocation, long-term assignments abroad, permanent resettling or family reunification [28]. Especially concerning migration from developing countries such as on the African continent, a widely spread point of view takes the stance that people are practically forced to migrate due to extreme poverty and conflict [8], underlined by media reporting of boat migration, often instrumentalized in politics [14]. This conventional narrative allows the conclusion that aid concentrated in the countries of origin has potential to curb migration [62]. This angle can however be challenged by including European demand for cheap labour and empirical evidence that most migrants "are not among the poorest" and "move on their own initiative" [10: 14]. This supports the conclusion that restrictive immigration legislation is ineffective as it feeds a never-ending spiral of migrants meeting the demand for unauthorized, cheap labour in Europe, which prompts stricter regulations to inhibit unauthorized labour and illegal migration, pushing migrants into permanent settlement, marginalization and again fostering unauthorized labor due to the more rigid immigration legislation which was developed due to illegal migration and unauthorized labor in the first place. This approach also questions the common narrative that increased local development will constrain migration as it will rather enable migration due to expanded availability of resources [9, 10].

Climate change has had global impact during the last decades [29]. The conventional narrative expects a significant amount of migrants induced by climate change [1, 16, 26, 50]. According to [68], effects of climate change on areas populated by humans can be assumed (such as flooding of river deltas or low lying regions), as well as impact on agriculture by desertification and increasing occurrence of natural disasters. While these effects may contribute to increased mobility they are very likely not the only reason for migration [19, 68]. While natural disasters, be they the result of climate change or not, certainly play a role in human mobility, displacement caused by such events might be of smaller magnitude than expected and of more temporary nature than previously assumed as shown by Gray et al. [19].

Security-wise, migration attracts increasing attention. Even before the so-called refugee crisis of 2015, EU states opined large-scale migration being a threat to their sovereignty, their national and regional borders, their economies and their societies. Most member states have reacted accordingly, tightening controls on irregular access to their territories and, in some cases, on legal channels [20]. Migration and especially irregular migration from sub-Saharan Africa and the Maghreb to Europe has been associated with security problems such as international crime and terrorism [35, 70].

2.1 Wicked Problems

The complexity inherent within the current humanitarian crisis associated with the mass migration can be framed in terms of a wicked problem. Tackling such wicked problems associated with human security is a challenging endeavour. Such wicked problems require thinking that is capable of grasping a systems view, including the interrelationships among the full range of causal factors underlying them.

The term 'wicked problem' emerged from the work of Rittel and Webber [56]. To them, wicked problems characterized a whole realm of social planning problems that precluded traditional linear, analytical approaches. Qualities of wicked problems include:

- *Wicked problems have many interdependencies and are often multi-causal.*
- *Attempts to address wicked problems often lead to unforeseen consequences.*
- *Wicked problems are often not stable.*
- *Wicked problems usually have no clear solution.*
- *Wicked problems are socially complex.*
- *Wicked problems hardly ever sit conveniently within the responsibility of any one organisation.*
- *Wicked problems involve changing behaviour.*

The wicked problem is characterized by complex causality [41] and the qualities of volatility, uncertainty, complexity and ambiguity (VUCA). Linear thinking is inadequate to problem frame such interactivity and uncertainty. The handling of wicked problems requires holistic rather than linear thinking. Within the context of humanitarian crisis domain [38], this approach is about thinking that is capable of grasping the big picture, including the interrelationships and the complex causality. Perspective, boundaries and interrelationships are key required elements to support the understanding of the dynamic nature of wicked problems.

3 Framing the Problem and Systems Thinking

Today we find ourselves confronted with problems of dynamic complexity and interdependency. Actions and interventions associated with such problems as disaster risk reduction [42], climate change [37], humanitarian relief [38], and non-traditional security [40] can have highly unpredictable and unintended consequences. Moving towards the development of solutions to these complex problem spaces depends on the lens we use to examine them and how we frame the problem.

Several events over the past number of years (such as that exhibited by the mass migration and humanitarian crisis in Europe, Middle East and North Africa) have exposed serious weaknesses pertaining to human security with respect to prevention and emergency-response capabilities in various countries. To address these complex human security challenges, it is argued by the Commission on Human Security (CHS), that a new paradigm of human security is required. It is recognized that:

> First, human security is needed in response to the complexity and the interrelatedness of both old and new security threats – from chronic and persistent poverty to ethnic violence, human trafficking, climate change, health pandemics, international terrorism, and sudden economic and financial downturns. Such threats tend to acquire transnational dimensions and move beyond traditional notions of security that focus on external military aggressions alone.
>
> Second, human security is required as a comprehensive approach that utilizes the wide range of new opportunities to tackle such threats in an integrated manner. Human security threats cannot be tackled through conventional mechanisms alone. Instead, they require a new consensus that acknowledges the linkages and the interdependencies between development, human rights and national security.

As argued in Goldin and Mariathasan [17] '…our actions are bound to have systemic consequences that we cannot foresee before they occur and often fail to understand afterwards; this is true for us individually, but potentially even more so for policy makers and institutions seeking to provide guidance and management in this highly complex environment'. Our linear mindset and reductionist approach to understanding complex problems fails. Systems thinking emerges as a new paradigm to support problem framing associated with human security. Jackson [30] defines systems thinking paradigm as '…a discipline for seeing the 'structures' that underlie complex situations, and for discerning high from low leverage change… Ultimately, it simplifies life by helping us to see the deeper patterns lying beneath the events and the details'.

As described in Masys [41], in this complex problem landscape, systems thinking is both a worldview and a process in the sense that it informs ones understanding regarding a system and can be used as an approach in problem solving [12]. 'Systems thinking' as discussed in Senge [58] emphasizes interconnectedness, causal complexity and the relation of parts to the whole, thereby challenging traditional linear thinking and simple causal explanations. Senge [58] describes systems thinking as 'a discipline for seeing wholes…a framework for seeing interrelationships rather than things, for seeing patterns of change rather than static snapshots'. The behaviour of the system is a result of the interaction and interrelationships that exists thereby acknowledging emergent behaviours and unintended consequences. Systems thinking purports that, although events and objects may appear distinct and separate in space and time, they are all interconnected. The butterfly defect, as described in Goldin et al. (2014) draws attention to the new nature of systemic risk, which is such that small perturbations now have much greater effects and permeate all dimensions of society. This is an underlying construct of complex causality that characterizes such societal issues as poverty, mass migration, and conflict. Problem framing through the application of systems thinking is certainly applicable to the wicked problem space of mass migration and humanitarian crisis.

3.1 Social Innovation

Social innovation is the process of developing and deploying effective solutions to challenging and often systemic social and environmental issues in support of social progress. In so doing, social innovation is aimed at improving human well-being and thereby connects to the human security domain. As described in the EU Guide to social innovation (2013), such societal challenges as aging of Europe, migration and sustainability have undergone a paradigm shift from a problem framing that focused on the narrow solution focus of economic actors to one that explores the full range of innovation opportunities and methodologies to influence policy. Societal problems such as poverty, social exclusion and migration requires problem framing beyond the linear approach. As described in Westley et al. [74: 136] '… when we think of a problem as complex, rather than complicated, the inquiry process unfolds in a radically different way. The starting assumption is that transformation, true novelty is possible. The status quo needs to be understood, but it is not perceived as a constraint. Resources are assumed to consist of not only what is already part of the system, but that which can be understood or created in relationship to each other'. An examination of human security and in particular the mass migration will be conducted through the lens of systems thinking and social innovation.

Social innovation may therefore act as a valid concept to addressing complex issues that have been targeted by short sighted and/or linear thinking before.

4 Discussion

The current mass migration and humanitarian crisis in Europe and North Africa has resulted in both regional and global impacts. Between January 2015 and February 2016, over 1.1 million people made their way to the EU, escaping conflict and poverty in their countries and seeking a better and safer life HPN [25]. Mass migration to Europe should not be a surprise. Such migration has taken place for decades and is often linked to humanitarian challenges elsewhere. The vast majority of migrants (88%) cited conflict, persecution and human rights abuses as the motivations for migration. Poverty and opportunities were also drivers that supported the mass migration. Goldin and Mariathasan [17] argue that '…Although laws, borders and restrictions separate countries, virtually all our activities and ideas have cross-border dimensions. Individual and local choices have global impacts and vice versa: what happens outside our borders has direct daily consequences for each of us every day. These connections are complex, frequently opaque and often beyond our control'.

As described in HPN [25] '…the flight to European shores reflected not only the pull of greater long-term security in Europe, but also the failure of the international humanitarian community to meet basic needs in other places'.

4.1 A Systems View: Why Has This Crisis Developed Now?

The failure of the international community to address conflicts, violence and human rights violations in countries of origin has been a key factor in the surge in irregular migration to Europe. As described in Metcalfe-Hough [45] 'there is no international political framework for ending the Syria conflict and no clear international strategy for addressing the related conflict in Iraq. Afghanistan is still far from any real stability despite years of international intervention, and the international community seems to be out of ideas on how to end the decades-long conflict in Somalia. Efforts to address the chronic poverty, inequality, weak governance and climate and environmental changes that constitute 'push' factors in many developing countries have also been inadequate'. Figure 1 captures the key influence factors pertaining to mass migration as described in Foresight [15].

The systems, sub systems and actors that comprise the complex human security 'ecosystem' embody an inherent interconnectedness between different levels, between environments, between ideas, between actions, and between intentions and actions thereby characterizing the problem space as highly dimensional, complex and volatile (subject to butterfly effects). Figure 1 highlights that addressing the migration crisis requires a systems view that recognizes the interconnected and interdependent factors that characterise the problem space. Employing a disaster forensics framework/approach [41] helps to reveal this inherent complexity to thereby support social innovation.

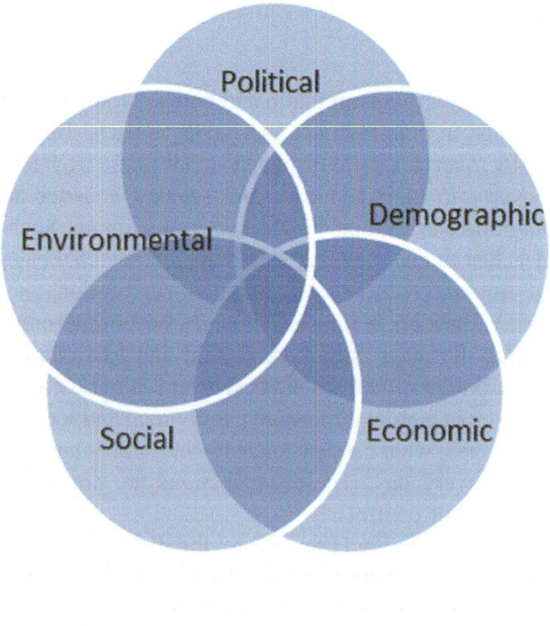

Fig. 1 Drivers of mass migration (themes derived from Foresight [15])

4.2 Applying Social Innovation

Social innovation emerges as an approach to better manage the complexity associated with mass migration. As described in Hagen-Zanker and Mallet [21] or Hamood [22], the European response to mass population movements has been coined by constraints, restriction and deterrence through measures such as militarization of frontiers, construction of barriers, externalization of borders and information campaigns in areas of origin aiming to discourage potential refugees and migrants to take on the journey to Europe. Newly arrived population in receiving nations was and is faced with tightened asylum laws and decreasing support by authorities. Actions taken by authorities appear focused on mere treatment of symptoms while neglecting a systemic view of the problem landscape.

The complexity associated with the problem space of human security makes interventions problematic. Approaching such a problem space with a linear mindset can lead to interventions that result in unintended consequences as described in Masys [38] and depicted in Fig. 2. In short, although each of our individual actions may be rational, collectively, they may lead to failure '…as complexity and integration grow, attribution becomes more difficult and the unintended or unknown consequences of actions increase. Failure to understand or even acknowledge the nonlinear and highly complex nature of global linkages on every level of governance leads to growing weaknesses and can paralyze decision making [17].

A key requirement for operationalizing social innovation is to discover and reveal the dimensions of a system. Ramalingam and Jones [53: 10] argue that '… any country can be seen as a system made up of a number of interconnected

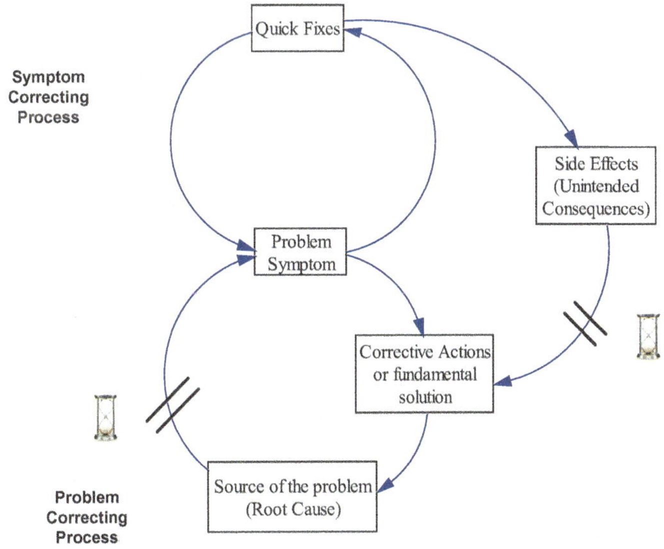

Fig. 2 Fixes that fail

elements and levels, from cities, towns, villages and communities all the way to individual citizens. In addition, human societies have social, cultural, physical, technical, economic and political dimensions [46], all of which are intricately intertwined. The degree of connectivity between these elements and dimensions has a crucial impact on how change happens within a given country'.

Social innovation in fact describes the entire process by which new responses to social needs are developed in order to deliver better social outcomes. This process is composed of four main elements:

- Identification of new/unmet/inadequately met social needs;
- Development of new solutions in response to these social needs;
- Evaluation of the effectiveness of new solutions in meeting social needs;
- Scaling up of effective social innovations [13].

Figure 3 highlights the process model and methodologies to support social innovation with regards to human security crisis. In scaling up social innovations in support of humanitarian relief associated with the migration crisis, a feedback loop is introduced that monitors weak signals and provides a continuous health check and response mechanism. Across the process model are the application of systems thinking, design thinking, developmental evaluation and reflective practices supporting the evolution and operationalization of social innovation.

A critical element is the requirement for a systems mapping of the human security landscape. This has been described in Masys [41]. A systems mapping reveals striking insights and solutions by helping to frame the problem across the human security landscape. As noted by Morgan [48]:

> 'First, people see the part to which they are connected but are largely unaware of the bigger system(s) that surrounds it. They miss their impact on others and others on them.
>
> Second, people tend to lack a time dimension. They see the present but not the past. They are intent on figuring out where a particular system should be in the future. They have some

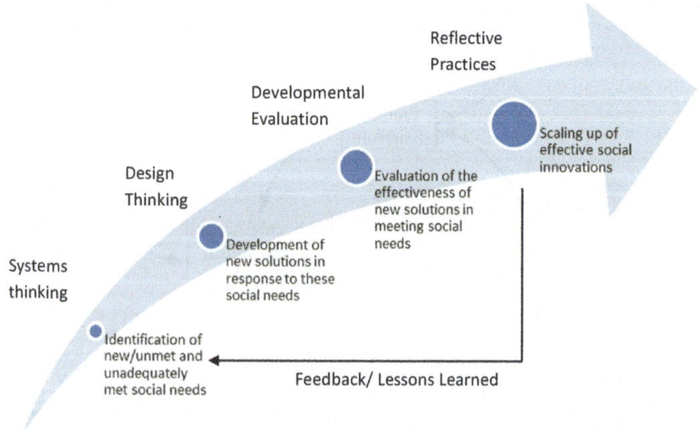

Fig. 3 Social innovation process

> interest in knowing where it is now. But they have little inclination to understand where it has been. They do not know the history of the present. '…everything that was left unprepared becomes a complex problem, and every weakness comes rushing to the forefront' [73].
>
> Third, participants miss - and in many cases mischaracterize - many of the key relationships that shape events.
>
> Fourth, people suffer from process blindness. They do not grasp the process dynamics, especially the deeper ones that are ongoing even within smaller systems. They suggest improvements which do not fit or even acknowledge the way the system actually works. They see individuals or events but not the processes of which they are a part'.

A social design and intervention strategy that focuses on the complex causal factors empowers people at local level to collaboratively invent solutions to economic and social problems. As noted by Patton [51] 'when a social innovation has a broad or durable impact, it will be disruptive and catalytic, meaning it will challenge the social system and social institutions that support the established order by affecting the fundamental distribution of power and resources and/or the basic beliefs that define the system, including the laws, policies and funding flows that govern it. ..To disrupt and change the broader system, there must eventually be a disruptive encounter with power, routine and beliefs'. This is a key paradigm shift in how we normally approach humanitarian crisis issues.

The concepts that are essential in problem framing within the context of human security are: perspectives, boundaries and entangled systems/interrelationships.

When operationalizing social innovation one must be cognizant regarding how boundaries drive how we see systems. Boundaries define who or what lies inside and what lies outside of a particular inquiry. The questions that shape our social intervention strategy are:

Perspectives

What are the different ways in which this situation can be understood?
How are these different understandings going to affect the way in which people judge the success of an endeavour?
How will it affect their behaviour and thus the behaviour of the system, especially when things go wrong from their perspective? With what results and significance?

Boundaries

What differences make a difference to the way in which a situation is understood or behaves?
Who or what is being excluded, marginalized or made a victim by the way in which this situation is being viewed or is operating?
What does this say about what is 'valued' by whom in this situation?
What are the consequences of boundary setting decision?

Interrelationships

What is the nature of the interrelationships within a situation?
What is the structure of these interrelationships?

What are the processes between them?
What are the patterns that emerge from those processes, with what consequences for whom? Why does this matter? To whom? In what context? [51]

Social innovation recognizes that 'action, reaction and interactions are dynamical, emergent, unpredictable and sometimes iterative. Operating under complex conditions requires ongoing adaptation' [51]. The dynamic complexity associated with human security landscape and humanitarian crisis interventions points to evaluation criteria that is rooted in building resilience and adaptability. Patton [51] argues that '...Under conditions of complexity, processes and outcomes are unpredictable, uncontrollable, and unknowable in advance. Uncertainties flow from turbulence in the environment, both evolutionary and transformational changes in the systems, and limits of knowledge'. This necessitates the requirement for reflective practices and organizational learning.

5 Self-organizing Organizations

While social innovation can be regarded as the fabric making up the dress of sustainable humanitarian assistance, self-organization (of groups) acts as the thread holding the entire construct together. Barton [2] describes self-organization as the development of a pattern in an open system without outside interference, as it progresses from chaos to order. Although numerous authors place self-organizing groups in a business setting [2, 27, 44, 61], a common advantage pointed out is the opportunity to act more independently and spontaneously in terms of coordination and reaction. Ramalingam et al. [54] contextualize more broadly: 'Self-organization is where macro-scale patterns of behaviour occur as the result of the interactions of individuals who act according to their own goals and aims based on their limited information and perspective on the situation.' Individuals, by adapting to a particular situation, interact and give rise to coping strategies, resulting in equilibrium between outside forces trying to organize the system and the reaction of agents within the self-organized system.

6 Fuelling Social Innovation: Self-organized Civil Society

In civil society, crises can prompt extraordinary willingness to help as could be observed in the wake of numerous events in the last decades (see for example Department of Manpower and Immigration, 1969 cited in Kaundert [31], Madokoro [36], Murber [49], Van Selm-Thorburn [72]. This enthusiasm is channelled by self-organizing relief efforts, such as Train of Hope in Europe during the events of 2015: a spontaneously established volunteer relief organization using social media to catalyze coordination and cooperation of volunteers for closing a gap in aid efforts authorities were leaving open [34, 59]. Civil society acts as a major resource in the management of humanitarian crises while being underestimated by authorities and legislation. By organizing itself spontaneously into effective organizations, civil society bears, in itself, the seed of contributing essentially to social innovation. It is therefore not to be seen as a mere entity to be regulated but rather as an equal partner for authorities [31]. While spontaneously established, self-organizing relief efforts reached equilibrium with conventionally founded organizations and authorities and succeeded in providing various services during the migration events of 2015.

In general, self-organization requires 'enabling approaches to empower actors at different levels of the team' [56]. This demands, as [56] conclude, the opportunity for involved individuals to choose self-organization as a tool for handling events as well as feedback mechanisms to derive lessons learned. Self-organization of relief in humanitarian assistance has, to various extents, always been part of relief efforts but with the background of the 2015 migration events in Europe it now appears to gain an identity of its own. Civil society's evolution, catalyzed by the digital age and social media, has leapt forward [6], not merely standing but also expanding its ground in the intricate, interconnected net of humanitarian assistance, public authorities and politics. Deriving and formulating lessons learned of self-organized efforts to facilitate their operationalization in the future constitutes a challenge as this type of aid cannot be commanded and controlled [31, 56] but only stimulated by an empowering and enabling legislative and socio-cultural environment.

In terms of systems thinking, this makes social innovation and self-organizing relief a wildcard, maybe more than other factors, as its influence on and it being influenced by other aspects varies with the particular set-up and development of self-organizing relief for the particular event. Organizational learning is a necessary requirement for self-organizing organizations as well as for public institutions: only through learning from the past, a potentially more efficient handling of the future becomes possible – a challenge that conventional, long established organizations often struggle with and that possibly shorter lived, atypical self-organizing groups have yet to master in their non-conventional structures.

7 Organizational Learning

Organizational learning is a critical element of strategic management and certainly resonates with the strategic interventions associated with social innovation in the human security landscape. The emergent notion of learning as the process of becoming a competent participant in a social and organizational process suggests that the traditional emphasis on the individual, the 'formal' team or the institutionalized organization as the key unit of analysis, may be rather less appropriate. Once we abandon or reject the assumption that organizations are homogeneous and functional units, we are left with the exciting and challenging task of making sense of, and describing, the work necessary for sustaining the process of collaboration.

We therefore ask how can co-ordination and learning be achieved in the presence of a variety of incoherent practices and thought systems? How can opportunities for innovation and change emerge in the course of this effort? And what are the conditions for boundary-crossing and boundary-restructuring? Learning may equally well be about how to negotiate current relationships, as it is about changing relationships.

8 Reflection on the Migration Crisis and Social Innovation

Apart from complex internal dynamics, the European external approach to address the destabilizing factors that cause political instability, economic deprivation and displacement seems to be inadequate and ineffective. Political, social and economic conditions in the major areas of origin of current migrants to Europe do not seem to be improving. Political efforts have been intensified to form a global coalition against the Islamic State after terror attacks in Paris in November 2015. Diplomatic efforts have been continuing to find a viable solution of Syrian crisis. However, political and social stability is unlikely to return in Syria in the immediate future. The crisis in Iran-Saudi relations over execution of shia cleric, Sheikh Nimr al-Nimr, would have implications on the Syrian crisis. European countries expressed their concern over growing hostilities between two Middle East countries. German Foreign Minister, Frank-Walter Steinmeier stated that Iran-Saudi Arabia tensions have resulted in strengthening the terrorists in the region and urged to act for resolving the military conflicts in the region. Economic deprivation and political suppression will continue to push people out of the African countries. Eritrea, South Sudan and Sudan remain highly violent; and poverty, drought and other natural calamities in Ethiopia, Kenya and other countries have exacerbated the humanitarian crises. After almost one and half decades of international engagement, peace and stability remain the major concerns in Afghanistan. The migrant crisis, therefore, has put internal as well as external challenges for the EU.

Opportunities for enabling, empowering approaches which may foster social innovation and self-organizing relief appear suffocated between discussions on refugee quotas for single EU countries, blame-shifting among various actors and 'a lack of self-reflection of the public sector' [18, 31]. Organizational learning, as well as communication between authorities and citizens, appear as definite challenges for established public structures. Bringing social innovation, self-organizing organizations and humanitarian aid into the picture, organizational learning from such an extensive, tightly-connected challenge would have to take place on various levels: the political large-scale, international level; the national political level and the grass-root organizational level. Issues of governmental and policy learning have been described as:

- using alibi policy as a symbolic means to prove that 'something is being done' without addressing the actual cause of the problem [4, 60]

- applying evidence-based ('instrumental') policy learning, which enables subjective selection of concerns to be addressed [11] and might result in 'cherry-picking' of evidence to back-up particular political positions [33]
- struggling for power while navigating among political processes to achieve a precise political goal [23, 43]

Likely, organizational learning in established relief organizations at grass-root level includes a political factor as well, due to direct or indirect financial sponsoring and resulting dependencies or because of political objectives an organization may be instrumentalized for. However, operational organizational learning will presumably have priority as it forms the core of grass-root organizations' business. Organizational learning in this context may have numerous definitions, the most fitting of which for this chapter's use being provided by [64] who define an organization's staff as being the most relevant knowledge repository. The key word in the above paragraph, however, is 'established relief organizations', meaning that permanent staff—voluntary or not—is available and that the organization in question is never completely shut down.

Self-organizing organizations, formed spontaneously and often temporarily, do not have permanent staff serving as 'an archive of knowledge' and run the risk of 're-inventing the wheel' or repeating other self-organizing groups' mistakes from the past, as their team composition varies more strongly than in established organizations. Here, the particular importance of public actors comes to the fore: as a top-down management of self-organized relief efforts is inherently impossible, it is policy makers' turn to distill lessons learned from relevant events, including interaction with self-organized groups, in order to be able to communicate needs and requirements from conventional relief actors to self-organized groups clearly and from the beginning on. This way, interfaces among traditional and non-traditional relief and authorities can be defined and established, and cooperation can be reinforced without relevant lessons learned being lost.

9 Conclusion

Facilitating the problem framing and solution navigation to support humanitarian relief pertaining to mass migration requires a better understanding of the complex problem space. Across the humanitarian relief domain from vulnerability analysis to building resilience, '...The most common source of mistakes is not the failure to find the right answers. It is the failure to ask the right questions. Nothing is more dangerous in business (or human security) than the right answers to the wrong questions' [55: 23]. Problem framing thereby emerges as the first step in support of social innovation interventions. As discussed, the complexity of the problem space in terms of interdependencies and interconnectivity across the drivers of mass migration necessitates a systems perspective. Disaster forensics described in [41] is a key approach to open the black box of mass migration drivers and interventions.

Social innovation in fact describes the entire process (as shown in Fig. 3) by which new responses to social needs are developed in order to deliver better social outcomes. Facilitating and fueling the emergence of self-organizing systems to support humanitarian assistance is a key requirement in navigating the solution space. This is about leveraging 'reflective practices' [42] in support of organizational learning and social innovation.

References

1. Baker A (2015) How climate change is behind the surge of migrants to Europe, TIME, Sep 07. Available online at http://time.com/4024210/climate-change-migrants/. Accessed 04 Apr 2017
2. Barton S (1994) Chaos, self-organization, and psychology. Am Psychol 49(1):5–14
3. Bilak A, Cardona-Fox G, Ginnetti J, Rushing EJ, Scherer I, Swain M, Walicki N, Yonetani M (2016) Global report on internal displacement. IDMC. http://www.internal-displacement.org/assets/publications/2016/2016-global-report-internal-displacement-IDMC.pdf
4. Bryson J, Crosby B, Middleton Stone M (2006) The design and implementation of cross-sector collaborations: propositions from the literature. Public Administration Review 66 (Special Issue):44–55. Available online at http://www.jstor.org/stable/4096569. Accessed 26 Mar 2016
5. Bundespressekonferenz (2015) Sommerpressekonferenz von Bundeskanzlerin Merkel, 31.8.2015. Available online at https://www.bundesregierung.de/Content/DE/Mitschrift/Pressekonferenzen/2015/08/2015-08-31-pk-merkel.html. Accessed 3 Jun 2017
6. Cairns R (2016) How social media, open source, and self-organisation are changing governance. The impact of migration from Syria. J Assoc Manage Educ Dev, e-O&P, 23(1, Spring: 12–20. Reproduced by kind permission of AMED www.amed.org.uk
7. Castles S, deHein H, Miller MJ (2014) The age of migration. international population movements in the modern world, 5th edn. Basingstoke: Palgrave Macmillan
8. Collier P (2013) Exodus: how migration is changing our world. Oxford University Pres, Oxford
9. deHaas H (2007) Turning the tide? Why development will not stop migration. Dev Change 38 (5):819–841
10. deHaas H (2008) The myth of invasion. The inconvenient realities of African migration to Europe. Third World Quartely 29(7):1305–1322
11. Duncan S (2009) Policy transfer: theory, rhetoric and reality. Policy Politics 37(3):453–458. Available online at http://www.ingentaconnect.com.ezproxy4.lib.le.ac.uk/search/articleoption1=tka&value1=policy+transfer+theory+rhetoric&pageSize=10&index=1. Accessed 04 Jul 2016
12. Edson R (2008) Systems thinking. Applied: a primer. ASysT Institute. http://www.anser.org/docs/systems_thinking_applied.pdf
13. EU (2013) Guide to social innovation. Feb 2013
14. Flahaux M-L, deHaas H (2016) African Migration: trends, patterns, drivers. Comparative Migration Studies 4(1):1–25. https://doi.org/10.1186/s40878-015-0015-6 Accessed 08 Oct 2017
15. Foresight: Migration and Global Environmental Change (2011) Final Project Report The Government Office for Science, London. http://podcasts.sciencforthepeople.ca/Castles_Uploads/UK%20Govt%20Scientist-migration-and-global-environmental-change.pdf

16. Freeman O (2014) Report on civil protection in diverse societies: migrants, asylum seekers and refugees inthe context of major risks prevention and management. Council of Europa, Strasbourg
17. Goldin I, Mariathasan M (2014) The butterfly defect: how globalization creates systemic risks and what to do about it. Princeton University Press, Princeton, NJ
18. Gratz W (2016) Das Management der Flüchtlingskrise. NWW Verlag, Wien
19. Gray CL, Mueller V (2012) Natural disasters and population mobility in Bangladesh. Proc Natl Acad Sci USA 109(16):6000–6005
20. Hagen-Zanker J, Mallett R (2015) Journeys to Europe: the role of policy in migrant decision-making. https://www.odi.org/sites/odi.org.uk/files/odi-assets/publications-opinionfiles/10297.pdf
21. Hagen-Zanker J, Mallett R (2016) On the limits of deterrence. Humanitarian Exchange 67:14–16
22. Hamood S (2006) African transit migration through Libya to Europe: the human cost. FMRS, AUC, Cairo
23. Heclo H (1974) Modern social politics in Britain and Sweden: from relief to income maintenance. Yale University Press, New Haven
24. Holborn LW (1956) The International Refugee Organization: a specialized agency of the United Nations, its history and work, 1946–1952. Oxford University Press, London
25. HPN (2016) Special feature: refugees and vulnerable migrants in Europe. Humanitarian Exchange Number 67 September 2016 http://odihpn.org/wp-content/uploads/2016/09/HE-67-FINAL.pdf
26. Illibauer C, Natschläger C (2017) Towards flexible business processes by supporting self-organizing groups. In: Piazolo F, Brehm L, Geist V, Schmidt R (eds) Information systems management and engineering. Springer, Cham, pp 57–68
27. Institute for the Study of Diplomacy (2017) New challenges to human security: environmental change and human mobility. Georgetown University, Working Group Report, Washington, DC
28. IOM (2004) Essentials of migration management: volume one: migration management foundations. International Organisation for Migration, Geneva
29. IPCC (2014) Summary for policymakers. In: Field CB, Barros VR, Dokken DJ, Mach KJ, Mastrandrea MD, Bilir TE, Chatterjee M, Ebi KL, Estrada YO, Genova RC, Girma B, Kissel ES, Levy AN, MacCracken S, Mastrandrea PR, White LL (eds.), Climate change 2014: impacts, adaptation, and vulnerability. part a: global and sectoral aspects. contribution of working Group II to the Fifth assessment report of the intergovernmental panel on climate change. Cambridge University Press, Cambridge, pp 1–32
30. Jackson MC (2003) Systems thinking: creative holism for managers. John Wiley & Sons Ltd, West Sussex, England
31. Kaundert M (2016) Which factors interfere with learning lessons from refugee crisis management? The case of Austria with a focus on refugee accommodation [Master's dissertation for University of Leicester, unpublished]
32. Krabbe J (1912) Toelating en uitzetting van vreemdelingen, Leiden, cited in Coleman N (2009) European readmission policy: third country interests and refugee rights. Koninklijke Brill NV, Leiden
33. Levy J (1994) Learning and foreign policy: sweeping a conceptual minefield. International Organization 48(2):279–312. Available online at http://www.jstor.org/stable/2706933. Accessed 04 Jul 2016
34. Lockwood S, Weaver CK, Debashish M, Simpson M (2015) The self organising of youth volunteers during the Rena oil spill in New Zealand. NZ J Mar Freshwat Res 50(1):28–41. https://doi.org/10.1080/00288330.2015.1063515 Accessed 28 May 2017
35. Lutterbeck D (2006) Policing migration in the mediterranean. Mediterr Politics 11(1):59–82
36. Madokoro L (2009) Good Material: Canada and the Prague Spring Refugees. Refuge 26(1): 161–171. Available online at https://refuge.journals.yorku.ca/index.php/refuge/article/viewFile/30618/28132. Accessed 16 Aug 2017

37. Masys AJ (2006) Understanding climate change through modelling and simulation: a case for verification, validation and accreditation. Published in the proceedings of the Climate Change Technology Conference, Ottawa, Canada. 9–12 May 2006
38. Masys AJ (2013) Human security—A view through the lens of complexity. In Gilbert T, Kirkilionis M, Nicolis G (eds), Proceedings of the european conference on complex systems 2012, Springer Proceedings in Complexity, pp 325–335
39. Masys AJ (ed) (2014) Networks and network analysis for Defence and Security. Springer Publishing
40. Masys AJ (ed) (2015) Applications of systems thinking and soft operations research in managing complexity. Springer Publishing
41. Masys AJ (ed) (2016) Disaster forensics: understanding root cause and complex causality. Springer Publishing
42. Masys AJ, Ray-Bennett N, Shiroshita H, Jackson P (2014) High impact/low frequency extreme events: enabling reflection and resilience in a hyper-connected World. 4th International Conference on Building Resilience, 8–11 September 2014, Salford Quays, United Kingdom. Procedia Economics and Finance 18 (2014) 772–779
43. May P (1992) Policy learning and failure. J Public Policy 12(4):331–54. Available online at http://www.jstor.org/stable/4007550. Accessed 05 Jul 2016
44. McClure BA (2005) Putting a new spin on groups. The science of Chaos. Lawrence Erlbaum Associates, New Jersey
45. Metcalfe-Hough V (2015) The migration crisis? Facts, challenges and possible solutions. ODI Briefing. https://www.odi.org/sites/odi.org.uk/files/odi-assets/publications-opinion-files/9913.pdf
46. Mittleton-Kelly E (2003) Ten principles of complexity and enabling infrastructures in complex systems and evolutionary perspectives of organisations: the application of complexity theory to organizations. Elsevier Press, London
47. Moore M, Westley F (2011) Surmountable chasms: networks and social innovation for resilient systems. Ecology and Society **16**(1):5. [online] URL: http://www.ecologyandsociety.org/vol16/iss1/art5/
48. Morgan P (2005) The idea and practice of systems thinking and their relevance for capacity development. European Centre for Development Policy Management. March 2005
49. Murber I, Fónagy Z (eds) (2006) Die Ungarische Revolution und Österreich 1956. Czernin, Wien
50. Myers N (2002) Environmental refugees: a growing phenomenon of the 21st century. Philos Trans R Soc B Biol Sci 357:609–613
51. Patton MQ (2011) Developmental evaluation: applying complexity concepts to enhance innovation and use. Guilford Press, New York
52. Pauver B, Twigg J, Sagramola S (2016) Migrants, refugees, asylum seekers: inclusion in disaster preparedness and response. Council of Europe, Lisbon
53. Ramalingam B, Jones H (2008) Exploring the science of complexity: ideas and implications for development and humanitarian efforts. Working Paper 285 ODI
54. Ramalingam B, Jones H, Reba T, Young J (2008) Exploring the science of complexity: ideas and implications for development and humanitarian effort, Working Paper 285. Overseas Development Institute, London
55. Ramirez R, Wilkinson A (2016) Strategic reframing: the oxford scenario planning approach. Oxford University Press, Oxford
56. Rittell HW, Weber MM (1973) Dilemmas in general theory of planning. Policy Sci 4:155–169
57. Schuler K (2016) Merkels neues wording. Zeit Online, Sept. 19. Available online at http://www.zeit.de/politik/deutschland/2016-09/fluechtlingspolitik-angela-merkel-cdu-horst-seehofer-csu/komplettansicht. Accessed 3 Jun 2017
58. Senge P (1990) The fifth discipline: the art and practice of the learning organization. Doubleday Currency, New York

59. Simsa R, Auf M, Bratke S-M, Hazzi O, Herndler M, Hoff M, Kieninger J, Meyer M, Mourad M, Pervan Al-Soqauer I, Rameder P, Rothbauer J (2016) Beiträge der Zivilgesellschaft zur Bewältigung der Flüchtlingskrise – Leistungen und Lernchancen. Kompetenzzentrum für Nonprofit Organisationen und Social Entrepreneurship, Vienna University of Economics and Business, Wien
60. Speer S (2011) Organised governmental learning: vocational education and training practices between peer review and peer learning. Research In Comparative and International Education 6 (3):330–340. Available online at http://rci.sagepub.com.ezproxy4.lib.le.ac.uk/content/6/3/330.full.pdf+html. Accessed 04 Jun 2016
61. Stacey RD (1992) Managing the unknowable. Strategic boundaries between order and chaos in organizations. Jossey-Bass, San Francisco
62. Stalker P (2002) Migration trends and migration policy in Europe. Int Migrat 40(5):151–179
63. Taylor I, Masys AJ (2017) Humanitarian. In: Masys AJ, Lin L (eds) Asia/Pacific security challenges: managing black swans an persistent threats. Springer Publishing
64. Toft B, Reynolds S (2005) Learning from disasters: a management approach. Palgrave Macmillan, Basingstoke
65. Tranos E, Gheasi M, Nijkamp P (2015) International migration: a global complex network. Environ Plan 42:4–22
66. UNDP (2015) Syria Alienation and Violence: Impact on Syrian Crisis Report. https://reliefweb.int/sites/reliefweb.int/files/resources/alienation_and_violence_impact_of_the_syria_crisis_in_2014_eng.pdf
67. UNDP (2016) Global project for electoral cycle support II, UNDP
68. UNDP (2017) Guidance note. Municipalities and people on the move. Cities' development policies for successful local management of migration and displacement. UNDP, New York
69. UNHCR (2014) World at War. http://www.unhcr.org/556725e69.pdf
70. UNODC (2006) Organized crime and irregular migration from Africa to Europe. United Nations Office on Drugs and Crime
71. Urry J (2002) The global complexities of September 11th. Theory, Culture and Society 19(4):57–69
72. Vanselm-Thorburn J (1998) Refugee protection in Europe: lessons of the Yugoslav Crisis. Kluwer Law International, The Hague
73. Weick KE, and Sutcliffe KM (2007) Managing the unexpected: resilient performance in an age of uncertainty, 2nd edn. Wiley, San Francisco, CA
74. Westley F, Zimmerman B, Patton M (2007) Getting to maybe: how the world is changed. Random House of Canada, Toronto
75. Whaples R (1995) Where is there consensus among american economic historians? The results of a survey on forty propositions. J Econ History 55(1):139–154. Available online at https://doi.org/10.1017/S0022050700040602. Accessed 3 Jun 2017

Complexity and Unintended Consequences in a Human Security Crisis: A System Dynamic Model of the Refugee Migration to Europe

Ivan Taylor and Anthony J. Masys

Abstract As described by UNHCR (The sea route to Europe: The Mediterranean passage in the age of refugees, 2015 [20]), 'Europe is living through a maritime refugee crisis of historic proportions. Its evolving response has become one of the continent's defining challenges of the early 21st century, with long-lasting implications for humanitarian practice, regional stability and international public opinion'. UNHCR's figures show that over one million people had reached Europe across the Mediterranean, mainly to Greece and Italy, in 2015. Of these, over 3700 were missing, believed drowned. This represents a regional human security crisis. Sagan (The limitation of safety—organizations, accidents and nuclear weapons. Princeton University Press, Princeton, New Jersey [17]) argues that 'Things that have never happened before happen all the time'. The mass migration of refugees should not have come as a complete surprise. The mass migration in Europe, North Africa and Middle East derives from a wicked problem space and is linked to humanitarian challenges elsewhere. Thus, the flight to European shores reflected not only the pull of greater long-term security in Europe, but also the failure of the international humanitarian community to meet basic needs in other places (Special feature: refugees and vulnerable migrants in Europe, 2016 [12]). This chapter explores the complexity of the refugee and humanitarian crisis in Europe and North Africa through the application of a System Dynamics model. Given the unintended consequences associated with policies and intervention strategies, the System Dynamics model examines the volatility of such interventions on migration.

Keywords Refugees · Humanitarian crisis · System Dynamics
Complexity · Systems thinking

A. J. Masys
University of South Florida, Tampa, FL, USA
e-mail: tmasys@health.usf.edu

I. Taylor (✉)
Policy Dynamics Inc., New Hamburg, ON N3A 2G5, Canada
e-mail: ivanwtaylor@gmail.com

A. J. Masys (ed.), *Security by Design*, Advanced Sciences and Technologies for Security Applications, https://doi.org/10.1007/978-3-319-78021-4_6

1 Introduction

The 1994 Human Development Report by the United Nations Development Programme (UNDP) defines human security as 'freedom from fear and freedom from want' highlighting seven key components (economic, food, health, environmental, personal, community and political security). These seven components do not exist as isolated domains but are rather interdependent and interconnected. This has a significant effect on the transmission of risk across borders. Masys et al. [9] argue that 'with our hyper-connected world underpinned by hyper or hybrid-risks, the impact of unexpected events such as floods, earthquakes, financial crisis, and cyber-attacks has revealed the fragility and vulnerabilities that lie within the social/technological/economic/political/ecological interdependent systems'. 'Global risks cannot be seen in isolation' [23]. The threats to human security are multiple, complex and interrelated in often mutually reinforcing ways. This certainly resonates with many of the drivers shaping the mass migration and contributing to the humanitarian crisis. The complex landscape associated with human security requires a holistic, systemic approach to understanding the underlying hazards, risk identification and management practices. Masys [8] describes the complex risk landscape that characterizes the Horn of Africa famine of 2011 and the subsequent humanitarian crisis. In so doing Masys [8] describes how risks are not confined to borders but have spill-over effects. The resulting humanitarian crisis thereby is characterized by a transboundary nature and complex causality [7], highlighting how human security challenges can emerge from feedback mechanisms, interdependencies and from linkages that are latent and often unrecognized (latent pathogens as described by Reason [13]).

Globally, we are in the midst of the highest levels of human displacement on record. According to the UNHCR [22] 'an unprecedented 65.6 million people around the world have been forced from home. Among them are nearly 22.5 million refugees, over half of whom are under the age of 18 …nearly 20 people are forcibly displaced every minute as a result of conflict or persecution'.

Of note is that in 2015, over 1,000,000 people had reached Europe across the Mediterranean, mainly to Greece and Italy. Of these, over 3700 were missing, believed drowned [2]. This represents a regional human security crisis. The human security crisis can be considered a wicked problem. The term wicked problem is applied to issues or challenges that defy and defeat traditional linear and analytical problem solving ([1, 15]. Wicked problems tend to be impossible to completely define, do not have clear end-points where they can be considered 'finished', are heavily dependent on stakeholder judgments and interpretation, defeat 'trial and error' approaches because each intervention changes the problem, and are unique in that there are no solution principles that can be drawn from other wicked problems [11].

An interesting characteristic of wicked problems is their dynamic nature. As described by Ritchy [14], wicked problems 'are sets of complex, interacting issues evolving in a dynamic social context. Often, new forms of wicked problems emerge *as a result* of trying to understand and treat one of them'. These characteristics lend themselves to the application of such problem structuring methods as System Dynamics [19].

2 System Dynamics

The fundamental ideas of System Dynamics were developed by Jay Forrester at MIT in the 1960s. The premise of Systems Dynamics is that the behaviour of systems, at whatever level, results from underlying structures of flows, delays, information and feedback relations. The outputs produced emerge as patterns of growth, decay, oscillation or chaotic behaviour that is often counter-intuitive [10].

2.1 Feedback

One of the central concepts behind the System Dynamics paradigm is feedback. Feedback can be thought of the situation in which information flows through the system and eventually returns in some form to its place of origin, creating a causation loop that can change future behaviour. Therefore, System Dynamics takes an endogenous view of systems as causally closed. There can be many causal loops in a model of a system. However, they come in two basic types: reinforcing loops in which a change in the variable travels through the system and eventually returns to cause a change in the same variable in the same direction; and, balancing loops in which a change in the variable travels through the system and eventually returns to cause a change in the same variable in the opposite direction. As their names suggest, reinforcing loops can cause the system to go out of control while balancing loops generally cause the system to settle into an equilibrium state.

2.2 Influence Diagrams

The feedback structure of a System Dynamics model is often portrayed using an influence diagram. The goal of an influence diagram is to communicate the complex nature of the systems model in relatively simple terminology. A subset of the system elements are shown in the diagram and they are connected by directed links or arrows (see Fig. 3). These links show the immediate influence of one system element on another but also shows the direction of that influence.

The connections between the elements can be traced using these relationships to determine if the loops are reinforcing or balancing. In principle, there can be many loops through the same system element, some reinforcing and some balancing. Therefore, it can be hard to determine the nature of the behaviour of the system elements simply from the influence diagram. This is one reason that a simulation model is necessary. Sometimes, the simulation model is based on the influence diagram which is generally more accessible to the stakeholders involved in the policy analysis. At other times, the simulation model is developed first by the analysts and then the influence diagram is used to communicate the nature of the simulation model to the stakeholders of the policy analysis so they can better understand the recommendations.

2.3 Simulation Models

A System Dynamics computer simulation model is usually developed in special purpose software. The special purpose System Dynamics software generally utilizes a graphical user interface that includes stock variables, flow variables and auxiliary variables. Stock variables are shown in the graphical user interface as boxes and represent states of the system. That is, concrete aspects of the system that can be seen and measured. Flow variables are shown in the graphical user interface as inflows and outflows connected to the boxes. These inflows and outflows are the only way in which the Stock variables change over time. The Flow variables can only be measured as changes over time. The Auxiliary variables are shown as labels that are connected to the Stock and Flow variables and each other using directed arrows. Auxiliary variables are generally considered information that is processed by the actors in the system.

One of the central features of a System Dynamics model is "unit consistency". That is, all of the variables must be identified in specific units and these units must be used in a manner that is consistent. For example, a Stock variable population would be measured in "people" and a Flow variable such as births or deaths would be measured in "people per year" or using some other unit of time. And a birth rate or death rate Auxiliary variable would be measured in "people per person per year". So by multiplying the birth rate in "people per person per year" by the population in "people", one would obtain the inflow of births in terms of "people per year"

In general, most System Dynamics models are deterministic rather than probabilistic and contain continuous variables rather than discrete variables. Another characteristic feature of a System Dynamics model is that the output is displayed in behaviour over time graphs. In these graphs, the variables of interest are shown on the vertical axis with time shown on the horizontal axis using continuous curves.

2.4 Mathematical Implementation of the System Dynamics Model

As described on the System Dynamics Society website [16], the System Dynamics computer simulation model is mathematically a system of highly coupled, nonlinear, first-order differential equations,

$$\frac{d}{dt}\mathbf{x}(t) = \mathbf{f}(\mathbf{x}, \mathbf{p})$$

where x is a vector of state variables, p is a set of parameters, and f is a nonlinear vector-valued function. This system of equations is generally solved using Euler integration by simulating time over discrete intervals of length 'dt' and stepping through the system of equations one 'dt' at a time. Each state variable is computed from its previous value and its net rate of change $x'(t)$: $x(t) = x(t - dt) + dt * x'(t - dt)$. The value of 'dt' is arbitrary but is selected to be small enough to have no discernible effect on the patterns of dynamic behavior in the model.

2.5 *Application to the Refugee Crisis*

It was felt that a System Dynamics model would be a good way to analyze various policies in the refugee crisis that is currently ongoing in the Middle East and Africa in which citizens, who are living in countries suffering from famine and civil war, are seeking to migrate to the more peaceful and developed nations of Europe.

2.6 *Discussion*

The humanitarian crisis and mass migration in Europe [6] is a human security issue that has resulted in both regional and global impacts. Between January 2015 and February 2016, over 1.1 million people made their way to the EU, escaping conflict and poverty in their countries and seeking a better and safer life [4].

Clayton and Holland [2] show that in 2015, in excess of 1,000,000 people had reached Europe across the Mediterranean, mainly to Greece and Italy. Of these, over 3700 were missing, believed drowned.

"From one death for every 269 arrivals in 2015, in 2016 the likelihood of dying has spiraled to one in 88," Spindler said [21]. On what's known as the Central Mediterranean route between Libya and Italy, "the likelihood of dying is even higher, at one death for every 47 arrivals," he said. But the crossing can be treacherous, with too many migrants—some fleeing war or persecution, others seeking a better life—crammed into what are often barely seaworthy boats [21].

The UNHCR [20] report highlighted six major findings regarding migration into Europe:

1. The majority of those taking the sea route to Europe are refugees, and their numbers continue to rise rapidly. Most people arriving by sea are fleeing from war, conflict or persecution at home, as well as deteriorating conditions in many refugee-hosting countries. EU States have a clear responsibility to offer them protection, and an obligation along with others to rescue people in danger at sea.

2. The number of deaths at sea rose to record levels in April 2015, then dropped significantly in May and June. While many factors contributed to the recent decline, improved European-led search-and-rescue operations beginning in May have had an immediate and positive impact. Yet the peak months still lie ahead.
3. There has been a major increase in refugees and migrants taking the 'eastern Mediterranean route' from Turkey to Greece. More than 85% of those arriving in Greece are from countries experiencing war and conflict, principally Syria, Afghanistan, Iraq and Somalia. From Greece, most move onwards across the Balkans to western and northern Europe. Italy remains the primary destination for Eritreans, Somalis and other people from sub-Saharan Africa.
4. As arrivals increase, reception capacity and conditions remain seriously inadequate. While conditions of reception in Italy vary a great deal, serious systemic gaps remain in Greece. The former Yugoslav Republic of Macedonia and Serbia collectively offer fewer than 3000 places of reception, significantly below the levels of arrivals (19,000 arrivals in the first weeks of June alone). This impacts people with special needs, including unaccompanied and separated children, increasing their vulnerability and risk of exploitation. This is an emergency situation, which requires urgent attention and far greater support for efforts to handle new arrivals. If this situation remains unaddressed, onward movement of refugees and migrants is likely to continue on a significant scale.
5. The number of refugees and migrants entering the western Balkans from Greece has already dramatically increased since the beginning of June, with over 1000 people entering every day. They face serious humanitarian and protection challenges linked to the hardship of the journey, the abuses of smugglers and criminal gangs, and the increasing tightening of the borders.
6. Countries of origin and the international community at large need to do better at preventing and resolving conflicts. Transit countries need to develop their asylum systems, including reception arrangements and identification processes.

The drivers of migration are well articulated in Foresight [3]. As depicted in Fig. 1, these drivers include: political, demographic, economic, social and environmental factors.

Intervention strategies described in Masys [7] and Kaundert and Masys [5] often do not meet the requirements of the problem space. This stems from the complexity and wickedness of such humanitarian crises. As described in Masys [7] and Kaundert and Masys [5] unintended consequences can emerge from intervention strategies creating greater 'heartache' as shown by the 'fixes that fail' influence diagram (Fig. 2).

This refugee crisis is not new. As noted by Sagan [17] 'Things that have never happened before happen all the time'. Salama et al. [18: 1804] recall 'the flight of 500,000–800,000 Rwandan refugees into the North Kivu region of eastern Zaire in 1994. … An average crude mortality rate of 20–35 deaths per 10,000 people per day was reported during the first month after the influx, during which an estimated 50,000 people died. These rates, which resulted from serial epidemics of cholera and shigellosis, were two-fold to three-fold the highest rates previously reported

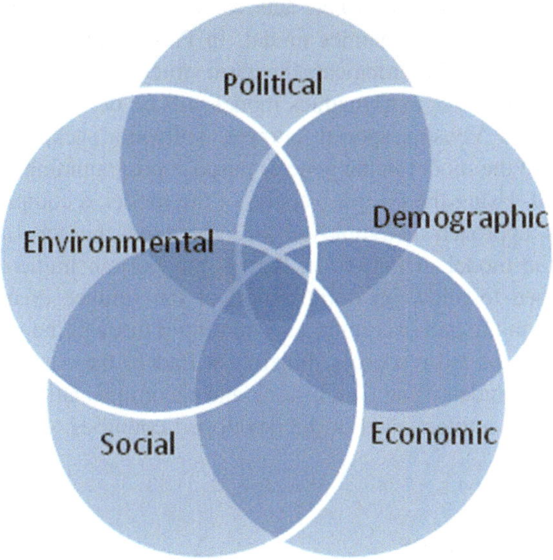

Fig. 1 Drivers of mass migration (themes derived from Foresight [3]

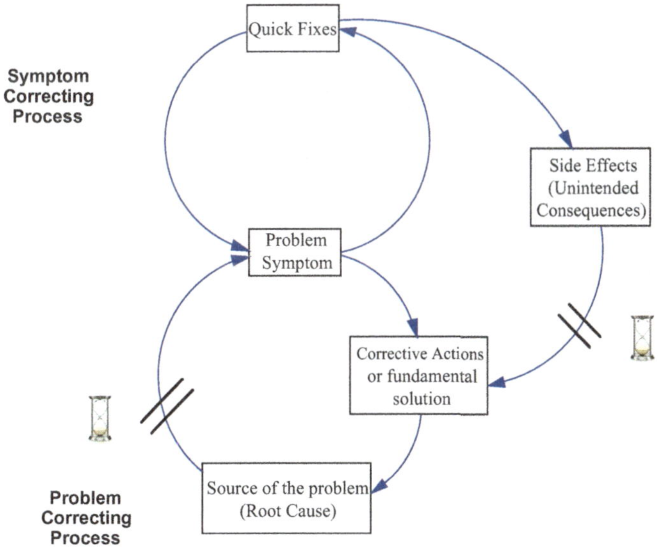

Fig. 2 Fixes that fail—unintended consequences

among refugees or internally displaced people (16–18 deaths per 10,000 people per day in Baidoa, Somalia in 1992)'. This crisis highlights a major concern in such refugee flights: the emergence of a public health crisis.

The humanitarian crisis described acts as the context and backdrop for the development of a System Dynamics model. In Part 3, interrelated influence diagrams were developed to communicate the basic structure and feedback loops in the model. In Part 4, the System Dynamics model will be described using a series of diagrams from the Vensim special purpose software package. In Part 5, an implementation of the model in the general purpose programming language Fortran will be described. Using the Fortran version of the model, a comprehensive design of experiments approach is employed to determine the most sensitive input parameters in the model in Part 6. In Part 7, the Fortran implementation of the model is modified to allow an optimization of the refugee crisis to reduce the avoidable deaths in the native country to a minimum through the most efficient and effective allocation of aid resources. Finally, in Part 8, the concluding material is provided which includes some conclusions and recommendations related to relief policies that can be implemented in the developed countries.

3 Influence Diagrams

We have developed two interrelated influence diagrams to describe many of the relationships involved in a refugee crisis. These influence diagrams express the relationships as items connected by causation arrows. A causation arrow means that a change in one item causes a change in another item. The arrow shows the direction of the change relationship. A change in the item at the tail of the arrow will cause a change in the item at the head of the arrow (i.e. one item will influence the other item). The arrows between two elements can be labeled with either an 's' or an 'o'. The 's' means that a change in the element at the tail of the arrow will cause a change in the element at the head of the arrow in the same direction. An 'o' means that a change in the element at the tail of the arrow will cause a change in the element at the head of the arrow in the opposite direction.

Figure 3 shows an influence diagram intended to show the movement of refugees between countries. The documentation of this influence diagram is provided at https://github.com/ivanwtaylor/Refugee-Crisis. Key parameters reflected in the influence diagram include the effects of health/disease, violence, poverty, population and humanitarian aid.

Figure 4 shows the influence diagram related to the distribution of humanitarian aid. We will describe the loops in this diagram in detail below. Notice that the humanitarian aid is connected to the movement of refugees in Fig. 3 using various shadow variables which are shown in light grey and surrounded by ' <>' brackets.

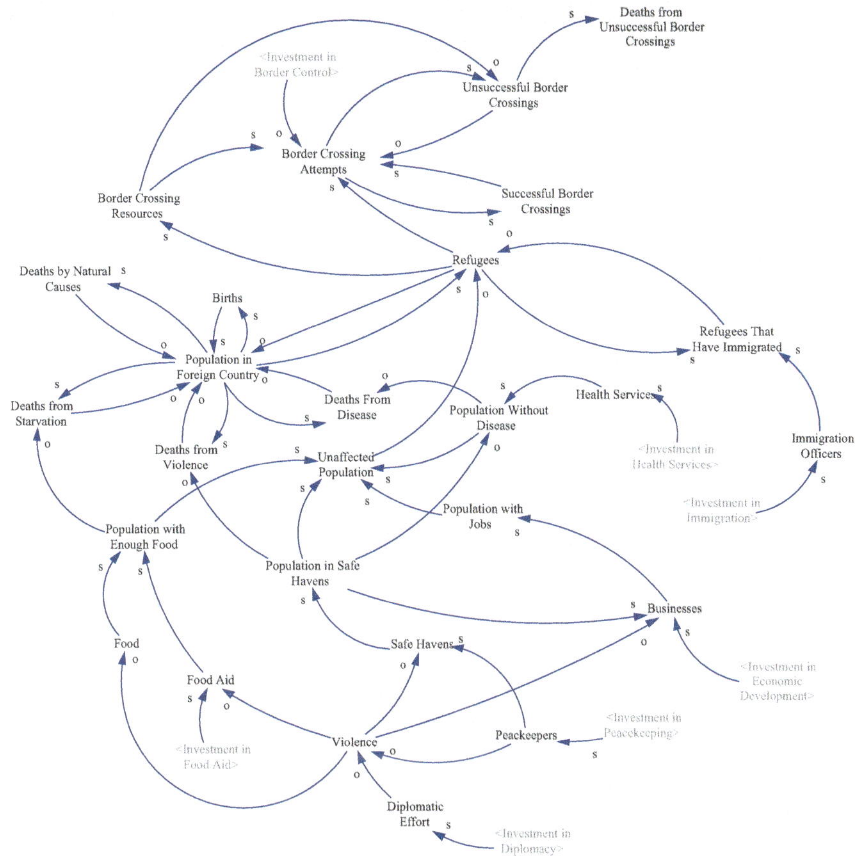

Fig. 3 The movement of refugees between two countries

3.1 Allocation of Investment in Humanitarian Relief

We assume the Investment Available is based on the number of vulnerable people. The Available Investment is then allocated to the various types of relief: Food Aid, Health Services, Peacekeeping, Diplomacy, Immigration, Economic Development and Border Control. The investment in these relief efforts is influenced in the 'same' direction by the Available Investment and in turn, the Investment in Food Aid, Health Services, Peacekeeping, Diplomacy, Immigration, Economic Development and Border Control will reduce the Available Investment. Therefore, they influence Available Investment in the 'opposite' direction.

We have a circular relationship, in that increased Available Investment will lead to increased Investment in Food Aid, Investment in Health Services, Investment in Peacekeeping, Investment in Diplomacy and Investment in Economic Development. These allocations will increase the Population with Enough Food,

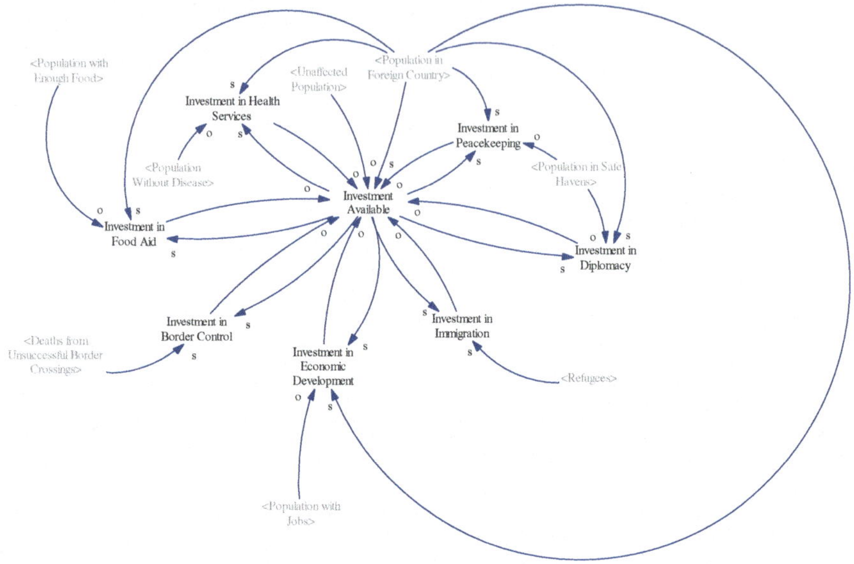

Fig. 4 The allocation of investment in the relief effort

the Population without Disease, the Population in Safe Havens and the Population with Jobs respectively which, in turn, will increase the Population Unaffected by famine, disease, war and poverty which will reduce the pressure to increase the Available Investment.

The Investment in Food Aid is determined by the Population in the Foreign Country minus the Population with Enough Food. So, the Investment in Food Aid is influenced by the Population in the Foreign Country in the 'same' direction and by the Population with Enough Food in the 'opposite' direction. Notice in Fig. 3, that the Investment in Food Aid influences Food Aid in the 'same' direction which influences the Population with Enough Food in the 'same' direction. So, we have a circular effect. The more Investment in Food Aid, the more Food Aid and the greater the Population with Enough Food but the greater the Population with Enough Food, the less the Investment in Food Aid.

The Investment in Health Services is determined by the Population in the Foreign Country minus the Population without Disease. So, the Investment in Health Services is influenced by the Population in the Foreign Country in the 'same' direction and by the Population without Disease in the 'opposite' direction. Notice in Fig. 3, the Investment in Health Services influences Health Services in the 'same' direction which influences the Population without Disease in the 'same' direction. So, we have another circular effect. The more Investment in Health Services, the more Health Services and the greater the Population without Disease but the greater the Population without Disease, the less the Investment in Health Services.

Both the Investment in Peacekeeping and the Investment in Diplomacy are determined by the Population in the Foreign Country minus the Population in Safe Havens. That is, the Investment in Peacekeeping and the Investment in Diplomacy are influenced in the 'same' direction by the Population in the Foreign Country and influenced in the 'opposite' direction by the Population in Safe Havens. Again, we have circular relationships. Increasing the Investment in Peacekeeping will increase the number of Peacekeepers which, in turn, will increase the number of Safe Havens. The increased number of Safe Havens will increase the Population in Safe Havens which will reduce the pressure to invest in peacekeeping. Similarly, an increasing Investment in Diplomacy will cause an increase in Diplomatic Effort which will decrease the amount of Violence which in turn will increase the number of Safe Havens and thereby the Population in Safe Havens which will reduce the pressure to invest in diplomacy.

The more refugees, the more Investment in Immigration that is demanded. So, the Investment in Immigration is influenced by the number of Refugees in the 'same' direction. Here the circular relationship means that the Investment in Immigration will cause an increase in the number of Immigration Officers and speed up the process of immigration for refugees which will decrease the number Refugees and thereby reduce the pressure to invest in immigration.

The Investment in Economic Development is based on the Population in the Foreign Country minus the Population with Jobs. That is, the Population in the Foreign Country influences the Investment in Economic Development in the 'same' direction and the Population with Jobs influences the Investment in Economic Development in the 'opposite' direction. The circular relationship here is that the increased Investment in Economic Development will cause an increase in the number of Businesses which will increase the Population with Jobs which will reduce the pressure to invest in economic development.

Finally, we have Investment in Border Control. Investment in Border Control is influenced by the number of Deaths from Unsuccessful Border Crossings in the 'same' direction. The circular relationship is that Investment in Border Control will try to reduce the number of Border Crossing Attempts which reduce the number of Unsuccessful Border Crossings which will reduce the number of Deaths from Unsuccessful Border Crossings which will reduce the pressure to invest in border control.

3.2 Discussion

From the discussion of the diagrams above, we can see that the refugee crisis is dominated by balancing feedback loops. Therefore, when we develop the quantitative System Dynamics model in the next part of this study, we can expect the system to reach an equilibrium after a certain period of time. However, because of delays in the humanitarian relief effort, the situation will probably become worse before it becomes better. Investments in border control may be able to reduce the number of deaths from unsuccessful unconventional border crossings but although very visible, these deaths are just the tip of the iceberg. They are vastly outnumbered by deaths from famine, disease and violence. We believe that the overall goal of reducing the total number of deaths caused by the crisis will be most effectively achieved by a rapid response by the other countries to reduce or eliminate the level of violence in the foreign country. Reducing the level of violence will reduce the likelihood of a famine, will increase the number of safe havens and ensure that the population can develop their economy effectively to create jobs. Increasing the number of safe havens will reduce the crowding in the safe havens and prevent the rapid spread of disease. All of these factors (food security, physical safety, health and jobs) will work together to alleviate the refugee crisis in the foreign country once the level of violence is reduced. The level of violence is affected by the response of the other countries in terms of peacekeeping and diplomacy. Diplomacy efforts can be mobilized quickly whereas peacekeeping would be expected to take a number of months to organize. Investments in food aid, health services, border control and immigration are necessary short-term responses to the refugee crisis until the level of violence is reduced. The investment in economic development aid is a necessary long-term response that will have a positive impact on the foreign country in that the population will not have to leave to create a better life for themselves.

The next step in this study is to develop a quantitative System Dynamics model using the special-purpose programming language Vensim. With this Vensim model, we should be able to create a simulation that reveals at least one way the refugee crisis could be resolved through the effective allocation of humanitarian aid in the short-term and in the long-term.

4 Stock and Flow Diagrams

In this section, the basic elements of the System Dynamics computer simulation are described. The computer simulation model is described as a sequence of interrelated sub-models. For each sub-model, a Stock and Flow diagram is provided using the special-purpose Vensim software package. This software can be used to envision and solve the mathematical system of first-order differential equations. In the diagrams, the state variables (Stocks) are shown as boxes. The flow variables are shown as in-flowing and out-flowing pipes with valves. The valves represent the control of the flows that change over time affecting the state variables. There are usually numerous arrows between the states and the flows and the auxiliary variables that represent the passing of information. The sub-models are inter-connected using shadow variables denoted using “<>” brackets and shown in light grey. These shadow variables are calculated in one of the other sub-models.

As was mentioned in Part 1, System Dynamics modeling requires that the mathematical formulae maintain unit consistency. One trick is often used to ensure this unit consistency, namely “effect” functions. We can think of one element in the model in some unit of measure affecting another element of the model in some other unit of measure through a non-linear effect function. These effect functions require a dimensionless input and usually translate this into a dimensionless output using a non-linear 'look-up' function. This dimensionless output is usually used as a multiplier. The effect function requires another trick to make the input dimensionless. For example, in the model, there is an Effect of Violence on Food Aid. The amount of Violence is measured in ongoing Battles. However, the effect function does not use the amount of Violence directly as an input because it is a dimensioned variable. Instead, the trick is to use the current amount of Violence divided by a value called the Normal Violence which is also measured in Battles. So, by dividing the current amount of Violence by the Normal amount of Violence, a relative value for the amount of Violence is obtained that is dimensionless. That is, the input is a value that indicates whether the amount of Violence is greater than Normal (i.e. relative value greater than 1.0), less than Normal (i.e. relative value between 1.0 and 0.0) or exactly Normal (i.e. relative value 1.0).

The look-up functions that implement these Effect functions are generally non-linear and can take on any desired shape. This is implemented in the software by specifying points on an x-y plot and joining these points by straight lines. The input values are shown on the x-axis and the output values are shown on the y-axis. Linear interpolation is used by the look-up function to translate the input value into an output value.

The formulae used in each of the sub-models is described in words below and the actual mathematical formulae, with their effect functions, initial values for the state variables, the normal values used in the effect functions and the other constants are provided at https://github.com/ivanwtaylor/Refugee-Crisis.

4.1 The Allocation of Investment Towards the Relief Effort

The first sub-model calculates the total amount of investment in relief, in millions of dollars per month, and then allocates that total amount of investment towards the various types of relief, namely, border control, food aid, health services, diplomacy, peacekeeping, immigration services and economic development (Fig. 5).

Total Investment per Month

The total amount of relief money per month is based on a normal amount of investment, in million dollars per month, multiplied by the effect of people in danger on total investment and the effect of deaths per month on total investment which are both calculated as dimensionless multipliers. The larger the fraction of the population in danger, the greater the total investment and the more deaths per month, the more total investment requested.

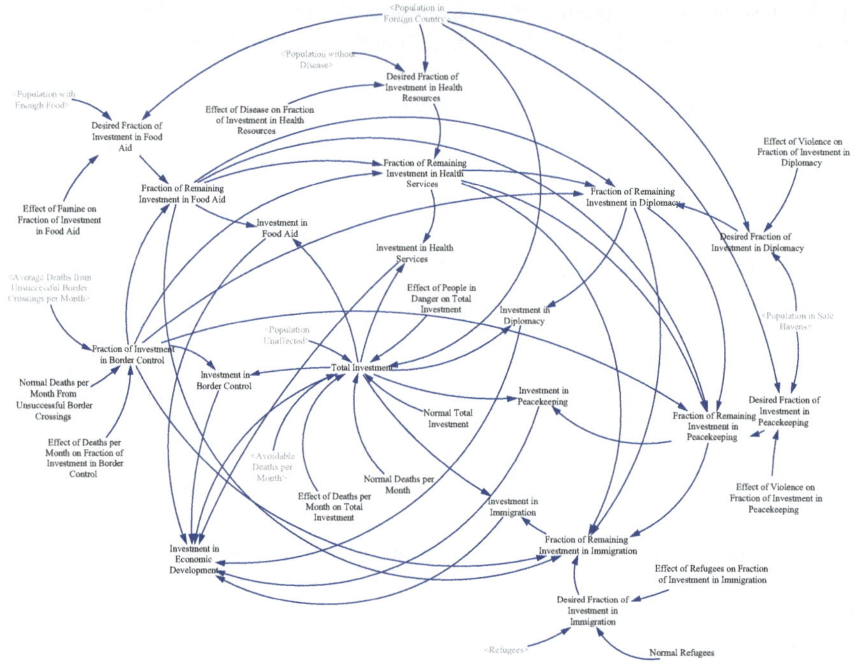

Fig. 5 Sub-model for the allocation of investment towards the relief effort

Investment in Border Control

The first allocation of money is to border control in millions of dollars per month. In this calculation, the fraction of total investment in border control is calculated based on an effect function which uses the average number of deaths caused by unsuccessful unconventional border crossings per month divided by the normal number of deaths per month caused by unsuccessful unconventional border crossings. So, the greater the relative number of deaths per month caused by unsuccessful unconventional border crossings, the larger the fraction of the total investment allocated to border control. This dimensionless fraction is multiplied by the total investment to determine the allocation of investment to border control.

Investment in Food Aid

The second allocation is to food aid in millions of dollars per month. This time, the desired fraction of total investment to food aid is calculated using the effect of famine on food aid. The input to the effect of famine is based on the fraction of the population without enough food which in turn is based on the total population of the country and the number of people with enough food. The larger the fraction of the population without enough food, the larger the desired fraction of total investment allocated to food aid. The actual fraction of total investment to food aid is the minimum of the desired fraction of investment to food aid and one minus the fraction of investment to border control. Then, the actual fraction of investment to food aid is multiplied by the total investment to determine the allocation of investment to food aid.

Investment in Health Services

The third allocation is to health services in millions of dollars per month. This time, the desired fraction of total investment to health services is calculated using the effect of disease on investment in health services. The input to the effect of disease on investment in health services is based on the fraction of the population with disease which in turn is based on the total population of the country and the number of people without disease. The larger the fraction of the population with disease, the larger the desired fraction of total investment allocated to health services. The actual fraction of total investment to health services is the minimum of the desired fraction of investment to health services and one minus the sum of the fraction of investment to border control and the fraction of investment to food aid. Then, the actual fraction of investment to health services is multiplied by the total investment to determine the allocation of investment to health services.

Investment in Diplomacy

The allocation of investment to diplomacy is based on the effect of violence. The desired fraction of investment to diplomacy is calculated as the effect of violence on the population which in turn is based on the fraction of the population affected by violence. That is, the total population in the country minus the number of people in safe havens divided by the total population in the country. The larger the fraction of the population at risk of violence, the greater the desired fraction of the total investment allocated to diplomacy. Again, the actual fraction of investment to diplomacy may not be the same as the desired fraction of the investment to diplomacy. The actual fraction of investment in diplomacy is the minimum of the desired fraction of investment to diplomacy and one minus the sum of the fraction of investment to border control, the fraction of investment to food aid and the fraction on investment to health services. The actual fraction of investment to diplomacy is then multiplied by the total investment to obtain the allocation of total investment to diplomacy in millions of dollars per month.

Investment in Peacekeeping

The fifth allocation is to peacekeeping. The desired fraction of total investment to peacekeeping is calculated using the effect of violence on investment in peacekeeping. The input to this effect is the fraction of people affected by violence which is based on the number of people in safe havens and the total population in the country. The total population in the country minus the number of people in safe havens is assumed to be the number of people at risk of violence. By dividing this value by the population in the foreign country, one gets the fraction of the population at risk of violence. The larger the fraction of the population at risk of violence, the larger the desired fraction of total investment allocated to peacekeeping. Then, the actual fraction of total investment to peacekeeping is the minimum of the desired fraction of investment to peacekeeping and one minus the sum of the fraction of total investment to border control, the fraction of total investment to food aid, the fraction of total investment to health services and the fraction of total investment to diplomacy. This actual fraction of investment to peacekeeping is multiplied by the total investment to obtain the investment in peacekeeping in millions of dollars per month.

Investment in Immigration

The allocation of investment to immigration services is based on the relative number of refugees fleeing the country. That is, the number of refugees fleeing the country is divided by the normal number of refugees and input to a function representing the effect of refugees on investment in immigration to obtain the desired fraction of total investment allocated to immigration services. The larger the relative number of refugees, the larger the desired fraction of total investment allocated to immigration. The actual fraction of investment allocated to immigration is the minimum of the desired fraction of investment to immigration and one minus the sum of the fraction of investment to border control, the fraction of investment to food aid, the fraction of investment to health services, the fraction of investment to diplomacy and the fraction of investment to peacekeeping. Then the actual fraction of investment allocated to immigration is multiplied by the total investment to obtain the investment in immigration in millions of dollars per month.

Investment in Economic Development

It is assumed that any remaining relief would be invested in economic development. So, the amount of investment in economic development is simply the total investment minus the sum of the investment in border control, the investment in food aid, the investment in health services, the investment in diplomacy, the investment in peacekeeping and the investment in immigration services.

4.2 The Movement of Refugees

Population in Foreign Country

The population in the foreign country is a stock variable, measured in families, as shown by the box in Fig. 6. This variable is given an initial value at the start of the simulation which represents an estimate of the number of families in the country before the crisis and then the population changes based on the in-flow and the out-flows. There is one flow variable into the stock variable (Net Birth Rate) and four flows out of it (Deaths Rate from Starvation, Deaths Rate from Violence, Deaths Rate from Disease and Net Additions to Refugees).

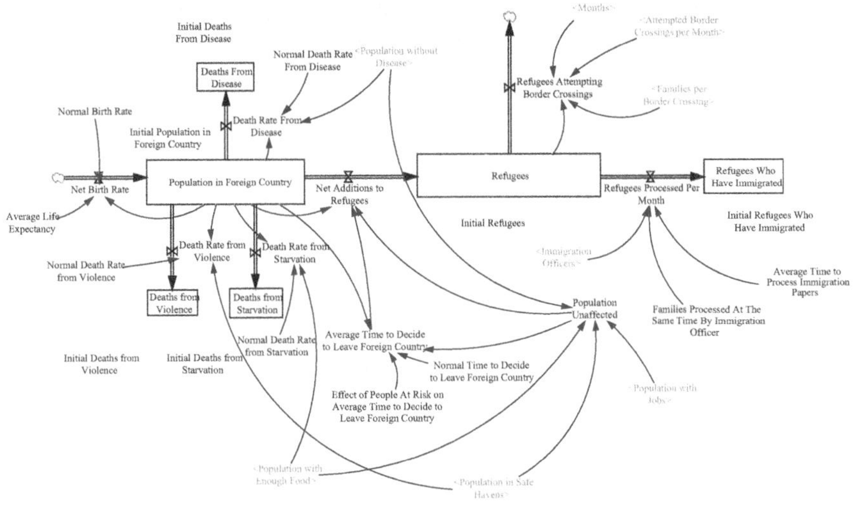

Fig. 6 Refugee movement sub-model

The net birth rate is the difference between births and deaths from natural causes. The number of births is calculated by multiplying the normal birth rate by the population in the foreign country. The number of deaths is calculated by dividing the population in the foreign country by the average life expectancy.

The death rate from starvation is calculated by multiplying the normal death rate from starvation by the population without enough food. The population without enough food which, in turn, is calculated as the population in the foreign country minus the population with enough food. The greater the population without enough food, the higher the death rate from starvation.

The death rate from violence is calculated by multiplying the normal death rate from violence by population at risk from violence. The population at risk from violence is calculated by subtracting the population in safe havens from the population in the foreign country. The larger the population at risk from violence, the higher the death rate by violence.

The death rate from disease is calculated by multiplying the normal death rate from disease by the population with disease. The population with disease which is calculated by subtracting the population without disease from the population in the foreign country. The larger the population with disease, the higher the death rate by disease.

Refugees will decide to leave the foreign country based on the impact of famine, disease, violence and poverty. The number of refugees deciding to leave the country is based on population affected by the crisis divided by the average time to decide to leave the country. The population affected by the crisis is calculated as the population in the foreign country minus the population unaffected by the crisis. The population unaffected by the crisis is the minimum of four values: the population with enough food, the population without disease, the population in safe havens and the population with jobs. The average time to decide to leave the country is calculated as the normal time to decide to leave the country times the effect of the fraction of people at risk on the time to decide to leave the country. The higher the fraction of people at risk, the lower the average time to decide to leave the country.

Deaths From Starvation

The deaths from starvation is also a stock variable, measured in families, which is initially set to zero and has a single in-flow of the death rate from starvation. The in-flow death rate from starvation is the same as the out-flow from the population in the foreign country described above.

Deaths From Violence

The deaths from violence is a stock variable, measured in families, which is also initially set to zero. It has a single in-flow of the death rate from violence. This in-flow is the same as the out-flow from the population in the foreign country.

Deaths From Disease

The deaths from disease is a stock variable, measured in families, which is also initially set to zero. It has a single in-flow of the death rate from disease. This in-flow is the same as the out-flow from the population in the foreign country.

Refugees

Refugees is a stock variable, measured in families, which is initially set to zero. It has one in-flow (net additions to refugees) and two out-flows (refugees processed per month and refugees attempting unconventional border crossings). The net additions to refugees is an in-flow to refugees that is the same as the out-flow from population in foreign country that was described above.

The refugees processed per month is calculated as the number of immigration officers times the number of families processed at the same time by an immigration officer divided by the average time to process immigration papers The greater the number of immigration officers, the greater the number of refugees processed per month.

The refugees attempting border crossings is calculated as the attempted border crossings per month multiplied by the families per border crossing. The attempted border crossings per month and the families per border crossing are calculated in the unconventional border crossings sub-model which will be described later.

Refugees Who Have Immigrated

The refugees who have immigrated is also a stock variable, measured in families, in which the initial value is set to zero and the single in-flow is the refugees processed per month which is the same as the out-flow from refugees.

4.3 The Food Supply

The food supply is a separate sub-model that includes a single stock variable for food, measured in food units which represents the amount of food required for one family for one month. It has two in-flows (food growth and food aid) and two out-flows (food consumption and food export) (Fig. 7).

Food growth is calculated by multiplying the food growth rate by the food. The food growth rate is based on the normal food growth rate multiplied by the one minus a value for famine which is a fractional value between zero and one. Famine is calculated based on the effect of violence which uses the current violence level divided by the normal level of violence required to destroy the harvest. The greater the relative violence level, the greater the level of famine.

Food aid is an in-flow which is calculated by multiplying the investment in food aid divided by the average cost of a food unit by the effect of violence on distribution of food aid. The effect of violence is calculated based on the current violence level divided by the normal violence level to disrupt distribution of food aid. Violence will impede the distribution of food aid. So the greater the relative level of violence, the less food aid that is provided.

Food consumption is calculated by dividing the food level by the normal food level and multiplying this by the average consumption per family and the population in the foreign country. Therefore, as the relative level of food declines the average distribution of food to the families will decline.

Food export is calculated as the food level minus the food consumption if this value is greater than zero.

Finally, the population with enough food is calculated as the food level divided by the average consumption of food per family.

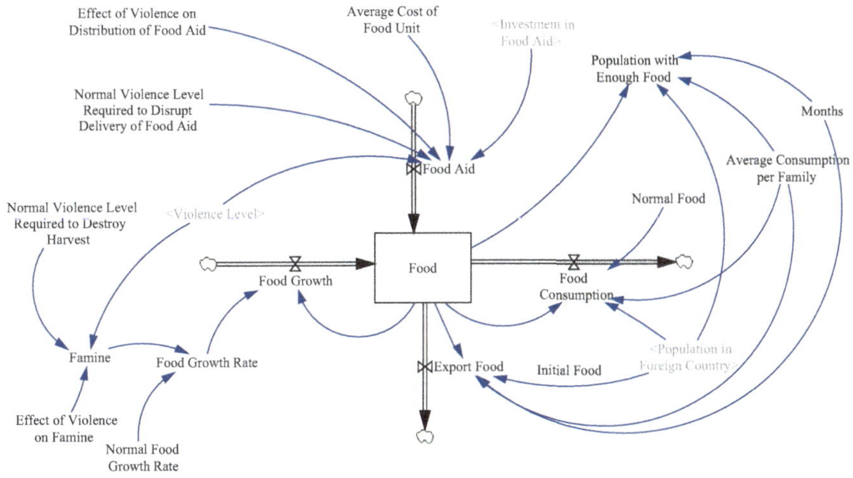

Fig. 7 Food supply sub-model

4.4 Disease

The disease model is primarily focused on the outbreak of cholera in the safe havens. There is an initial number of families effected by a cholera outbreak. There is an in-flow of families being infected with the disease and an out-flow of families being cured of the disease (Fig. 8).

The in-flow of infections per month is based on the population with disease and the normal infection rate. These are multiplied together and then the effect of crowding comes into play. The effect of crowding on disease rate is a multiplier that is high when there is a large population in the safe havens compared to the capacity of the safe havens and lower when there is smaller population in the safe havens compared to the capacity of the safe havens.

The out-flow of cured families per month is calculated as the number of medical practitioners times the people a medical practitioner can handle divided by the average time to cure the disease.

Then the population without disease is the population in the foreign country minus the population with disease.

4.5 The Safe Havens

Safe Havens is a stock variable, measured in havens, which is initially set to a value that ensures that the entire foreign country consists of safe havens based on the average population per safe haven. It has one in-flow (safe havens created) and one out-flow (safe havens destroyed) (Fig. 9).

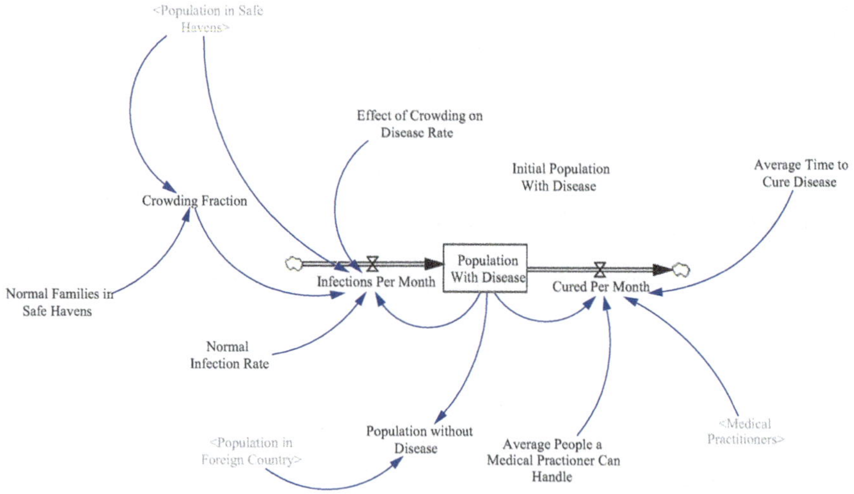

Fig. 8 Disease sub-model

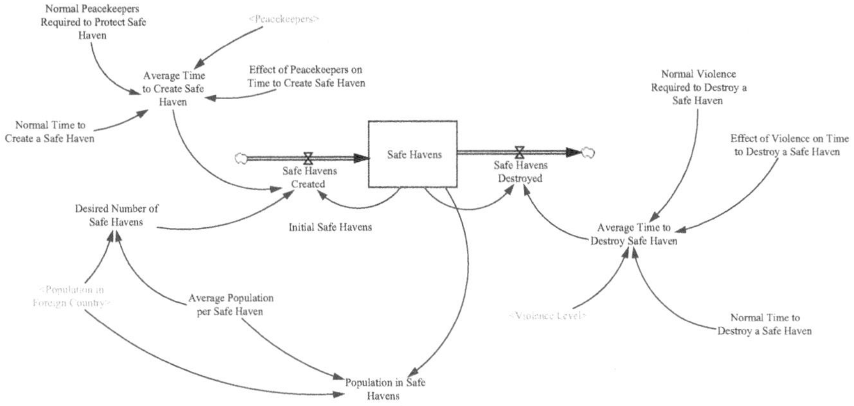

Fig. 9 Safe havens sub-model

Safe havens created is calculated by finding the desired number of safe havens which is the population in the foreign country divided by the average population per safe haven. Then the current number of safe havens is subtracted from the desired number of safe havens and divided by the average time to create a safe haven to obtain the safe havens created per month. The average time to create a safe haven is calculated as the normal time to create a safe haven times the effect of peacekeepers on the time to create a safe haven which uses the number of peacekeepers divided by the normal number of peacekeepers.

The safe havens destroyed is calculated as the current number of safe havens divided by the average time to destroy a safe haven. The average time to destroy a safe haven is calculated by multiplying the normal time to destroy a safe haven by the effect of violence on the time to destroy a safe haven where this is based on the current violence level divided by the normal violence level required to destroy a safe haven. The higher the level of violence, the less time it will take to destroy a safe haven.

4.6 *Violence*

Violence, measured in ongoing battles, is set to an initial value of one. Then, it changes based on the net change in violence. The net change in violence is calculated as the desired level violence minus the actual level of violence divided by the average time to change the violence level. The desired level of violence is the current violence level times the violence escalation rate minus the sum of the peacekeepers divided by the number of peacekeepers to reduce a battle and the diplomatic effort divided by the diplomatic effort to reduce a battle (Fig. 10).

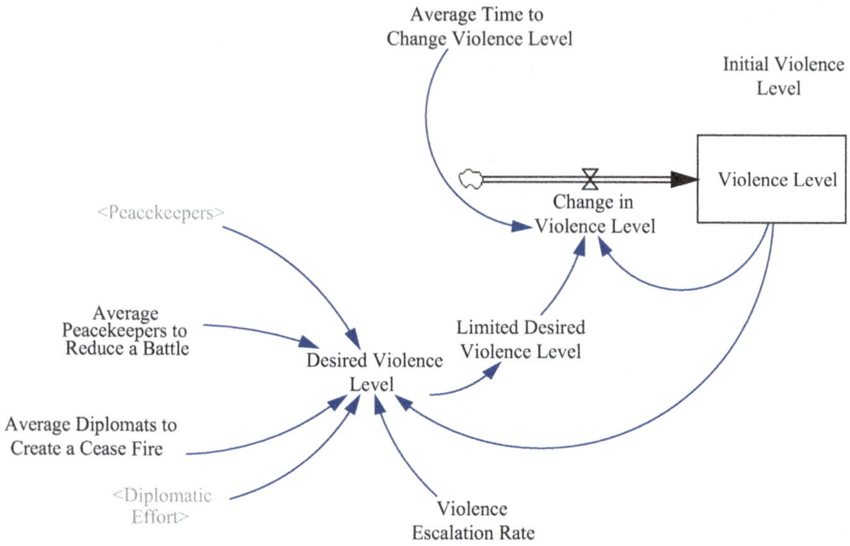

Fig. 10 Violence sub-model

4.7 Peacekeepers

The level of peacekeepers is initially set to zero then has a flow called the change in peacekeepers which represents both an in-flow and a potential out-flow. The change in peacekeepers is simply the desired number of peacekeepers minus the current level of peacekeepers divided by the average time to deploy peacekeepers. The desired number of peacekeepers is simply the investment in peacekeeping divided by the average monthly cost of a peacekeeper (Fig. 11).

The sub-models for diplomatic effort, health services, immigration officers and border control resources are similar to the peacekeepers sub-model. In each case, a desired level of diplomats, medical practitioners, immigration officers and border control ships is calculated by dividing the associated investment by the corresponding average cost. Then the change in the stock is based on the difference between the desired level and the current level of the stock divided by the average time to change the stock. They are all set to an initial value of zero at the start of the simulation.

4.8 Economic Activity

The economy sub-model has one stock variable, called 'businesses', which is initially set to a value so that the entire population of the foreign country has a job. Then the stock variable, businesses, changes based on the one in-flow (businesses created) and the one out-flow (businesses destroyed).

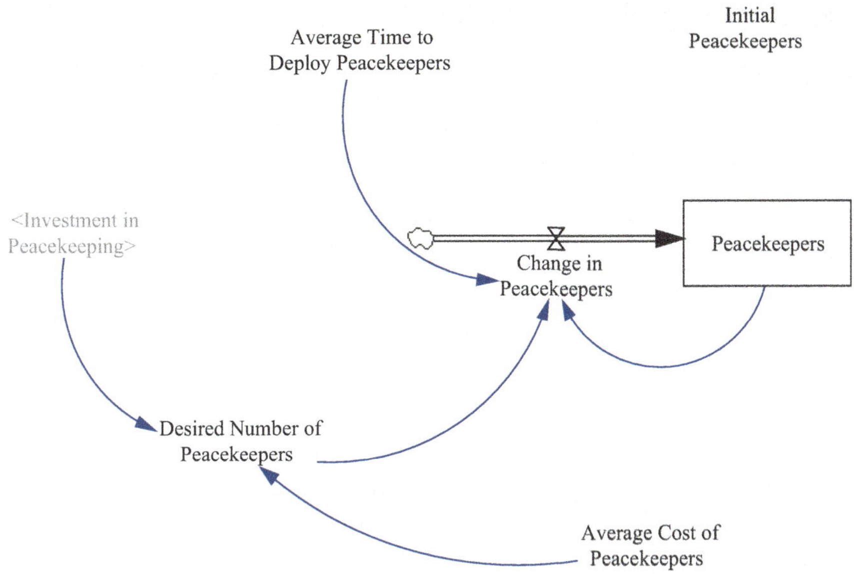

Fig. 11 Peacekeepers sub-model

Businesses created is positive if the desired number of businesses is greater than the current number of businesses and is equal to the investment in economic development divided by the average cost of creating a business. Business destroyed is equal to the number of businesses divided by the time to destroy a business. The average time to destroy a business is equal to the normal time to destroy a business multiplied by the effect of violence on the time to destroy a business which, in turn, is based on the current level of violence divided by the normal level of violence required to destroy a business. The greater the relative violence level, the smaller the time to destroy a business (Fig. 12).

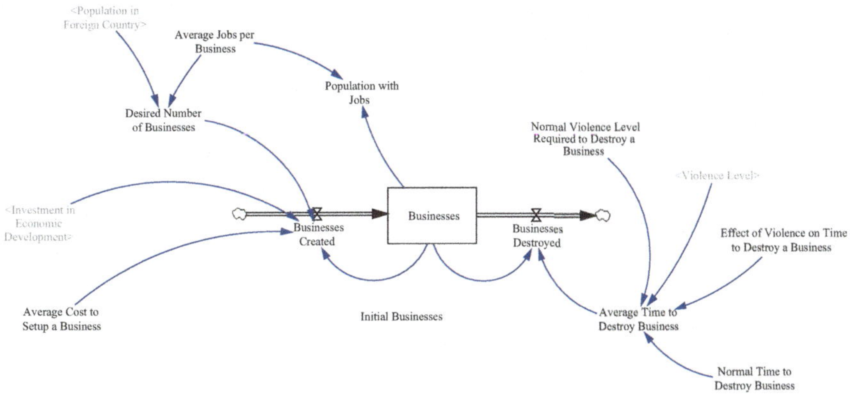

Fig. 12 The economy sub-model

4.9 Border Crossing Resources

The level of border crossing resources is initially set to 100 boats then has an flow called the change in border crossing resources which represents both an in-flow and a potential out-flow. The change in border crossing resources is simply the desired number of border crossing resources minus the current level of border crossing resources divided by the average time to add border crossing resources. The desired number of border crossing resources is the normal border crossing resources times the effect of refugees on border crossings resources which in turn is based on the number of refugees divided by the normal number of refugees. The greater the relative number of refugees, the greater the increase in border crossing resources (Fig. 13).

4.10 Unconventional Border Crossings

The unconventional border crossings sub-model consists of three stock variables: successful border crossings, unsuccessful border crossings and deaths from unsuccessful border crossings.

The central variable in this sub-model is attempted border crossings per month which is based on the normal attempted border crossings per month multiplied by a number of effects. The effect of border crossing resources on attempted border crossings is based on the current border crossing resources divided by the normal

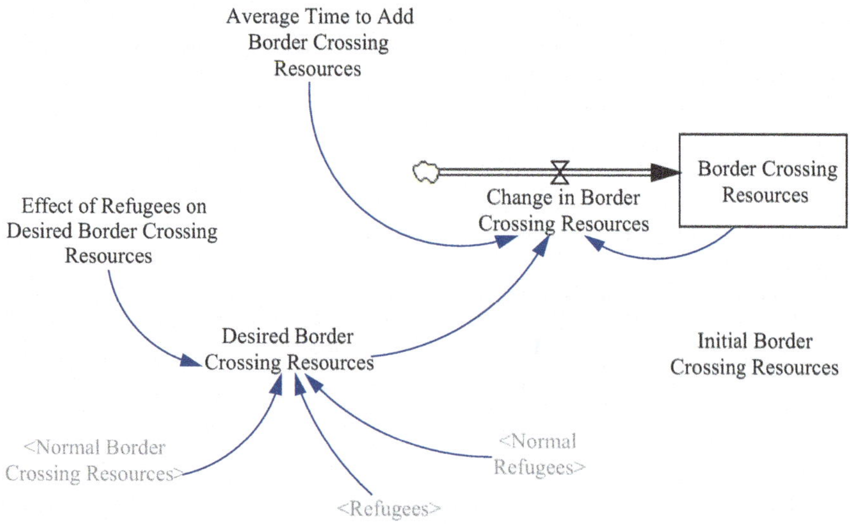

Fig. 13 Border crossing resources sub-model

border crossing resources with the greater the relative amount of border crossing resources, the more attempted border crossings per month. The effect of refugees on attempted border crossings is based on the number of refugees divided by the normal number of refugees with the larger the relative number of refugees, the larger the number of attempted border crossings per month. The effect of border control on attempted border crossings is based on the current level of border control resources divided by the normal border control resources and, in this case, the greater the relative border control resources, the smaller the number of attempted border crossings per month. The effect of successful border crossings on attempted border crossings is based on the average number of successful border crossings per month divided by the normal number of successful border crossings per month, with a larger relative number of successful border crossings encouraging a greater number of attempted crossings per month. The effect of unsuccessful border crossings on attempted border crossings is based on the average number of unsuccessful border crossings per month divided by the normal number of unsuccessful border crossings per month. In this case, unsuccessful border crossings will discourage future border crossings. So, the greater the relative number of unsuccessful border crossings, the smaller the number of attempted border crossings per month (Fig. 14).

Unsuccessful Border Crossings

The stock variable unsuccessful border crossings has one in-flow, unsuccessful border crossings per month. The unsuccessful border crossings per month is calculated as the attempted border crossings per month multiplied the fraction of border crossings that are unsuccessful. The fraction of border crossings that are unsuccessful is calculated by multiplying the normal fraction of border crossings that are unsuccessful by the effect of border crossing resources on the fraction of border crossings that are unsuccessful which, in turn, is based on the current border

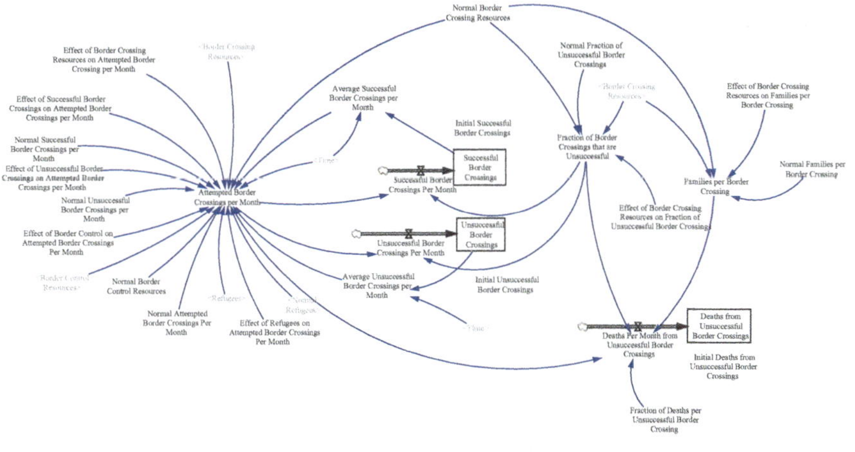

Fig. 14 Unconventional border crossings sub-model

crossing resources divided by the normal border crossing resources. In this case, the smaller the relative amount of border crossing resources, the more likely the border crossing will be unsuccessful.

Successful Border Crossings

The stock variable successful border crossings has one in-flow, successful border crossings per month. The successful border crossings per month is calculated as the attempted border crossings per month multiplied one minus the fraction of border crossings that are unsuccessful.

Deaths from Unsuccessful Border Crossings

The number of deaths from unsuccessful border crossings is one of the most visible aspects of the refugee crisis. The deaths per month from unsuccessful border crossings is the product of the attempted border crossings per month, the fraction of border crossings that are unsuccessful, the families per border crossing and the fraction of deaths per unsuccessful border crossing.

The families per border crossing is calculated by multiplying the normal number of families per border crossing by the effect of border crossing resources on families per border crossing which, in turn, is based on the current border crossing resources divided by the normal border crossing resources. In this case, the smaller the border crossing resources, the larger the number of families per border crossing. That is, the boats are assumed to be more crowded.

4.11 Avoidable Deaths

The final sub-model is called avoidable deaths (Fig. 15). The avoidable deaths per month is calculated as the sum of the average deaths by violence per month, the average deaths by starvation per month, the average deaths from disease per month and average deaths from unsuccessful border crossings per month. These values, in turn, are based on the stock variables calculated in other sub-models (deaths from violence, deaths from starvation and deaths from disease in the refugee movement sub-model, and deaths from unsuccessful border crossings in the unconventional border crossings sub-model).

Also, the cumulative deaths is calculated as the sum of the deaths from violence, deaths from starvation, deaths from disease and deaths from unsuccessful border crossings.

4.12 Results from the Baseline Run

The constants, initial values and normal values, as well as the effect functions, have been estimated in a rough fashion to obtain a baseline run. One of the characteristics of System Dynamics modeling is the belief that it is better to provide a rough

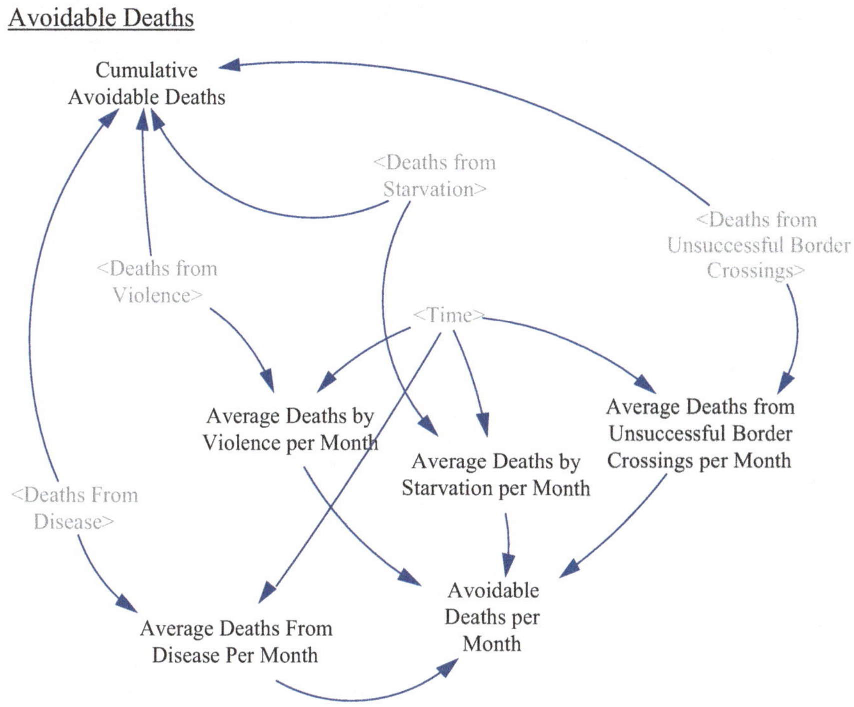

Fig. 15 Avoidable deaths sub-model

estimate of an important variable than to ignore the factor by leaving it out of the model. Thus using these rough estimates, the following graphs show the results of the baseline run. More details are provided at https://github.com/ivanwtaylor/Refugee-Crisis.

In Fig. 16, one can see the nature of the refugee crisis. The number of refugees increases quite a bit over the first two and a half years of the crisis, then declines as the immigration processes kick in and eventually appears to level off as the crisis is resolved. In the meantime, the refugees who have immigrated increases steadily over the five year period reaching over 80 thousand families.

In the first months of the crisis, the number of avoidable deaths per month is dominated by deaths from starvation that begins to decline as food aid arrives (see Fig. 17). The avoidable deaths caused by disease also increases quickly at the beginning but levels off after 18 months and declines slowly as the health practitioners are deployed. The avoidable deaths caused by violence increases quickly half way into the crisis as the gap between the number people in the foreign country and the number of people in the safe havens increases (i.e. the number of people at risk of violence increases) but levels off and starts to decline after 36 months and declines slowly as the peacekeepers are deployed and the diplomatic effort starts to

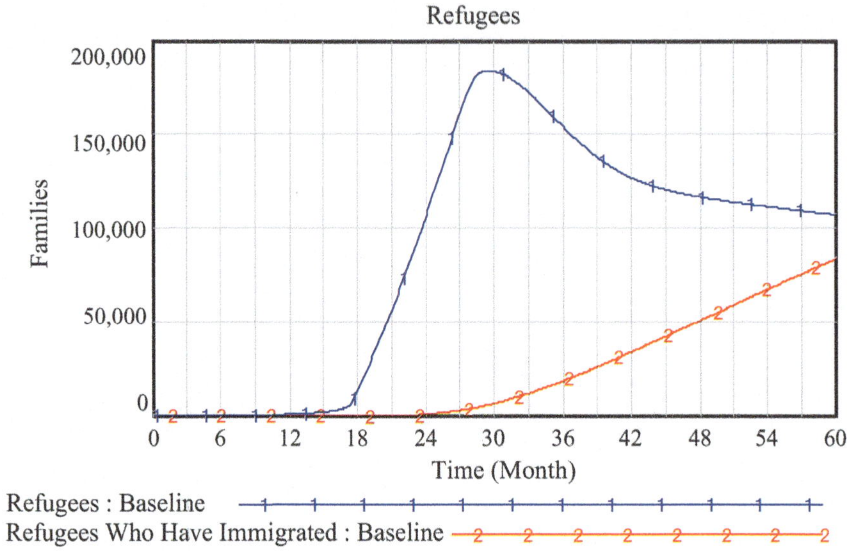

Fig. 16 The nature of the refugee crisis

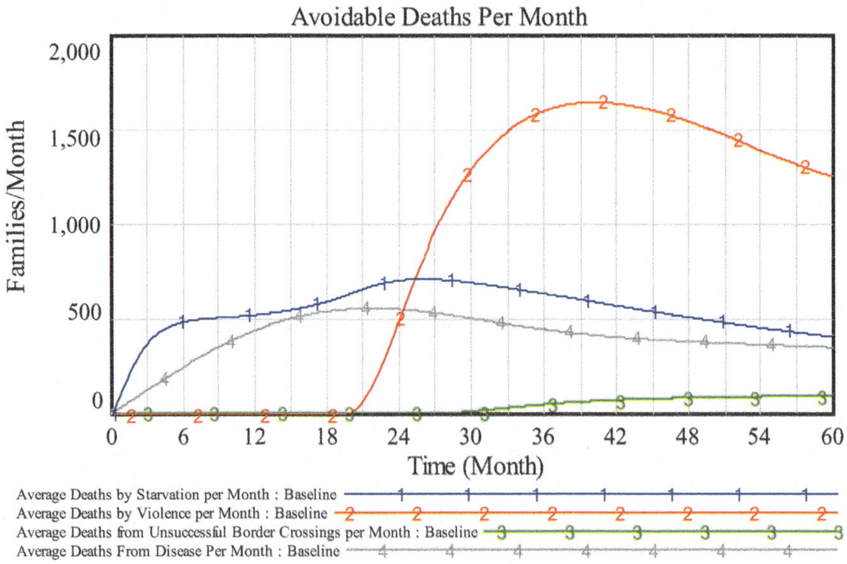

Fig. 17 Average avoidable deaths per month

take effect. The avoidable deaths caused by unsuccessful border crossings is much smaller than the other avoidable deaths but seems to be increasing slightly throughout the period under examination.

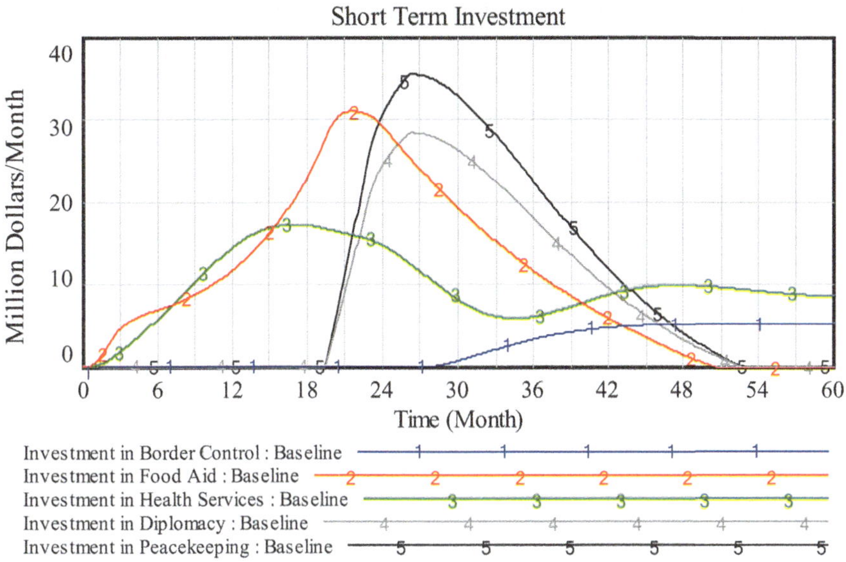

Fig. 18 Allocation of aid

Figure 18 shows the allocation of investment between the various short-term forms of relief. Investment in food aid increases quickly, peaks and then begins to decline as the situation stabilizes. Investments in health services also increases quickly peaks and then declines. However, there appears to be the beginnings of an oscillation effect with health services. Investments in peacekeeping and diplomacy take about 18 months to get going and then rapidly increase, peak and decline to stabilize at a relatively low level. The investment in border control also takes a while to come into effect, then increases and stabilizes at a modest level.

4.13 *Discussion*

These results above can be, in no way, considered definitive since the constants, the normals and the effects are only estimated in a rough manner. We will now examine the sensitivity of the results to these estimated values.

The special purpose Vensim software package is useful for building and testing a prototype model. The software has also provided a relatively easy way to visualize the System Dynamics model using stock and flow diagrams. However, the Vensim software is unwieldy when it comes to conducting analysis if one wants to change a number of variables at once which is what is desired when a detailed sensitivity analysis is conducted or when an optimization process is used.

Therefore, the model will be converted into a more versatile general purpose programming language, namely Fortran. In Part 5 of this study, the Fortran version of the refugee crisis model will be documented. Then in Part 6, a detailed sensitivity analysis is conducted using a design of experiments technique. This is followed in Part 7 with an optimization of the model which uses a genetic algorithm to minimize the number of avoidable deaths.

5 Fortran Version of the Model

The model was converted into a computer program using the general-purpose programming language Fortran which is documented in https://github.com/ivanwtaylor/Refugee-Crisis.

To ensure that the Fortran version of the model is a true implementation of the Vensim prototype model, the results from both models were imported into an Excel spreadsheet. Then for each variable at each step time, the results were compared by subtracting the Fortran results from the Vensim results and seeing if the difference was approximately zero. Once this was achieved, the Fortran version of the refugee crisis model was sufficiently accurate to conduct further analysis.

5.1 Sensitivity Analysis

As noted in Part 4, the input parameters used in the baseline run of the model were simply rough estimates of the actual values. It is important to determine which of these parameters need to be the most accurately estimated to obtain reasonable results. Therefore, a sensitivity analysis will be conducted using a technique called design of experiments. This technique ensures that the greatest amount of results are obtained with a reasonable number of simulation runs (experiments). The Fortran version of the model can be easily modified to show the results in terms of a dependent output value and a series of independent values for the input variables.

5.2 Optimization of the Model

The Fortran program also can be modified to conduct an optimization in which an output value can be maximized or minimized by changing the parameters. The optimization of the model will be conducted using a tool from artificial intelligence called a genetic algorithm.

6 Design of Experiments

Design of experiments, when applied to computer simulations, is an extremely efficient method of conducting sensitivity analysis. Sensitivity analysis attempts to determine the effect of changes in the input parameters to the computer simulation to the outputs of the simulation. The goal is to determine which parameters are most important in predicting the results and which are less important. As we have seen in Part 4 of this report, there are many input parameters to our refugee crisis model that need to be estimated. Which of these parameters have the most impact on the results is therefore a relevant question and determining this in the most efficient manner will be useful. By efficient, we mean minimizing the number of simulation runs required to get the maximum information about the parameters' sensitivity.

6.1 Changes to the Fortran Program

Each of the input parameters will be randomly assigned either a high value or a low value. The high value will correspond to twice the original input value that was used in the baseline run. The low value will correspond to half of the original input value that was used in the baseline run. After all of the 98 independent values are assigned either high or low values, the simulation program is run. Then a line is written to the output file which starts with the dependent value for the cumulative avoidable deaths followed by the 98 independent variables. This output file is written as a comma separated variables file that can be read by the statistical software package R. The documentation of the revised program along with the results of the design of experiment is provided at https://github.com/ivanwtaylor/Refugee-Crisis.

6.2 The R Commands

The statistical package R is a command line program. The first thing that must be done is to read the comma-delimited file into memory. Then we run a linear regression model on the data using the independent variables to predict the dependent variable. Then we can display the results of the linear regression.

6.3 Interpreting the Preliminary Results

The statistical significance of the independent variable is determined by the P-value of the linear regression. The values shown in Table 1 represent those independent variables that have a statistically significant impact on the dependent variable 'cumulative avoidable deaths'. The larger the absolute value of the coefficient, the greater the impact. The sign of the coefficient suggests the direction of the effect. So a positive value, implies that the higher the independent variable, the higher the dependent variable and a negative value implies that the higher the independent variable, the lower the dependent variable.

Table 1 Significant independent variables

Independent variable	Coefficients
Initial population in foreign country	365,088
Normal death rate from violence	208,327
Normal time to create a safe haven	81,361
Average consumption per family	69,794
Normal peacekeepers required to protect safe haven	66,971
Average cost of peacekeepers	66,033
Normal deaths per month	53,611
Initial violence	48,620
Violence escalation rate	38,060
Average time to deploy peacekeepers	37,494
Average cost of a food unit	33,474
Normal death rate from starvation	31,015
Effect of people at risk on average time to decide to leave foreign country	24,660
Effect of violence on fraction of investment in peacekeeping	23,295
Normal time to decide to leave foreign country	22,296
Average life expectancy	19,227
Effect of violence on fraction of investment in diplomacy	−16,255
Normal food growth rate	−19,681
Effect of violence on time to destroy a safe haven	−26,379
Normal violence required to destroy a safe haven	−42,768
Normal food	−45,672
Average population per safe haven	−106,248
Initial safe havens	−106,993
Normal total investment	−107,149

So from the results in Table 1, we can see the statistically significant independent variables sorted from most impact in increasing the cumulative avoidable deaths to the most impact at decreasing cumulative avoidable deaths. The initial population of the foreign countries has the most impact, causing the cumulative avoidable deaths to go up because there would be more deaths from violence, deaths from starvation, deaths from disease simply because there would be more people in the foreign country to be effected by the crisis. The next highest impact variable is the normal death rate from violence which increases the cumulative avoidable deaths for obvious reasons. The impact drops off considerably with the remaining independent variables.

An increase in the normal time to create a safe haven would cause the cumulative avoidable deaths to increase because it would likely affect the ability of safe havens to respond to an increase in the level of violence. An increase in the average consumption of food per family would cause the cumulative avoidable deaths to increase because of increased pressure on the food supply to avoid deaths from starvation. An increase in the normal peacekeepers required to protect safe havens would cause the cumulative avoidable deaths to increase because the actual number of peacekeepers required to protect safe havens would have to increase to keep the relative number of peacekeepers required to protect safe havens the same (that is, the relative number of peacekeepers required to protect safe havens is the actual number of peacekeepers available divided by the normal peacekeepers required to protect safe havens). An increase in the average cost of peacekeepers would increase the cumulative avoidable deaths because the same investment in peacekeeping would provide fewer actual peacekeepers to reduce violence, support food aid distribution and protect safe havens. Increasing the normal deaths per month would cause an increase in the cumulative avoidable deaths because the relative deaths per month would decline causing the total investment in aid to decrease. An increase in initial violence or an increase in the violence escalation rate would have an obvious effect of cumulative avoidable deaths by increasing the level of violence in the country causing the destruction of safe havens and thus more people at risk of violence. Increasing the level of violence would also result in more deaths from starvation because the food growth would be more likely effected by famine and the distribution of food aid would be disrupted. An increase in the average time to deploy peacekeepers would result in a slower reaction time to increases in the violence level allowing the violence level to increase further than it would have otherwise. Increasing the average cost of food would cause more deaths from starvation because the investment in food aid could buy less food for distribution to the people of the country in need. Increasing the normal death rate from starvation would have an obvious effect on cumulative avoidable deaths. An increase in the effect of people at risk on average time to decide to leave the country and the normal time to decide to leave the country would result in greater cumulative avoidable deaths because more people would stay in the country as the crisis became greater causing more deaths from violence, starvation and disease. An increase in the average life expectancy would increase the cumulative avoidable deaths because fewer people would die of natural causes and more people would be around to die of avoidable causes.

Now we will look at the independent parameters that would cause a decrease in the cumulative avoidable deaths. If the normal total investment in aid goes up, then the cumulative avoidable deaths goes down considerably. This is a good sign because it means that aid from foreign countries can have a considerable effect on reducing the number of deaths in the foreign country. If the initial number of safe havens and/or the average population per safe haven goes up, there will be less families affected by violence and considerably less cumulative avoidable deaths. An increase in the normal food would decrease the cumulative avoidable deaths because then the relative amount of food consumption would decrease and there would be fewer people dying of starvation. Increasing the normal violence required to destroy a safe haven will mean the relative violence required to destroy a safe haven will decrease or increasing the effect of violence on the time to destroy a safe haven will mean that the time to destroy a safe haven will increase resulting in fewer safe havens destroyed and more people protected from violence. Increasing the normal food growth rate would have an obvious effect on the deaths from starvation. Increasing the effect of violence on the fraction of investment in diplomacy will cause more investment in diplomacy that would have the effect of reducing the violence level.

6.4 Discussion

In our design of experiments, we looked at all of the input parameters used in the System Dynamics model of the refugee crisis to determine which input parameters had the most impact on the cumulative avoidable deaths. The parameters shown in Table 1 will need to be estimated more carefully than some of the other input parameters in the model because they have the most impact on the results.

It is interesting to note, that only 24 independent variables have a statistically significant impact on the cumulative avoidable deaths in the model. Therefore, the parameter estimation process can be greatly reduced while still obtaining accurate projections.

7 Optimization of the Model

The Fortran program can be modified to conduct an optimization in which an output value can be maximized or minimized by changing the parameters. The optimization of the model will be conducted using a tool from artificial intelligence called a genetic algorithm. The parameters of the model are considered to be a vector of genes in an organism. The program starts with a random set of vectors representing a population of organisms. A subset of this population is assumed to survive based on the quality of their output. Then this fittest population will breed and produce offspring which are a mix of their genes (parameter values). There also will be the possibility that a small number of these offspring will have random mutations of their genes. Then the most well adapted offspring (those with the highest quality results) will form a new generation of the population. The program will stop once a goal is achieved or when a number of generations have been simulated.

We will vary the 26 parameters that the aid agencies have some control over and try to find a set of values that minimizes the cumulative avoidable deaths at the end of the five year period.

7.1 Results of the Optimization

The optimization process is one of diminishing returns. That is, the greatest amount of improvement is found in the first few generations and as the number of generations gets larger the relative improvements becomes less and less. Figure 19 is a

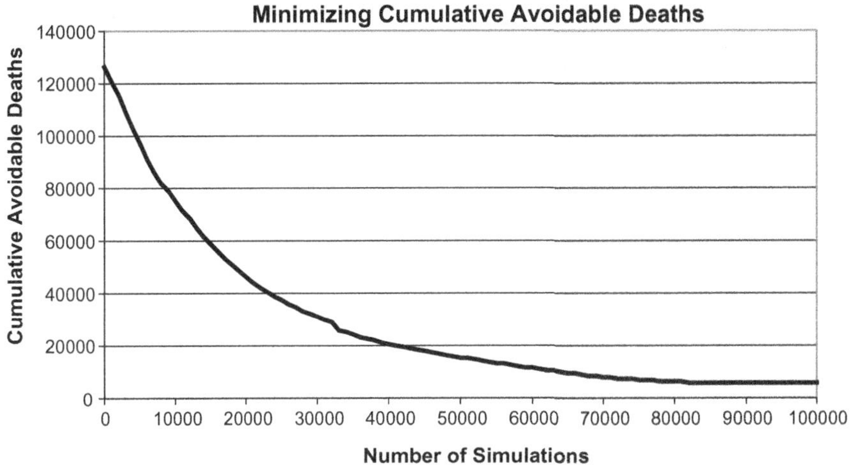

Fig. 19 Typical results from the genetic algorithm optimization

typical run of 100,000 generations. The improvement goes from over 120,000 cumulative avoidable deaths in the five year period in the baseline run to approximately 20,000 cumulative avoidable deaths in the five year period after 40,000 generations of the genetic algorithm. This is followed by a further reduction to approximately 5500 cumulative avoidable deaths over the five year period by the 100,000th generation.

Table 2 shows the parameters that were used in the baseline run for the 26 variables that are under the control of the aid agencies along with the same parameters for the optimal run found using the genetic algorithm. We can see with slight changes to these controllable parameters, we could potentially save over 120,000 lives.

There is a random element in the genetic algorithm. So we ran the algorithm a number of times and compared the results. After 100,000 generations, the results in terms of cumulative avoidable deaths was around 5500 no matter the random number generator seed. Also the parameter values were approximately the same for all of the runs. So we believe we have a robust solution to the refugee crisis.

The massive reduction in cumulative avoidable deaths is created by changes in all of the parameters under the control of the aid agencies because they are all highly connected. Going from the top to the bottom of this table, we see that the border control resources would need to be used efficiently with the normal border control resources less than 5. Similarly, the peacekeepers required to protect a safe haven would need to be used efficiently and be less than 500. The time to create a safe haven would need to be less than 4 months to reach the optimal results. The normal total investment would need to be at least 35 million dollars per month.

The diplomatic service would need to be effective with the average diplomats to create a cease fire as 2 and the average peacekeepers would need to be effective at

Table 2 Parameters of baseline run and optimal values compared

Variable name	Baseline run	Optimal value
Cumulative avoidable deaths	126,443	5544
Normal border control resources	10	4.48
Normal peacekeepers required to protect safe haven	1000	421.71
Normal time to create a safe haven	24	3.59
Normal total investment	20	35.89
Average diplomats to create a cease fire	10	1.87
Average peacekeepers to reduce a battle	1000	648.41
Average people a medical practitioner can handle	100	180.89
Average time to cure disease	2	1.06
Average time to deploy border control resources	6	3.51
Average time to deploy peacekeepers	12	5.45
Average time to employ diplomats	6	4.78
Average time to hire medical practitioner	6	4.16
Average time to process immigration papers	12	3.69
Average time to train immigration officers	12	5.43
Families processed at same time by immigration officer	5	16.23
Effect of border control on attempted border crossings per month	1	0.46
Effect of deaths per month on fraction of investment in border control	1	0.47
Effect of deaths per month on total investment	1	0.48
Effect of disease on fraction of investment on health services	0.2	0.11
Effect of famine on fraction of investment in food aid	0.2	0.02
Effect of peacekeepers on time to create safe haven	1	0.39
Effect of people in danger on total investment	0.1	0.06
Effect of refugees on fraction of investment in immigration	10	3.31
Effect of violence on distribution of food aid	1	3.81
Effect of violence on fraction of investment on diplomacy	0.2	0.03
Effect of violence on fraction of investment in peacekeeping	0.4	0.12

reducing battles with the average peacekeepers required to reduce a battle being 650 soldiers. We will also need to have effective peacekeepers creating safe havens by having the mid-point of the effect function as 0.4. Medical practitioners would have to be effective as well with capability to handle 180 patients simultaneously and the average time to cure disease of 1 month.

The reaction times would need to all be low. The average time to deploy border control resources would need to be 3.5 months. The average time to deploy peacekeepers would need to be 5.5 months. The average time to employ diplomats would need to be 4.75 months. The average time to hire medical practitioners would need to be 4 months. The average time train immigration officers would need to be 5.5 months and the immigration officers would need to be efficient with their average time to process immigration papers at 3.5 months and the number of families they process at the same time as 16.

Then we would need to have effective border control with the effect function mid-point of 0.5 and the responsiveness of the number of deaths per month from unsuccessful border crossings to investment in border control would need to have the mid-point of the effect function as 0.5. We would need to have total aid investment responsive to the average avoidable deaths per month with the effect function mid-point as 0.5. We would need to have the health services investment process responsive to the disease rate with the effect mid-point as 0.1 and the food aid responsive to famine by having the mid-point at 0.02. We need to have highly effective peacekeepers with their effect on time to create a safe haven mid-point as 0.4.

We need to be highly responsive to the number of people in danger by having the mid-point of the effect function as 0.06 and the responsiveness of immigration on the number of refugees by having the mid-point of the effect function as 3. We need to have the responsiveness of food aid distribution to the level of violence with the effect function mid-point as 4 and the responsiveness to violence in the investment in diplomacy and peacekeeping by having the mid-point values as 0.03 and as 0.1 respectively.

The graphs below show the final result of the genetic algorithm compared to the results of the baseline run.

Figure 20 shows the small number of cumulative avoidable deaths in the optimal run compared to the baseline run.

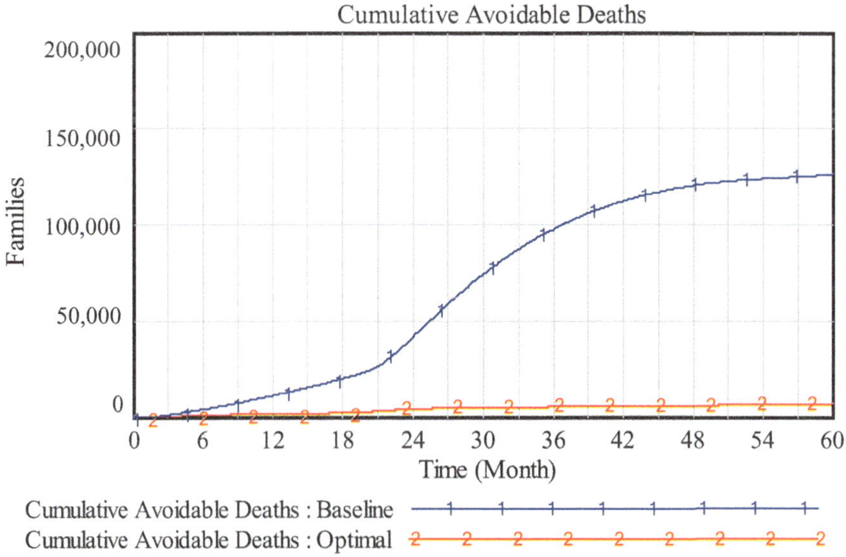

Fig. 20 Comparison of cumulative avoidable deaths results

We can see from Fig. 21 that the total investment over the period of five years is not that much different between the baseline run and the optimal run. The difference obviously is that the investment in aid is more efficiently used in the optimal run.

By comparing Figs. 17 and 22 we can see that the avoidable deaths per month are an order of magnitude lower in the optimal case compared to the baseline case.

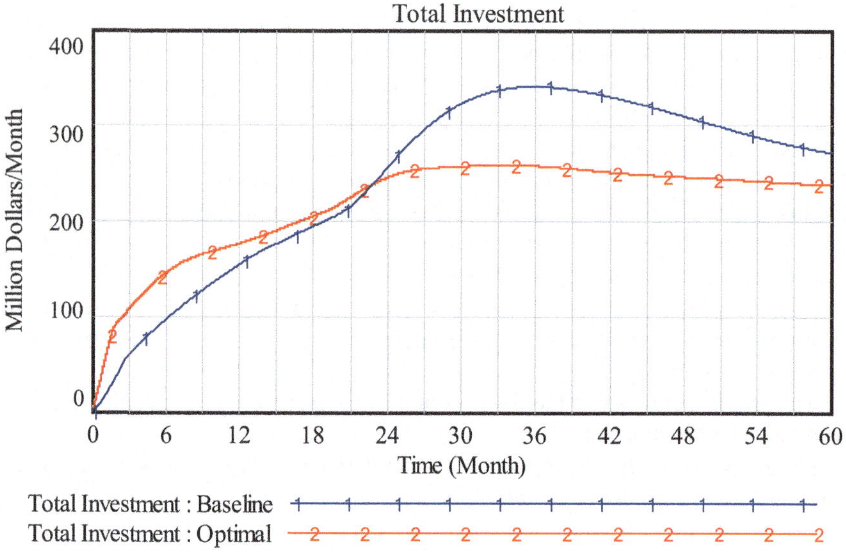

Fig. 21 Comparison of total investment results

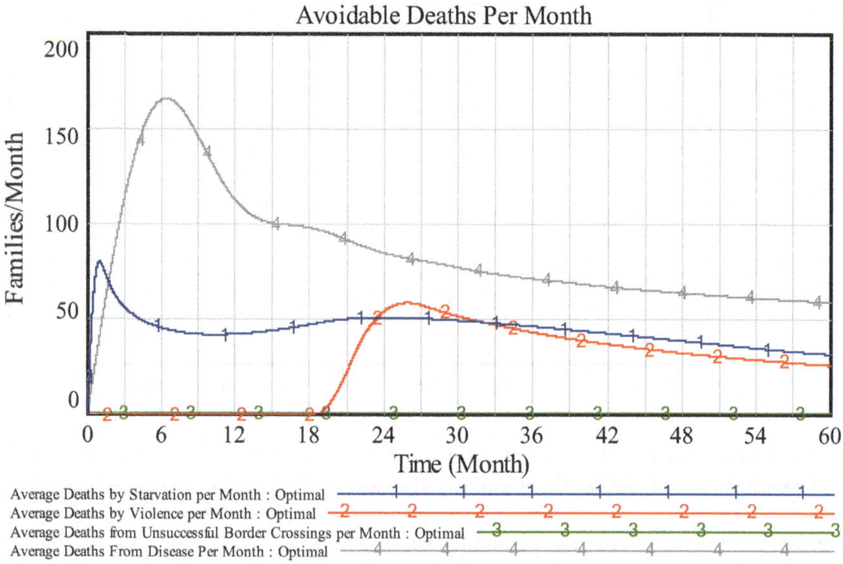

Fig. 22 Average avoidable deaths per month for the optimal run

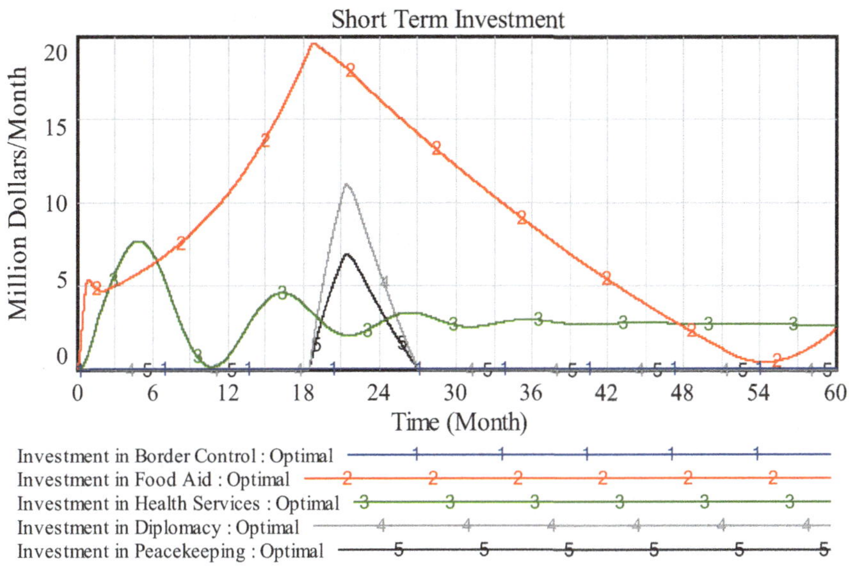

Fig. 23 Short-term investment results in the optimal run

The short-term investment in aid seems to be much lower per month in the optimal case compared to the baseline case as shown in a comparison of Figs. 18 and 23.

The details of the genetic algorithm optimization are provided at https://github.com/ivanwtaylor/Refugee-Crisis.

8 Unintended Consequences and the Need for Further Work

8.1 Unintended Consequences

In this part, we will discuss some of the potential unintended consequences that are contained in the System Dynamics model and some of the potential unintended consequences that still need to be modelled.

We have seen the potential of System Dynamics modelling to provide insight into the complex problem of the refugee crisis. We have found a potential positive unintended consequence; namely, if we can minimize the deaths from violence, starvation and disease, and improve the immigration process and create jobs through economic development, this will also resolve the problem of deaths caused by unconventional border crossings. However, it is not clear how the aid nations could influence some of the variables which would need to be modified to make these positive changes. On the other hand, we have seen that the direction of the changes have been indicated even if the aid nations can not arrive at the ideal values.

There are a number of other possibly negative unintended consequences that are readily apparent from the System Dynamics model's balancing feedback loops. For example, the investment in peacekeeping and diplomacy might reduce the level of violence in the short-term but once the level of violence is reduced, there will be less pressure on the aid nations to invest in further peacekeeping and diplomacy. Since violence is assumed to increase without interventions, the level of violence will start to rise again in the longer-term once the pressure is off the aid countries to provide peacekeeping and diplomatic effort. This will cause an oscillation in violence levels decreasing and increasing indefinitely without ever completely resolving the violence and the refugee problem in the long-term.

Similarly, investment in food aid may reduce the number of people without food in the short-term but then there will be less pressure on the aid nations to provide further food aid. Once the violence level increases again, the ability to produce food in the country will be reduced causing the number of people without food to increase again. This would create another oscillation effect with food aid in the long-term.

All of the balancing feedback loops in the model would produce the same type of oscillating result.

When this chapter was reviewed by a refugee from Yemen,[1] he noted a number of negative unintended consequences that were not modelled. First, there is the potential of food aid being abused. He has seen refugees lie about the size of their families to obtain more food than their family needs and then take the extra food aid to the market and sell it to obtain extra cash. This has the unintended consequence undermining the producers of food in the country and ruining their ability to make a living by providing food for their fellow citizens thereby creating long-term problems of food insecurity.

Figure 24 shows an influence diagram of the unintended consequences of food aid. We can start with the food needed. The more food needed, the more potential money that can be earned by refugees selling food aid. The more potential money earned, the higher the stated need for food. The higher the stated need for food, the more food aid delivered to a family of refugees. The more food aid delivered to a family of refugees, the larger the gap between the food aid delivered and the food needed by the family. The larger the gap between the food aid delivered and the food needed by the family, the more food is available to sell in the market and the more money earned by the refugee family. However, the more food in the market sold by refugees, the less food sold in the market by farmers. The less food sold by farmers, the less money earned by farmers and the less farmers' profit. The less farmers' profit, the less food produced and the less farmers' food sold in the market. This vicious cycle repeats until the farmers are driven out the market by refugee

[1]Private discussion Ivan Taylor and Ali Sheikh, 28 Oct 2017.

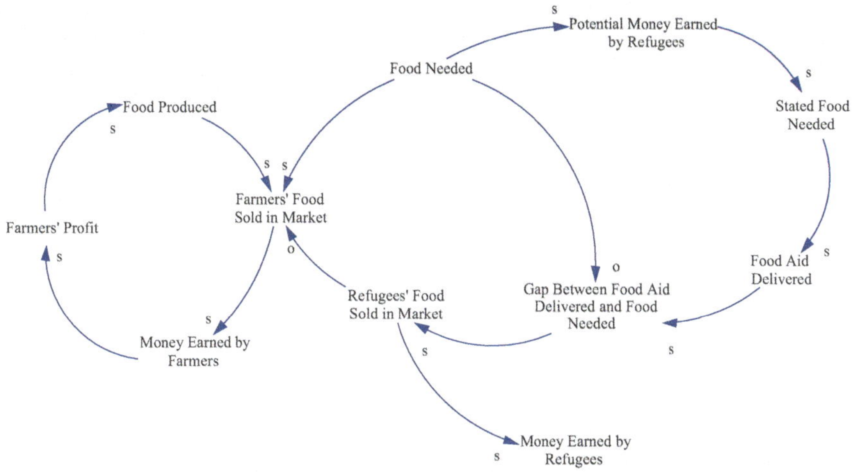

Fig. 24 Model of unintended consequences of food aid

families overstating their food need and selling the extra food which will make the refugees completely dependent on food aid to meet their true needs in the long-term.

Another negative unintended consequence noted, that was not modelled, was caused by the payment of wages to peacekeepers in the country. Peacekeepers from many African nations would get paid at a much higher rate in the peacekeeping role than they would be paid as soldiers in their own country. Therefore, it may not be in their best interest to resolve the violence quickly. They may even instigate more violence to keep getting paid as peacekeepers.

Figure 25 shows an influence diagram of the unintended consequences of payments to peacekeepers. We can begin from the desired compensation. The desired compensation and the compensation are tied to the compensation gap in the logical way. The compensation gap is connected to the effort to reduce violence by the peacekeepers in such a way that if the compensation gap gets larger, the effort to reduce the level of violence gets smaller and vice versa. The more effort to reduce the violence, the lower the level of violence and the less need for peacekeepers. The less need for peacekeepers, the fewer peacekeepers in the country and the less compensation. Therefore, reducing the level of violence increases the compensation gap and reduces the effort of the peacekeepers and in this way, we can see that the peacekeepers have limited incentive to reduce the level of violence.

These negative unintended consequences are the result of second-order effects that were not directly modelled in our original System Dynamics model. They indicate the need for more work on this model to ensure the desired results are obtained and sustained.

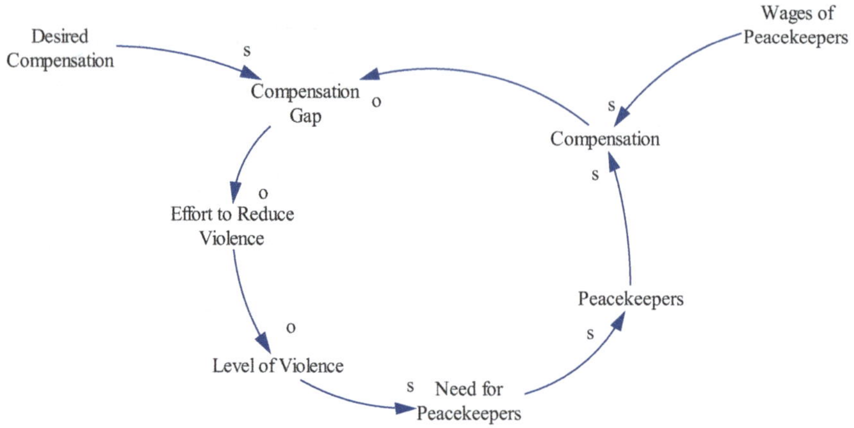

Fig. 25 Model of unintended consequences involving peacekeeping

9 Summary and Conclusions

Globally, we are in the midst of the highest levels of human displacement on record. According to the UNHCR [22] 'an unprecedented 65.6 million people around the world have been forced from home. Among them are nearly 22.5 million refugees, over half of whom are under the age of 18 … nearly 20 people are forcibly displaced every minute as a result of conflict or persecution'. The threats to human security are multiple, complex and interrelated in often mutually reinforcing ways. This certainly resonates with many of the drivers shaping the mass migration and contributing to the humanitarian crisis. The complex landscape associated with human security requires a holistic, systemic approach to understanding the underlying hazards, risk identification and management practices. In Part 2 of this report, we introduced the concept of System Dynamics modeling that we felt would be useful for modeling the refugee crisis in Northern Africa. We developed a System Dynamics simulation of the refugee crisis based on a series of Influence Diagrams. The Influence Diagrams were discussed in Part 3 of this report. To run this simulation, we needed to estimate more than 100 parameters. We did this is in rough manner to obtain a baseline run. The philosophy of System Dynamics modeling is that one should estimate the parameters of the model in a rough manner rather than ignore the factor altogether. A prototype System Dynamics model was developed in the special-purpose language Vensim. This was useful for testing the concepts and documenting the sub-models in stock and flow diagrams and equations in English language expressions. This was discussed in Part 4 of this report. Then the model was converted to the general-purpose programming language Fortran and tested against the Vensim version of the model. The Fortran version of the model was discussed in Part 5 of this report. The Fortran version of the model was useful for conducting sensitivity analysis on the parameters. The detailed sensitivity analysis of the parameters allowed us to determine which parameters needed to be estimated with greater care to produce accurate projections and which were less important in the model. This sensitivity analysis was conducted using a technique from statistical analysis called design of experiments. We found that a subset of 24 variables were the most important to estimate accurately while the remaining variables could be estimated in a rough manner because they had less impact on the model projections. This sensitivity analysis is documented in Part 6 of this report. Finally, in Part 7 we discussed how we were able to minimize the number of avoidable deaths over the five year period of study by varying the parameters under the control of the aid agencies. We found that by using information on the crisis for a rapid response in terms of aid and by using this aid in a highly effective manner, there is the potential to save many thousands of lives.

References

1. Ackoff R (1994) Systems thinking and thinking systems. Syst Dyn Rev 10(2–3):175–188
2. Clayton J, Holland H (2015) Over one million sea arrivals reach Europe in 2015. http://www.unhcr.org/afr/news/latest/2015/12/5683d0b56/million-sea-arrivals-reach-europe-2015.html
3. Foresight (2011) Migration and global environmental change: final project report. The Government Office for Science, London https://www.gov.uk/government/uploads/system/uploads/attachment_data/file/287717/11-1116-migration-and-global-environmental-change.pdf
4. HPN (2016) Special feature: refugees and vulnerable migrants in Europe. Humanitarian Exchange Number 67 September 2016 http://odihpn.org/wp-content/uploads/2016/09/HE-67-FINAL.pdf
5. Kaundert M, Masys AJ (2017) Mass migration, humanitarian assistance and crisis management: embracing social innovation and organizational learning. In: Masys AJ (ed) Security by design. Springer Publishing, Berlin
6. Masys AJ (ed) (2016) Disaster forensics: understanding root cause and complex causality. Springer Publishing, Berlin
7. Masys AJ (2013) Human security—a view through the lens of complexity. In: Gilbert T, Kirkilionis M, Nicolis G (eds) Proceedings of the European conference on complex systems 2012. Springer Proceedings in Complexity, pp 325–335
8. Masys AJ, Ray-Bennett N, Shiroshita H, Jackson P (2014) High impact/low frequency extreme events: enabling reflection and resilience in a hyper-connected world. In: 4th international conference on building resilience, 8–11 September 2014, Salford Quays, United Kingdom. Procedia Economics and Finance vol 18, pp 772–779
9. Mingers J, White L (2010) A review of the recent contributions of systems thinking to operational research and management science. Eur J Oper Res 207:1147–1161
10. Navarro J, Hayward P, Voros J (2008) How to solve a wicked problem? Furniture foresight case study. Foresight 10(2)
11. ODI (2016) Special feature: refugees and vulnerable migrants in Europe. Humanitarian Exchange Number 67 September 2016. http://odihpn.org/wp-content/uploads/2016/09/HE-67-FINAL.pdf
12. Reason J (1990) Human error. Cambridge University Press
13. Ritchey T (2013) Wicked problems: modelling social messes with morphological analysis. Acta Morphologica Generalis AMG 2(1)
14. Rittell HW, Weber MM (1973) Dilemmas in general theory of planning. Policy Sci 4:155–169

15. SDS (2017) The System dynamics society—What is SD. http://www.systemdynamics.org/what-is-s/
16. Sagan S (1993) The limitation of safety—organizations, accidents and nuclear weapons. Princeton University Press, Princeton
17. Salama P, Spiegel P, Talley L, Waldman R (2004) Leassons learned from complex emergencies over the past decade. Lancet 364:1801–1813
18. Sterman J (2000) Business dynamics: system thinking and modeling in a complex world. McGraw-Hill
19. UNHCR (2015) The sea route to Europe: The Mediterranean passage in the age of refugees. http://www.unhcr.org/5592bd059.pdf
20. UNHCR (2016) Mediterranean death toll soars, 2016 is deadliest year yet. http://www.unhcr.org/news/latest/2016/10/580f3e684/mediterranean-death-toll-soars-2016-deadliest-year.html
21. UNHCR (2017) Figures at a glance. http://www.unhcr.org/figures-at-a-glance.html
22. WEF (2015) Global Risks 2015 10th Edition: Insight Report

Cyber-Physical Systems Governance: A Framework for (Meta)CyberSecurity Design

Polinpapilinho F. Katina and Charles B. Keating

Abstract This chapter discusses a framework for improving security of cyber-physical systems through purposeful design, execution, and evolution of metasystem functions. State actors (i.e., government agencies), non-state actors (i.e., for-profit and non-profit organizations), and their systems operate under highly emergent and complex conditions. Under these conditions, system performance is not always deducible from the constituent systems. Moreover, such systems are often interdependent and dynamically interacting with other systems such that the state of each system is influenced by and is influencing states of the interconnected systems. In these conditions, leaving a system to develop through processes of 'accretion' (ad hoc evolution of a system) or 'self-organization' (totally unconstrained evolution of a system) might increase the probability of missing performance expectations. In contrast, 'purposeful design' is an invitation for a more determined engagement in system development to increase the probability of producing expected and desired performance. In this chapter, emerging research in *Complex System Governance* (CSG) is suggested as an emerging field to direct more purposeful design for systems. Specifically, a CSG enabled framework for security design focused on Cyber-Physical Systems (CPS) is provided. The framework, grounded in Systems Theory and Management Cybernetics, emphasizes more holistic design for integration, coordination, communication, and control for development of CPS.

P. F. Katina (✉) · C. B. Keating
National Centers for System of Systems Engineering,
Engineering Management and Systems Engineering,
Old Dominion University, Norfolk, VA, USA
e-mail: pkatina@odu.edu; pkatina@uscupstate.edu

C. B. Keating
e-mail: ckeating@odu.edu

P. F. Katina
Division of Natural Sciences and Engineering,
University of South Carolina – Upstate, Spartanburg, SC, USA

A. J. Masys (ed.), *Security by Design*, Advanced Sciences and Technologies for Security Applications, https://doi.org/10.1007/978-3-319-78021-4_7

Keywords Complex System Governance (CSG) · Cyber-Physical Systems (CPS) Management cybernetics · MetaCyberSecurity · Metasystem · Systems theory Vulnerability assessment

1 Introduction

In 2016, an attacker accessed the Society for Worldwide Interbank Financial Telecommunication (SWIFT) system and instructed the Federal Reserve Bank of New York to transfer over US$900 million from Bangladesh Bank's account to different accounts in the Philippines. While US$850 million, were prevented by SWIFT, US $101 million went through. Of these, US$20 million, traced to Sri Lanka, were recovered. US$81 million, transferred to the Philippines, were laundered through casinos and some were then forwarded to Hong Kong [1]. This attack on Bangladesh bank's SWIFT system highlights vulnerabilities in financial and banking systems.

Unfortunately, such vulnerabilities are not limited to banks. Such vulnerabilities are becoming common to other systems, including, among others, energy (i.e., electricity generation, transmission and distribution, oil and gas production, transport and distribution), telecommunication systems, water supply (i.e., drinking water, waste water, and sewage), agriculture (i.e., food production and distribution), public health (i.e., hospitals, ambulances), and transportation systems (including aviation, highways, maritime, railway networks, and freight and pipeline systems). It is well-established that such systems "…are so vital that their incapacity or destruction would have a debilitating impact on the defense or economic security of the United States [or any other nation for that matter]" [2: p. 37, 347]. Gheorghe and his colleagues [3] go on to suggest that such systems are "critical because they provide services that are vital to one or more broad governmental or societal functions or attributes. This can be related to survivability of citizens as far as the safety of their life is concerned, or to their quality of life" (p. 5). And yet, there appears to be an increase in prevalence of threats and attacks[1] on such systems with high potential impact [1, 4]. To this end, Rasmussen and Batstone [5] assertion that: "the frequency and magnitude of organizational failures and the subsequent impacts are increasing at an alarming rate" [5: p. ii] stands true, decades later.

These events are responsible, in part, for increasing interest in understanding operating states of infrastructure systems. This understanding includes concepts of 'risks' and 'vulnerabilities' and means of increasing infrastructure 'resilience' as well as 'deterring' threats. A key aspect of related research is classification of threats. In this case, threat classification is focused on the grouping of different risks based on several different measures. These measures might include potential cost or likely impact, likelihood of occurrence, and required countermeasures. At the very

[1]See for example, http://www.swissre.com/; http://globalriskinsights.com/; https://www.weforum.org/reports/the-global-risks-report-2017.

least, one ought to consider three possible classifications of threats: *natural*, *technical*, and *malicious*. A natural threat can include events such as earthquakes, floods, hurricanes, fires, and extreme heat. Occurrence of these events can lead to incapacity and disruption in services of critical systems (e.g., banks and healthcare systems), property loss, and even death. A technical threat involves accidents and human error in engineered systems—including elements of ubiquitous computing and information and telecommunication technologies, which are increasingly embedded in many aspects society. These systems can fail, unintentionally, from our inability to achieve 100% reliability and is evidenced by accidents and human error. A malicious threat, on the other hand, involves intentional failure of systems and can include acts of terrorism, insiders, sabotage, and state-sponsored threats [1].

In each classification, given a system of interest, specific threats can be identified and means for dealing with such threats can then be developed. Furthermore, incentives can be established for enhancing capability for detecting threats, risks, and vulnerabilities, preventing threats and attacks, protection against attacks, developing approaches to mitigate effects of attacks, and means to recovering from attacks. Obviously, such activities involve decisive efforts. However, such efforts must be undertaken in recognition of several realities involving the operating landscape and nature of systems in the 21st century [6]. Table 1 is drawn to elaborate on the landscape of realties for systems the 21st century.

The grim reality could be that these conditions will escalate, accelerate, and most likely get worse before stabilizing or getting better. In response, our systems (including state actors such as government agencies and non-state actors such as for-profit and non-profit organizations) frequently confound, disappoint, and often fail to achieve desired performance levels, irrespective of increased investment of resources or well-intentioned rhetoric.

With respect to system development, we argue that most systems generally take one of three primary developmental paths: accretion, self-organization, or purposeful design. System development by accretion involves leaving the system to evolve in an 'ad hoc' or haphazard way. This is similar to the forming of stellar systems where systems and their parts (including process) accumulate into 'massive objects' by gravitationally attracting more matter in ad hoc fashion such that the whole is fragmented with little to no sense [7, 8]. In contrast, self-organization involves leaving the system to 'self-organize', which involves the structure and resulting behavior developing with minimal oversight and constraint [9]. The result is a system that evolves virtually unfettered, consuming minimal energy to achieve its state. The final system development perspective involves purposeful design, execution, and evolution of a system—with implications for advancing the state of the system. Table 2 provides a summary of the three contrasting forms of system development.

It should be evident that using 'accretion' and 'self-organization' as system development approaches are not preferred to effectively deal with the realities of the 21st century. They are simply not sustainable or consistent. Using these approaches, one might 'get lucky.' More often, however, 'we get what we get' and are often disappointed, particularly as systems and their problems become more complex. System development by accretion or self-organization represent an absence of

Table 1 Five realities for 21st century systems

The reality	Description	Example
Ambiguity	Increasing lack of clarity in understanding and interpretation of both the system and context within which they exist	Difficulties in establishing the boundary conditions for what is included and excluded in a system for analysis purposes (e.g., which stakeholders might be considered most relevant may vary)
Complexity	A high number of systems so intricate and interconnected that complete understanding, prediction, control, or explanation is impossible	A transportation system might include any or all of automobiles, trains, planes, water craft, and pedestrians. The ability for complete understanding is unattainable given the magnitude of elements and interactions
Emergence	Unpredictability of events and system behaviors from constituent elements. Such behaviors might be known and experienced after occurrence of events. However, there is lack of ability to predict such behavior	When a catastrophic weather event occurs, the ability to predict direct and indirect consequences (e.g., traffic flows, flooding damage, causalities, etc.) is doubtful
Interdependence	There is mutual influence among systems through which the state of each system influences and is also influenced by the state of interconnected systems	Number of internal and external relationships in a Smart Grid. These relationships (i.e., links) expose Smart Grids to threats originating from interdependent systems
Uncertainty	Presence of incomplete knowledge of situations casting doubt for decisions and actions. Decisions and actions are taken without full knowledge of consequences	Deployment of new communication technologies to monitor emergence response resulting in bypassing traditional command structure

Table 2 Three contrasting forms of system development

	System development type		
Characteristic	Accretion	Self-organizing	Purposeful
Structure/behavior/performance	Fragmented	Emergent	Designed
Development energy consumed	Medium	Low	High
Primary focus for improvement	Isolated/piecemeal	Self-balancing/Laisse Faire	Holistic/integrated
Design preference	ad hoc	Unfettered	Intertional

purposeful system design. The absence of purposeful system design and development can be evidenced by deficiencies spanning the holistic spectrum of technology, organizational, managerial, human, social, information, policy, and political dimensions. However, as noted in Table 2, although purposeful system development is preferred, it does take the greatest resource expenditures to achieve.

As we consider the development of Cyber-Physical Systems, our objective for this chapter is to advocate for purposeful design consistent with the emerging Complex System Governance (CSG) perspective and approach to system development. To achieve this objective, the remainder of this chapter is structured as follows: Sect. 2 provides a review of literature related to frameworks for dealing with hazards, risks and vulnerabilities. Emphasis is placed on Cyber-Physical Systems (CPS). Section 3 establishes foundations of CSG as a system design framework. The aim is to provide the conceptual foundations of the proposed framework of CSG for Cyber-Physical System development, including CSG grounding in General Systems Theory (i.e., the set of laws, principles, and concepts that govern all complex systems) and Management Cybernetics (i.e., the system science of effective organization/structure) and associated Metasystem functions. In Sect. 4, a CSG-based framework for CPS is articulated. This framework is developed in direct response to better assert development of MetaSecurity issues (i.e., security at the meta-level) in CPS. The chapter concludes with a set of implications of research for practitioners and potential research directions to accelerate CPS development through inclusion of CSG.

2 Literature Review

The terms 'risk' and 'vulnerability' are different, even though they are related. Risk is typically defined in terms of *probability of occurrence* of an event and magnitude of the resulting *consequences* [10]. However, there are different perspectives on the concepts of risk, including, for instance, risk being associated with determination of event sequences and evaluation of event frequencies and probabilities in industrial settings [11]. In systems analysis, risk is associated with probabilities of unknown outcomes and uncertainties [12]. In systems engineering, risk is related to technical factors in the systems life cycle [13]. Risk is also related to one's perception and relative to one's position in an organizational setting [14]. Holton [15] and Knight [16] limit presentation of risk to terms of uncertainty.

A consistent among these definitions, arguably, is the implication that risk is that which occurs without one's planning, anticipation, or intent. Of interest, however, is the manner of analysis associated with risk—risk assessment. In risk assessment, one might select a particular stress (i.e., threat or hazard) of concern and then seek to identify its consequences on the system of interest [17]. In such assessments, the objective can be a general classification of risk as 'acceptable' or 'unacceptable' and representing results in a risk matrix. This is where concepts such as ALARA (i.e., *As Low as Reasonably Achievable*), play a major role in taking every reasonable effort to reduce exposures to hazards and to limit the potential for harm.

Vulnerability, like risk, has many definitions, accepted to various degrees with no one universally accepted definition [17–19]. Einarsson and Rausand [20] as well as Holmgren et al. [21] and Turner et al. [22] suggest that vulnerability should be seen in terms of system properties that may weaken (or limit the ability of a system to survive and perform its mission) a system in the presence of threats that can originate both within as well as outside the system boundaries. A closely related term is *degree of vulnerability*. In Song's [17] research, the difference between vulnerability and degree of vulnerability is made more obvious: vulnerability is a system's susceptibility and resilience to hazards while degree of vulnerability is a numerical index of vulnerability based on different criteria, usually in the range of 0–100%. In the present discussion, *Quantitative Vulnerability Assessment* (QVA) might take a central role in expressing vulnerability in numbers, defensible in the so-called scientific approach to vulnerability assessment [19]. QVA is a result of a warranted analogy with *Quantitative Risk Assessment* (QRA), a term coined in decades past by risk and safety communities and focused on the numerical articulation of risk.

To make a distinction between risk vulnerability more apparent, at least in the manner of analysis, Song [17] offers the following explanation: in *risk assessment*, one might select a particular stress (i.e., threat or hazard) of concern, and then seeks to identify consequences for a variety of system properties. In *vulnerability assessment*, one selects a particular system (or component) of interest and examines how it can be affected by a variety of stressors. In risk, such analysis will involve identification of risks and reduction of consequences while in vulnerability, one identifies vulnerabilities and focuses on their reduction [23]. In Song's clear-cut words, vulnerability describes "inherent characteristics of a system that create the potential for harm but are independent of the risk of occurrence of any particular hazard" [17: p. 19]. Arguably, risk and vulnerability are related, assessment-wise. In assessing risk, one might discover hazards and means of affecting the system. At the same time, one discovers vulnerabilities (to hazards) in the system and develops or enhances system properties against the discovered hazards.

Keeping in mind the concepts of risk, risk assessment, vulnerability, and vulnerability assessment, a simple but relevant activity might be the identification of frameworks for dealing risk and vulnerability in complex systems. In the present case, a complex system is defined as 'a system composed of a group of related units (subsystems), for which the degree and nature of the relationships is imperfectly known' [24]. Under this definition, Katina [25] suggests that 'it is difficult to predict overall system behavior even if the behaviors of subsystems are predictable.' This notion is grounded in Guckenheimer and Ottino's [26] characteristics of complex systems: (i) many interacting parts, (ii) emergent behavior, (iii) adaptation and change, and (iv) systems uncertainty. Any approach for dealing with such systems requires going beyond traditional approaches based in the reductionist paradigm (understanding by braking a system into increasingly smaller elements). In this case, reductionism is closely aligned with the traditional scientific method, which holds that a complex organism is nothing but the sum of its parts, and therefore they can be reduced to constituent elements [27–29].

However, reductionism has been rejected by prominent researchers in favor of the 'science of wholes' approach typified in the movement of General Systems Theory [27, 28, 30–46]. In the view of von Bertalanffy [45], the "method [traditional scientific approach] worked admirably well insofar as observed events were apt to be split into isolable causal chains, that is, relations between two or a few variables. It was at the root of the enormous success of physics and the consequent technology. But questions of many-variable problems always remained" (p. 409). In so far as dealing with complex problems of the 21st century, Hammond, [27] notes that "problems confronting humanity at this stage in our history (poverty, violence, crime, environmental degradation and nuclear weapons…terrorism) are systemic and cannot be understood or resolved in isolation…we needed a more ecological or systemic world, based on an understanding of our fundamental interconnectedness and interdependence, with each other and with all of life" (p. 430).

And while a General Systems Theory (GST) has yet to emerge [25, 47, 48, 49], the aspects of GST describing isomorphic concepts, laws, principles, and theorems applicable to different systems is prevalent in literature [29, 38, 47, 50–55]. Consequently, system practitioners and theorists have developed systems theory-based frameworks and methodological approaches that can be used to understand complex systems including their structures and behaviors [25, 56, 57]. In this case, a methodology includes theoretical underpinnings and is used to "refer to methods for exploring and gaining knowledge about systems" [58: p. 3] and is defined as "…procedures for gaining knowledge about systems and structured processes involved in intervening in and changing systems" [58: p. 134]. This definition is further elaborated by Jackson [56], who suggest a methodology as a source of "guidance given to practitioners about how to translate [systems] philosophy and [systems] theory of an approach into practical application" (p. 51). There is no shortage of methodological approaches underpinned by GST as depicted in Table 3. The proponents associated with each methodology are discussed elsewhere (e.g., see Jackson [56]; Katina and Calida [57]). While there is overlap among these methodologies, each methodology is purposely designed and offers utility in the context of problematic situation as well as purpose of the analysis [57, 59].

In view of the purpose of this exploration, purposeful design for security in critical Cyber-Physical Systems (CPS), there remains a scarcity of methodological approaches addressing hazards, risks and vulnerabilities associated with operations in CPS. A critical CPS was previously defined as "a system of systems that provides essential goods and services necessary for public well-being with the aid of control systems in the form of information and telecommunications" [60: p. 174]. In this construction, critical CPS encompasses aspects of hardware, workforce, environment, facilities, business and organizational entities as well as complex interactions essential for the maintenance of vital societal functions [60, 61]. Certainly, these systems include, among others, health, safety, security, economic or social well-being of people. The destruction or disruption of these systems would have a significant impact on their users and associated interdependent systems.

The conceptions above are drawn from research in *critical infrastructures*—especially the emerging ideas related to CPS. For simplicity, critical infrastructures are "…systems and assets, whether physical or virtual, so vital to the United States [or any nation] that the incapacity or destruction of such systems and assets would

Table 3 A summary of GST-based methodologies

Classification	Methodology	Description
Hard systems	Systems analysis	This methodology brings to bear tools of 'modern science' and technology, searching for regularizes in system behavior and to provide evidence related to different aspect of the system, including cost, benefits, and consequences various possible responses to the problem at hand
	Systems engineering	Focused on defining technical and business customer needs with the goal of producing quality products that meet user needs. This approach emphasizes the life cycle associated with systems
	Operations research	Commonly associated with determining a maximum or minimum variable (e.g., profit, performance, yield, loss, risk). Such application areas as inventory, resource allocation, waiting-time, replacement, and processes, operations research was developed to deal with issues by the application of mathematical modeling and analysis
Soft systems	Systems dynamics	Concerned with limits of growth and understanding of the system structure using feedback loops as the main determinants of system behavior. Mathematical in nature, this approach involves examination of system boundaries, the network of feedback loops, variables of 'rates' or 'flows', 'levels' or 'stocks,' and leverage points impacting system outputs
	Organizational cybernetics	Embodies the idea that organizations are black boxes characterized by complexity, self-regulation and probabilistic behaviors. Central to this approach is the Viable System Model, which is based on a neurocybernetic model, consisting of five essential subsystems that provide for continued system viability (existence). VSM is a model rather than a methodology, as it does not have a clear set of prescribed phases for deployment
	Strategic assumption surfacing and testing	Grounded on the premise that the formulation of a right solution to the right problem requires uncovering critical assumptions underlying policy, planning, and strategy. This enables management to compare and contrast and gain new insights based on making explicit and challenging potentially faulty assumptions inherent in formulation of complex system problems
	Interactive planning	Focused on creating a desired future by designing desirable present conditions. It is made up two parts: *idealization* (focused on the desirable state) and *realization* (understanding what is necessary to achieve that state). These parts are through six interrelated phases

(continued)

Table 3 (continued)

Classification	Methodology	Description
Soft systems	Soft systems methodology	Emerged as a response to a recognized need for methods that can be used to intervene in 'ill-structured' problem situations. This approach emphasizes the importance of learning about systems while still focusing on 'goal-seeking' endeavors that answer 'what' should be done and 'how' it should be done. Additionally, this methodology places emphasis on understanding the 'soft' aspects of system issues (e.g. socio-political considerations) ignored in systems engineering
	Systems of systems engineering methodology	Intended to provide a high-level analytical structure to explore complex system problems related to the integration of multiple and disparate systems. To enhance our understanding of complex systems, SoSEM invests heavily in the understanding and problem framing of issues related to multiple system integration
	Critical systems heuristics	Focused on addressing 'unfairness in society.' It promotes emancipatory systems thinking for planners and citizens alike. The approach emphasizes guided (heuristic question based) reflection (critical inquiry) from a holistic (systems) perspective
	Organizational learning	Concerned with single-loop and double-loop as essential learning approaches to detect and correct errors (undesirable variabilities) in a system. Learning is based on contrasting 'expected outcomes' with the 'obtained outcomes' to identify variance. Contrasting these outcomes involves learning based on correcting errors through single-loop (improving compliance to existing system) or double-loop (modifying organizational (system) norms, policies, and objectives) [110]
	Sociotechnical systems	Concerned with engaging in both social system influences, technical system influences, and their influence on one another. The overall system is modified based on conducting a joint optimization of the technical and social systems
	Total systems intervention	A meta-methodology that emerged out of the recognition of strengths of capabilities of individual systems approaches, the need for pluralism in systems thinking, and calls for emancipatory ideas in systems thinking. The approach is based on engaging in three interrelated aspects: (1) creativity—targeted to systemic framing of the situation/system, (2) choice—selection of systems approach or combination of approach(es) based on the framing, and (3) implementation—deployment of the approach to improve the system
	Complex system governance	An approach based on the design, execution, and evolution of nine metasystem functions that provide for the control, communication, coordination, and integration of a complex system

have a debilitating impact on security, national economic security, national public health or safety, or any combination of those matters" [62]. This definition can be applied to CPS when one considers goods and services that CPSs offer, especially in the context of operational environments characterized by high levels of ambiguity, complexity, emergence, interdependence, and uncertainty. The following summary statements follow the characterization of CPS found in Alur [63] and Katina et al. [60, 61] research:

- CPS must respond to changes in their environment on a continuous basis via input and feedback from output parameters to provide appropriate responses. The appropriate responses involve maintenance of critical service reliably.
- CPS depend on concurrency principles in which multiple events can occur simultaneously. These events include, among others, computations or different interacting systems and events.
- CPS use feedback loops to correct deviations necessary to maintain consistency in the path forward. In this case, feedback loops are part of control systems essential to CPS. Control systems (e.g., supervisory control and data acquisition: SCADA) operate with coded signals over system communication channels.
- The design of CPS considers real-time performance and computation as well as reporting of results. In fact, the criticality of CPS calls for real-time monitoring of system products (i.e., goods and services) essential for public well-being. Monitoring might involve measuring system performance. Short and long-term responses to indicators might then include approaches (e.g., for prevention, protection, mitigation, among others) in response to threats such that the system is made more resilient.
- A general consensus is that CPS are designed for 'safety-critical' applications where operational errors can lead to intolerable consequences (e.g., loss of life). In these cases, it is paramount to establish a system works as-correctly-as-possible prior to deployment.

Perhaps a key advantage of CPS is the increasing ability to remotely sense for risks, creating the opportunities for better decisions and actions related to systems and their environment [63, 64]. However, this capability comes at a price. First, there is a need for computing elements (including software) to coordinate and communicate with sensors. Second, sensors must be able to monitor cyber and physical aspects of the system and the environment based on predetermined indicators. Thus, these relationships create co-dependence between cyber systems and physical systems, producing a tight coupling of CPS at different scales—both temporal and spatial. This tight coupling is becoming prevalent in many systems, including, automatic pilot avionics, autonomous automobile, medical monitoring, process control, robotics, smartphones, and smart grids. For instance, in the transport case, individual automobiles and road infrastructure systems are able to communicate with other automobiles sharing real-time data (e.g., traffic flow, location, and other issues). This information could then be used in accident prevention and congestion avoidance.

However, such potential advantages must be addressed within the context of increasing prevalence of attacks and threats to infrastructure systems and the potential for catastrophic consequences [1, 4, 65–67]. These concerns are not new. Examples include, among others, recent power blackouts in Brazil [68], several industrial episodes [69], SQL Slammer worm [70], StuxNet computer worm [71], and water breach [72].

The European Parliament has noted that while CPS technologies might have "started out as a military application for tactical intelligence gathering, drones are increasingly applied in humanitarian operations and civil surveillance tasks such as border patrol…[such technologies have become] cheap to buy and easy to operate, thus criminals (including terrorists) could use these to deliver lethal effect" [66: p. 67].

In terms of vulnerability, [1: p. 22] recognizes that "cyber attacks are not necessarily sophisticated or inevitable and are often the result of exploited—but easily rectifiable and, often, preventable—vulnerabilities." That is, there is a need to consider the nature of our systems (CPS) as well as those who (i.e., people) operate such systems. Certainly, these efforts include assessing vulnerabilities in CPS—from individual smartwatches to industrial-level Internet-of-Things and national and international systems. This also entails consideration of behavioral factors influencing safety culture associated with CPS usage. For the United Kingdom (UK), this has translated into a vision of creating a 'nation that is secure and resilient to cyber threats; a nation that is prosperous and confident in the digital world' [1]. Implementation of such a vision includes deterrence, defense, developing, international action and cross-cutting influence strategies.

Suffice to say, "cyber-physical systems are prone to failures and attacks on their physical infrastructure, and cyber attacks on their data management and communication layer" [73: p. 2715]. The previous examples serve to illustrate the need for enhancing capabilities in detection, prevention, protection, and recovering from incidents that might affect CPS operability. A consideration for classification of threats as 'natural,' 'technical,' 'malicious,' or a 'combination' thereof, might be a good starting point. Strategic approaches for dealing with CPS-related hazards (i.e., risk and risk assessment approaches) and system development (i.e., vulnerability and vulnerability assessment) are necessary.

The U.K. has promoted the vision to be: 'secure and resilient to cyber threats, prosperous and confident in the digital world' and developed strategies to support that vision [1] as depicted in Table 4.

Obviously, attaining the articulated vision, requires purposeful design involving, for instance, 'creating stronger information sharing networks' and 'creating deterrence measures.' These activities must be coupled with increasing *cyber hygiene*—being aware of technical vulnerabilities and avoiding the legacy and unpatched systems, *sufficient training and skills*—having the requisite knowledge to meet cyber security needs across different systems, and knowing the hacking tools and resources that could be used in a cyber-attack [1].

Methodologically, there is currently a lack of theoretically-grounded holistic approaches to guide practitioners dealing with emerging hazards and vulnerabilities

Table 4 U.K.'s national cyber security strategic outcomes

Strategic approach	Strategic outcome
Deterrence	Ensuring the U.K. has the capability to effectively detect, investigate, and counter cyber-threat activities of their adversaries
	Ensuring that impact of cybercrimes on the U.K. and its interests is significantly reduced. Developing means to ensure that cybercriminals are deterred from targeting the U.K.
Defense	The U.K. has the capability to manage and respond effectively to cyber incidents to reduce the harm caused to the U.K. while effectively countering cyber adversaries
	Our partnerships with industry for active cyber defense result in large scale phishing and malware attacks that are no longer effective
	The U.K. is more secure as a result of technology products and services having cyber security designed into them and activated by default
	Government networks and services will be as secure as possible from the moment of their first implementation. The public will be able to use government digital services with confidence, and trust that their information is safe
	All organizations in the U.K., large and small, are effectively managing their cyber risk, are supported by high quality advice designed by the NCSC [National Cyber Security Centre], underpinned by the right mix of regulation and incentives
Developing	There is the right ecosystem in the U.K. to develop and sustain a cyber security sector that can meet our national security demands
	The U.K. has a sustainable supply of home grown cyber skilled professionals to meet the growing demands of an increasingly digital economy, in both the public and private sectors, as well as defense
	The U.K. is universally acknowledged as a global leader in cyber security research and development, underpinned by high levels of expertise in U.K. industry and academia
	The U.K. government is already planning and preparing for policy implementation in advance of future technologies and threats and is 'future proofed.'
International action and influence	The threat to the U.K. and our interests overseas is reduced due to increased international consensus and capability aimed towards responsible state behavior in a free, open, peaceful and secure cyberspace
Cross-cutting	U.K. Government policies, organizations and structures are simplified to maximize the coherence and effectiveness of the U. K.'s response to the cyber threat

to CPS [60, 61, 74]. Three aspects elaborate this deficiency. First, there is a recognition that traditional probabilistic approaches might not be sufficient to address emerging threats as they are not based on 'experiences' [75]. Second, the intricate interconnections among the different systems might raise questions based on 'analysis' that suggests a system is, or can be, clearly isolated [19, 76].

This is especially the case since some of our systems "have become hopelessly interconnected and overcomplicated, such that in many cases even those who build and maintain them on a daily basis can't fully understand them any longer" [31: p. 2]. Finally, there is a need to consider limitations of present methods [73] in addressing CPS needs. While current approaches are designed to address issues in CPS, they tend to focus on a limited set of elements of the problem domain. For instance, monitoring, using a specific approach such as graph theory, which offers limited utility in a CPS problem space that is much more holistic in nature and exists beyond graph theory capabilities to adequately address. Thus, there remains a need for much more evolved and holistic approaches to deal with CPS issues. This need serves as the subject of the remainder of this chapter. In particular, we focus on Complex System Governance as one approach that might help in fulfilling this need.

3 Foundations of CSG Governance Framework

The need for more theoretically-grounded holistic approaches to CPS is described in the preceding section. This section provides the necessary underpinnings of Complex Systems Governance (CSG) as a theoretically-grounded and holistic approach to critical CPS, with an emphasis on security[2]. CSG is an emerging field, representing an approach to improve system performance through purposeful design, execution, and evolution of essential metasystem functions [77–82]. CSG emphasizes communication, control, coordination, and integration (C3I) of complex systems though effective performance of metasystem functions necessary. The approach to CSG finds its roots at the intersection (i) *General Systems Theory*, (ii) *Management Cybernetics*, and (iii) *Governance.*

General Systems Theory (GST) emerged in the early 20th century as an approach to discovering trends in various disciplines [29]. One of the main objectives of GST was to provide an alternative to the reductionist approach closely aligned with the scientific method, which holds that a complex organism is nothing more than the sum of its parts [25]. In postulating GST, von Bertalanffy's objective was to bridge the gap that exists in different disciplines via the discovery of principles and laws common across disciplines [29]. He notes,

> A consequence of the existence of general system properties is the appearance of structural similarities or isomorphisms in different fields. There are correspondences in the principles that govern the behavior of entities that are, intrinsically, widely different [29: pp. 32–33].

[2]Security is taken in its broadest sense to include protection and deterrence, defense, developing international action and influence for cyber systems and interdependent systems in the cyberspace against cyber-threats—internal and external to the system of interest.

The traditional scientific mindset has, notes Hammond [27] is:

> ...rooted in the mechanistic worldview we inherited from the scientific revolution of the seventeenth century...we needed a more ecological or systemic world, based on an understanding of our fundamental interconnectedness and interdependence, with each other and with all of life. (p. 430)

In essence, GST is related to ideas of looking at the world through the lens of 'wholes,' 'having irreducible properties,' 'environment,' 'centralization,' 'self-organization,' and 'holarchy of nature' [28]. These ideas call to mind the need to see organizations, relationships, and interrelations among systems.

There are three related but different terms associated with GST: systems science, systems technology, and systems philosophy. Systems science places emphasis on the "scientific exploration and theory of 'systems' in various sciences (e.g., physics, biology, psychology, social sciences), and general systems theory as the doctrine of principles applying to all (or defined subclasses of) systems" [45: p. 414]. Therefore, this aspect of GST deals with knowledge of the connected 'wholes' as opposed detailed and isolated systems.

Systems Technology is the aspect of GST dealing with "problems arising in modern technology and society, including both 'hardware' (control technology, automation, computerization, etc.) and 'software' (application of systems concepts and theory in social, ecological, economical, etc., problems)" [45: p. 420]. This viewpoint of GST suggests that the world's most pressing issues involving, for instance, pollution, economies, health, politics, and international conflicts, cannot be solved in isolation from one another [45, 53, 83]. In fact, increasing levels of ambiguity, complexity, emergence, interdependence, and uncertainty suggest a holistic (systemic) and interdisciplinary approaches to complement traditional scientific-based approaches. There is no shortage of methods that can be used to gain knowledge and intervene in behaviors of complex systems (see e.g., Hieronymi [51]; Katina [84]). Consequently, the systems technology element of GST is concerned with the development of unique sets of methods and tools enabling discovery of problematic situations, understanding, and bringing about positive change.

A third element of GST is *systems philosophy*. In this case philosophy addresses the philosophical underpinnings (i.e., paradigm) within which GST operates. These underpinnings include (i) systems ontology, (ii) systems epistemology, and (iii) the nature of man. *Systems ontology* deals with how an observer views reality with two main opposing extremes (realism and nominalism) characterizing reality. Realism is focused on the perspective of reality being independent of the observer. In contrast, nominalism is rooted in the notion that reality is a 'construction' of the observer and thus does not exist independent of the observer. Epistemology deals with how we obtain and communicate knowledge with two main opposing extremes (positivism and anti-positivism) characterizing knowledge. Positivism holds that the knowledge is objective, reducible, and concrete. In contrast, anti-positivism holds that knowledge is subjective, irreducible, and fallible. The third element addresses the nature of human beings in which people can be viewed as deterministic or voluntaristic in nature. Deterministic view suggests that there is a predetermined outcome that is deducible from cause-effective relationships. The voluntarism perspective suggests that there is 'free will' with respect to determining relationships—this exist beyond predetermined outcomes or absolute cause-effect relationships. The concepts of systems philosophy, ontology, and nature of man are discussed elsewhere [38, 85, 86].

The proceeding discussion is provided for two reasons: First, there is not one common and widely accepted definition of GST. Second, GST emerged to complement and in contrast to the classical scientific approach, with an emphasis on wholes as opposed to parts. Interestingly, although GST was initially developed in mathematical terms and emphasizing isomorphic relationships [29], Hammond [27] notes that "much of his [von Bertalanffy's] writing reflects a deeper concern with the mechanistic and reductionist orientation of then current models in biology and psychology" [27: p. 436]. This is supported by the work of von Bertalanffy [45] who later stated that "classical science in its various disciplines, such as chemistry, biology, psychology, or the social sciences, tried to isolate the elements of the observed universes…We have learned, however, that for an understanding not only the elements but their interactions as well are required—say, the interplay of enzymes in a cell, the interactions of many conscious and unconscious processes in the personality, the structure and the dynamics of social systems, and so forth" (pp. 414–415).

Again, while a universally accepted GST has yet to emerge, the aspect of GST describing isomorphic concepts, laws, principles, and theorems applicable to different systems is evident in systems literature [25, 38, 47, 50, 51, 87, 88]. In the work of Adams et al. [47] and Whitney et al. [55], thirty (30) propositions—inclusive of laws, principles, and theorems—are proposed as means to investigate situations from a systems view-point along seven (7) axioms. These axioms include centrality, context, design, goal, information, operational, and viability. Most recent research suggests over 80 principles of systems theory [51]. CSG relies on this view of GST, where principles are used as a basis for governance of complex systems with an emphasis on 'integration' and 'coordination.' In this case, integration is concerned with means for continuous maintenance of system integrity. This integrity requires a dynamic balance between autonomy of constituent system entities and integration of those entities to form a coherent whole. Coordination accounts for interactions (relationships) with system entities, other systems, and the environment such that a system does not experience excessive oscillation.

Cybernetics was originally defined as the science of control and communication in animals and machines [89]. Since that time, and to extend beyond traditional management thinking, organizational cybernetics has come to signify a 'science of effective organization.' This approach embodies the idea that an organization can be seen as black boxes characterized by complexity, self-regulation and probabilistic behaviors [56]. Central to this approach is Stafford Beer's Viable System Model (VSM), which is based on a neurocybernetic model, consisting of five essential subsystems that are essential to maintain the continued existence (viability) of a system [90–92].

Using principles of communication and control, Stafford Beer developed the VSM, with supplement developments from Espejo and Harnden [93], and Keating and Morin [94]. Beer's VSM, as elaborated by other related works, envisioned the necessary and sufficient subsystems of productive (S1), coordination (S2), operations (S3), monitoring (S3 Star [*]), system development (S4), learning and transformation (S4*), and system policy and identity (S5) as essential functions necessary for organizational viability (continued existence) under turbulent environmental conditions. Recent and emerging research have extended VSM functions. In particular, the emerging field of *Complex System Governance* (CSG), is evolving as an approach to improve system performance by purposeful 'design, execution, and evolution of the metasystem functions necessary to provide control, communication, coordination, and integration of a complex system' [81]. CSG offers an extended view of system viability functions. Figure 1 provides CSG's overall description [82] while Table 5 is based on Keating and Bradley's [95] research into the revised set of functions for CSG. The term 'metasystem' is deliberately used to suggest functions/roles that are beyond the purview of individual systems [96–98]. In this table, individual systems can be represented as M1. However, M1 is not listed as the concern for CSG is as the control and communication beyond individual systems (i.e. the metasystem level).

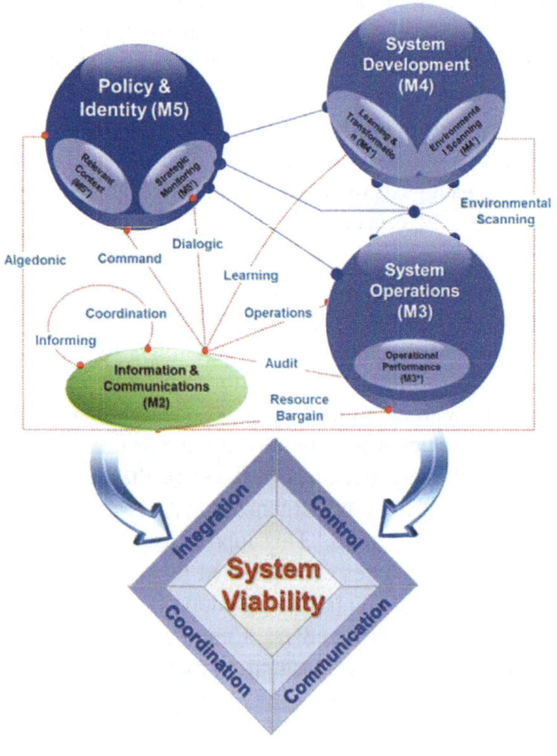

Fig. 1 CSG metasystem functions

Table 5 Metasystem functions in CSG Framework

Ares of concern	Metasystem function	Primary role
System identity	M5: Policy and identity	Focused on overall steering and trajectory for the system in the fulfillment of its mission. Maintains identity and balance between current and future focus
	M5*: System context	Focused on the specific context within which the metasystem is embedded. Context is the set of circumstances, factors, conditions, or patterns that enable or constrain execution of the system
	M5′: Strategic system monitoring	Focused on oversight of the system performance indicators at a strategic level, identifying performance that exceeds or fails to meet established expectations
System development	M4: System development	Maintains the models of the current and future system, concentrating on the long-range development of the system to ensure future viability
	M4*: Learning and transformation	Focused on facilitation of learning based on correction of design errors in the metasystem functions and planning for transformation of the metasystem
	M4′: Environmental scanning	Designs, deploys, and monitors sensing of the environment for trends, patterns, or events with implications for both present and future system viability
System operations	M3: System operations (M3)	Focused on the day to day execution of the metasystem to ensure that the overall system maintains the established performance levels and fulfills its missions
	M3*: Operational performance	Monitors system performance to identify and assess aberrant conditions, exceeded thresholds, or anomalies
System information	M2: Information and communications	Designs, establishes, and maintains the flow of information and consistent interpretation of exchanges (communication channels) necessary to execute metasystem functions

There exist several unique advantages for using concepts of management cybernetics in understanding complex situations. For instance, Jackson [56] establishes:

- It enables the establishment of a clear identity for the system/organization in question with an achievable and understood purpose within the environment. As the environment changes, the organization needs to 'reinvent and reconfigure itself.'
- The VSM approach (and subsequently, the CSG approach) offers an easy route for developing a shared understanding and language for dealing with organizational issues (including pathologies) affecting system viability and performance.

- There is a solution to the perennial problem of centralization versus decentralization. The parts can be given autonomy and empowered without threatening managerial control and organizational cohesion. In this case freedom and control are complementary rather than opposites.
- The establishment of identity and system purpose enables consistent direction and execution of the individual systems (M1s) and their support systems to enable the realization of a viable and effective organization.
- In the VSM, decision-making and control are spread throughout the structure of the constituent systems and integrating metasystem. This helps to amplify leadership (influence) throughout the system of interest.

The above points are meant to amplify general utility associated with Beer's VSM. In addition to the perspective and formulation from the VSM, the concept of metasystem within the emerging CSG perspective [6] is amplified by the following critical elaborations:

- The metasystem is considered as logically 'separate' from the systems that it seeks to integrate (this is consistent with Beer's articulation of the metasystem in relationship to the autonomous systems that it seeks to integrate).
- Beer's [90] original articulation of viability functions includes five functions. Recent research expands these functions to nine functions and places emphasis on metasystem functions.

In summary, management cybernetics informs the *communication* and *control* essence for CSG. In this case, 'communication' involves the flow, transduction, and processing of information within and external to the system. Consistency in communications enables thinking, decisions, actions, interpretations, and knowledge creation for a system to be made from a common frame of reference with respect to the system. In CSG, the primary function of 'control' is to provide the minimal constraint necessary to ensure continued system performance and behavior, while supporting the highest degree of autonomy for governed entities.

Closing out the CSG framework is the matter of governance. In Greek language, governance is synonymous with kybernetikos which translates to the art of steering [99]. The equivalent term in Latin, *guberneles*, relates to *gubemator* or to govern. Schneider and Kenis [100] have extended governance to include societal control and self-regulation. Moreover, Schneider and Bauer [101] espouse that "if a 'problem' is defined as the difference between a preferred state and an undesired status quo, the function of governance is 'problem -solving' in the sense of moving to desired states" (p. 11). In this case, Katina [25] suggests that governance is related to 'regulation' that leads to the realization of desired short and long-term goals. This view of governance reverberates throughout several governance perspectives (see Table 6).

It is obvious that several of these perspectives are driven by the nature of the system in question. In CSG, the interest in governance involves the three essential aspects, including direction, oversight, and accountability. First, governance involves sustaining a coherent identity and vision that supports consistent decision,

Table 6 Perspectives in governance

Governance type	Description	Proponents
Process-centric	A governing arrangement where one or more public agencies directly engage non-state stakeholders in a collective decision-making process that is formal, consensus-oriented, and deliberative and that aims to make or implement public policy or manage public programs or assets	Ansell and Gash [111: p. 544]
	…social turbulence kept within bounds, and change steered in desired directions… preserves order and continuity, but not necessarily the maintenance of the status quo	Dunsire [112: p. 18]
Structure-centric	…the totality of conceptual ideas about these Interactions…(these in relation to the act of governing)	Kooiman [113: p. 79]
	…the activity of coordinating communications in order to achieve collective goals through collaboration	Willke [114: p. 10]
State-centric	…the process through which state and nonstate actors interact to design and implement policies within a given set of formal and informal rules that shape and are shaped by power…	The World Bank [115: p. 3]
Hybrid	…the reflexive self-organization of independent actors involved in complex relations of reciprocal interdependence, with such self-organization being based on continuing dialogue and resource-sharing to develop mutually beneficial joint projects and to manage the contradictions and dilemmas inevitably involved in such situations	Jessop [116: p. 142]
	…interdependence between organizations… continuing interactions between network members, caused by the need to exchange resources and negotiate shared purposes…game-like interactions, rooted in trust and regulated by rules of the game negotiated and agreed by network participants…a significant degree of autonomy; they are self-organizing	Rhodes [117: p. 1246]
Corporate governance	…the system of checks and balances, both internal and external to companies, which ensures that companies discharge their accountability to all their stakeholders and act in a socially responsible way in all areas of their business activity	Brennan and Solomon [118: p. 890]
New public management	…the means for achieving direction, control, and coordination of wholly or partially autonomous individuals or organizations on behalf of interests to which they jointly contribute	Lynn et al. [119: p. 235]

(continued)

Table 6 (continued)

Governance type	Description	Proponents
Public policy	…the ways in which stakeholders interact with each other in order to influence the outcomes of public policies	Bovaird [120: p. 220]
	…the processes and institutions, both formal and informal, that guide and restrain the collective activities of a group	Keohane and Nye [121: p. 12]
International security	…the emergence and recognition of principles, norms, rules and behavior that both provide standards of acceptable public behavior and that are followed sufficiently to produce behavioral regularities	Keohane and Nye [122]
Social and political	Governance denotes the structures and processes which enable a set of public and private actors to coordinate their interdependent needs and interests through the making and implementation of binding policy decisions in the absence of a central political authority	Krahmann [123: p. 11]
	…arrangements in which public as well as private actors aim at solving societal problems or create societal opportunities, and aim at the care for the societal institutions within which these governing activities take place	Kooiman [124, p. 139]
Earth	…the interrelated and increasingly integrated system of formal and informal rules, rule-making systems, and actor-networks at all levels of human society (from local to global) that are set up to steer societies towards preventing, mitigating, and adapting to global and local environmental change and, in particular, earth system transformation, within the normative context of sustainable development	Biermann et al. [125]

action, interpretation, and strategic priorities. Second, governance involves providing for control, communication, and integration of system and their parts/entities—this is the essence of the oversight aspects of governance. Finally, governance involves accountability for system development. Accountability is focused on ensuring efficient utilization of resources, monitoring performance, and exploration of aberration conditions. Together, the triad of governance (direction, oversight, and accountability) provide essential underpinnings for further elaboration of the CSG paradigm.

The preceding section laid out the foundations of CSG, anchored in the underlying conceptual framework provided by GST, metasystem functions that enable system viability (management cybernetics), and the need for governance functions that achieve direction, oversight, and accountability. Applications of this theoretically-grounded and holistic approach have thus far been limited to curriculum development [102], DoD system acquisition [103, 104], and the water utilities domain [105]. Recent research (see Katina et al. [60, 61]) has extended CSG in the domain of critical infrastructure with a focus on CPS and the need to design and perform diagnosis based of metasystemic functions. This suggest CSG applicability in different domains and problem systems. In the following section, the CSG-based framework is applied to CPS with focus on security.

4 Cyber-Physical Systems Governance Framework

CPS was previously defined as a system-of-systems providing essential goods and services necessary for public well-being with the aid of control systems in the form of information and telecommunications. CPS, similar to other critical infrastructures, must operate under conditions characterized by high degrees of ambiguity, complexity, emergence, interdependence, and uncertainty. We suggest that these conditions can mask risks and vulnerabilities influencing system performance. With the view in mind of risk assessment (i.e., discovering hazards how they affect the system) and vulnerability assessment (i.e., discovering vulnerabilities (to hazards) in the system and developing/enhancing system properties against hazards), this section is developed to suggest the applicability of CSG as a framework for addressing the issue of security in CPS. Security, viewed from a CSG perspective, suggests that establishment and execution are designed, executed, and evolved from the higher level of 'meta-security' system governance provided by CSG. This view of CPS security encompasses aspects of risk assessment (external) and vulnerability assessment (internal) for hazards affecting a system of interest, with emphasis being placed on integration and coordination (stemming from systems theory) and communication and control (based in management cybernetics). Table 7 present the current state of CSG functions and their projection for MetaCyberSecurity of CPS.

There are five (5) primary conclusions with respect to CSG functions and their implications for security in critical CPS:

- *CONCLUSION 1*: In the CSG framework presented for CPS, the cyber- and physical systems are taken as individual systems (M1s) that must be integrated to form a 'unit-whole' (i.e. the governing metasystem for the CPS). While each system has its dedicated and prescribed functions, the unit-whole is expected to perform functions and operate as an integrated CPS, beyond the capacity of individual constituent systems forming the CPS. It is at this intersection of cyber- and physical that issues such as meta-security must be addressed. In this case meta-security addresses issues (risks and vulnerabilities) above-and-beyond those which individual systems may be ill equipped to either recognize or respond to.
- *CONCLUSION 2*: The CSG framework provides a means to address issues above-and-beyond individual systems constituting an extended CPS. This suggests the consideration elements of integration and coordination provided for the integrated CPS based in systems theory (i.e., a consideration of principles

Table 7 CSG framework for security in cyber-physical systems

CSG areas of concern	Metasystem function	Primary role	MetaCyberSecurity implications for CPS
CPS system identity	M5: Policy and identity	Focused on overall steering and trajectory for the system in the fulfillment of its mission. Maintains identity and balance between current and future focus	Establishing and maintaining critical CPS identity though dealing with current landscape of threats while projecting the system to deal with future threats
	M5*: System context	Focused on the specific context within which the metasystem is embedded. Context is the set of circumstances, factors, conditions, or patterns that enable or constrain execution of the system	Exploration of contexts of operation for the CPS in question. This includes understanding set of cyber-related factors enabling or constraining the system execution
	M5′: Strategic system monitoring	Focused on oversight of the system performance indicators at a strategic level, identifying performance that exceeds or fails to meet established expectations	Exploration, development, and assessment of strategic cyber-performance indicators. While indicators might vary from system to system, the lack of such indicators ensures that indicator-based analysis and development cannot be undertaken
CPS system development	M4: System development	Maintains the models of the current and future system, concentrating on the long-range development of the system to ensure future viability	The maintenance of current performance in the midst of rapid changes in cybersecurity threats suggests a continual capacity to identify threats and make modifications. Additionally, identifying future threats can suggest changes in the system trajectory to get ahead of future threats
	M4*: Learning and transformation	Focused on facilitation of learning based on correction of design errors in the metasystem functions and planning for transformation of the metasystem	Ensuring that incident events (cyber hazards, threats, risks, and vulnerabilities) are analyzed for any insights that can be used to correct design and increase system resilience

(continued)

Table 7 (continued)

CSG areas of concern	Metasystem function	Primary role	MetaCyberSecurity implications for CPS
	M4′: Environmental scanning	Designs, deploys, and monitors sensing of the environment for trends, patterns, or events with implications for both present and future system viability	Threats are constant in CPS. However, the nature, source, and timing are emergent. Effective environmental scanning for CPS threats as well as potential opportunities stemming from the environment require an active scanning design to permit timely identification, assessment, interpretation, response, and response outcomes evaluation
CPS System operations	M3: System operations (M3)	Focused on the day to day execution of the metasystem to ensure that the overall system maintains the established performance levels and fulfills its missions	Ongoing performance of a CPS must provide for integration to continuously achieve the higher-level system purpose. Provision of sufficient controls to ensure continuity in this achievement
	M3*: Operational performance	Monitors system performance to identify and assess aberrant conditions, exceeded thresholds, or anomalies	System performance in accordance with agreed upon outputs/outcomes requires monitoring. This monitoring of the CPS must provide rigorous and routine identification of the level of ongoing performance (good or bad) of the system
CPS System information	M2: Information and communications	Designs, establishes, and maintains the flow of information and consistent interpretation of exchanges (communication channels) necessary to execute metasystem functions	CPS protects the flow of information and communications for a system. The design of the metalevel flow of information and communications necessary to permit the CPS to continue to perform its mission. This is achieved by the active design, execution, and evolution of information and communication at the metasystem level for CPS, which is different than the information flow at lower levels of an operational set of CPS constituents

governing all systems) and management cybernetics's (i.e., a means to developing effective systems) provisions of communication and control.

- *CONCLUSION 3*: The CSG metasystem functions can offer utility for CPS during assessment in three primary ways. First, is the assessment of the metasystem functions and how the CPS is designed for and executing these functions. This assessment takes that form of evaluation of the metasystem functions (i.e., existence of metasystem functions and degree to which they are effectively performed) for a given critical CPS. Second, is the evaluation of the effectiveness of specific mechanisms of a particular CPS that are used to achieve each of the CSG metasystem functions (i.e., MetaCyberSecuirity of the CPS[3]). This evaluation is intended to gather data and information on the specific mechanisms associated with metasystem functions. Such mechanisms include, for instance, mechanisms for monitoring system performance and assessing aberrant conditions for M3*. In this case, one looks for mechanisms and their effectiveness (i.e., a mechanism for monitoring/assessing aberrant conditions might exist but not be effective in assessing conditions affecting performance). Obviously, the nature of the specific mechanisms for execution of CSG functions might depend on the nature of the CPS and its operations. Additionally, the mechanisms may be dynamic, adding, deleting, or modifying over shifting time and circumstances.
- *CONCLUSION 4*: Finally, there is a matter of governance—the achievement of direction, oversight, and accountability for a CPS.
  - Each CSG function, and its relation to MetaSecurity, involves sustaining a coherent identity and vision which supports consistent decision, action, interpretation, and cyber and physical strategic priorities. These priorities might involve implementation of long and short resilience-based policies that position the system to deal with current and future risks and vulnerabilities.
  - Providing for control, communication, and integration of the CPS system (metasystem and constituent system levels). This ensures that the constituent system/parts/entities have sufficient autonomy to effectively deal with risk and vulnerabilities, dissemination of data and information regarding risks and vulnerabilities. Also important is the CPS establishing a level of balance between this autonomy of constituents and integration of the overall system. Finally, governance involves accountability for CPS development, which involves ensuring efficient utilization and effective targeting of resources to deal with priority risks and vulnerabilities, monitoring hazards and threats

[3]Cyber security is typically concerned with the protection of internet connected systems (to include hardware, software and associated infrastructure), the data on them, and the services they provide, from unauthorized access, harm or misuse [1]. In this case, there is no distinction between issues (i.e., risks, threats, and vulnerabilities) at the system level and the metasystem level. MetaCyberSecurity attempts to address system and metasystemic issues recognizing that a system (including CPS-related systems) does not operate in isolation. This approach enables the consideration of internal and external system risks and vulnerabilities regardless of cause—malicious, technical, or natural.

with the greatest potential to affect performance, and exploration of the nature of aberration hazards and threating conditions as they unfold as well as those that may present future-based threats.

- *CONCLUSION 5*: In the presented framework, the concept of MetaCyberSecurity encompasses aspects of risk and its assessment involving external issues. This includes, but is not limited to, malicious (e.g., data breach), technical (e.g., lack of updating patches), and natural (e.g., solar flares) affecting the system as well as vulnerability (again malicious, technical or natural) and its assessment involving the internal nature of the system and hazards affecting the system.

It is obvious that risks and vulnerabilities in critical CPS involves more than having technical capability to address technical risks and vulnerability. In designing a CPS purposefully, there is need for consideration of several factors that characterize the present and future conditions of the CPS problem domain. Among these are ambiguity, complexity, emergence, system interdependence, and uncertainty. As an example, approaching risk and its assessment in an isolated system might involve identification of hazards and how they might affect the system. However, in taking an interdependent systems perspective, risk assessment must go beyond the immediate system of interest. Such an approach involves dealing with risks of the immediate system of interest as well as risks which might emanate from the interaction of a system with other interdependent systems. In effect, this entails consideration beyond the isolated boundaries of a 'single' CPS operating independently. Additionally, complete, absolute, or infallible knowledge of the domain within which a CPS operates is a fiction. Thus, this perspective recognizes the ambiguity, complexity, emergence, and uncertainties associated with current systems and their operating environment. This is suggested by Arbesman [31] as the lack of understanding brought about through interconnectedness and overcomplication.

5 Research Implications and Directions

Arguably, CSG and its reference model (metasystem functions) are in the embryonic stages of development and applications. The attribution of CSG to MetaCyberSecurity is targeted to enable practitioners in critical CPS to better deal with current and emerging risks and vulnerabilities stemming from the problems associated with a landscape marked by increasing levels of ambiguity, complexity, emergence, system interdependence, and uncertainty from which CPS must operate. Essentially, CSG helps in the articulation of the nature, role, and functions for system viability—offering a compelling response to avoid negative implications of *system drift (the undirected development of a system)*. As a matter of illustration, just as a powerless ship will drift along its intended course subject to uncontrollable currents, so too can our systems (e.g. CPS) experience system drift resulting from development by processes of *accretion* or *self-organization*. System drift symbolizes a system that is subject to the unintended consequences that accrue in the absence of a purposefully executed design. In the end, system drift describes a condition all too familiar to practitioners who must navigate systems through increasingly complex turbulent environments, while confronting seemingly intractable issues on a daily basis. CSG is a coherent response to system drift which we suggest might be characteristic of modern CPS. Simply, drifting systems/programs (e.g. CPS) fail to deliver anticipated value, much less produce high performance. Examples of drifting systems/programs are numerous (e.g., see Bradley et al. [103]) and we might suggest that current developments of CPS are following this 'system drift pattern'. However, drifting systems take on a new meaning when it comes to critical infrastructure systems whose failure can have significant impact on health, safety, security, economic or social well-being. We have presented CSG as an alternative set of lenses from which to view program deficiencies including risk and vulnerabilities (in the security paradigm) that affect system performance. While one cannot provide assurances that CSG framework will absolutely preclude risks and vulnerabilities in CPS, the establishment the framework and the evaluation of CSG functions along with mechanisms (and their effectiveness) might provide CPS owners and practitioners insights into potential cyber-related risks and vulnerabilities and a chance for development of more insightful countermeasures.

The development and application of CSG for CPS is in its infancy. Further research and development directions for CSG in relationship to research in CPS security is envisioned as following the research and development model provided in Fig. 2. This model suggests a close coupling of four primary elements to holistically engage CSG development and deployment to improve practices in the MetaCyberSecurity field. The basis of this model suggests that CSG research is the system-science based engineering of technologies supporting application development that advances practices related to design, execution, and development of CPS areas of (i) science based in systems theory, (ii) engineering of technologies, (iii) application development, and (iv) practice.

The systems-based science forms the broad foundation of the multiple different fields that explore the phenomena associated with explaining the behavior and performance of systems. For CSG, the systems science basis is grounded in GST and systems theory and management cybernetics. In this case, system science provides the conceptual underpinnings for all derivative developments including 'technologies' involving methods and tools for dealing with risks and vulnerabilities in CPS. Thus, systems science provides a conceptually rigorous theoretical underpinning that grounds all further development and applications—including CPS.

Based upon the underlying systems science, engineering involves the development of implementing technologies. These technologies are developed as CSG supporting artifacts (e.g., tools, techniques, software), grounded in systems science and addressing a targeted aspect of design, execution, or development of complex systems. This is the attempt to provide CPS practitioners with system science derived/grounded artefacts to perform CSG functions more effectively.

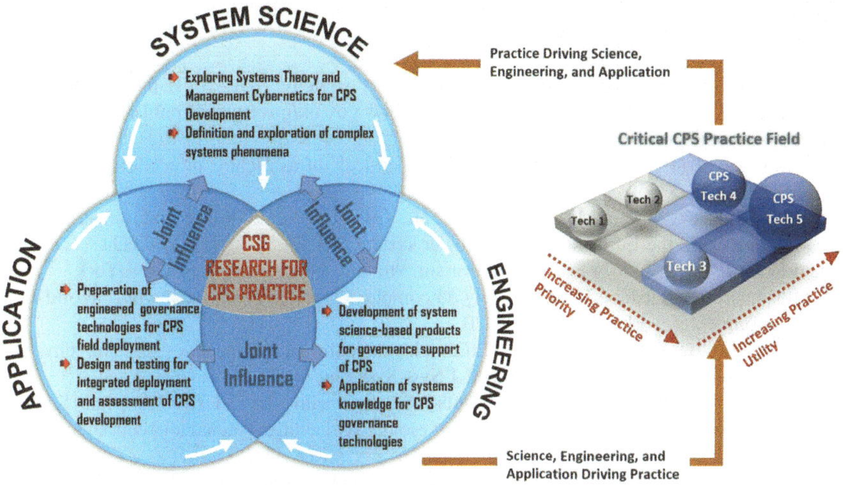

Fig. 2 CSG research and development framework

Application is focused on development of the particular methods, processes, and methodologies that bridge the divide between the engineering of technologies and preparation for deployment in practice. The application emphasis is the appropriate preparation of technologies for deployment in practice based settings for CPS. This also entails integration with other technologies and methods/methodologies as well as providing for effective deployment within operational/practice contexts.

Ultimately the beneficiary of the systems science, engineering, and application triad is the practice field where deployment is targeted. This is where the different technologies, as deployed through application design, are targeted to enhance practices related to better design, execution, and development of complex systems and their problems. This calls for case demonstration of deployable applications where the issue of security cannot be achieved separate from the world of practice. The pragmatic focus of the cyber and physical systems and the need to address security concerns necessitates the development practical tools, techniques, and technologies. It is though these needs that practitioners can realize utility of CSG framework in dealing with current and emerging risks and vulnerabilities in CPS. Moreover, such risks are not restricted to purely cyber concern, but rather may be external to cyber considerations but still generate significant impact concerns. Such concerns can include system pathology [25, 51, 106–109] originating with violation of system governing principles and having significant impacts on performance of CPS.

References

1. HM Government (2016) National cyber security strategy 2016–2021, p 84. UK Cabinet Office, London. Retrieved from https://www.gov.uk/government/publications/national-cyber-security-strategy-2016-to-2021
2. Clinton WJ (1996) Executive order 13010: critical infrastructure protection. Fed Reg 61 (138):37345–37350
3. Gheorghe AV, Masera M, Weijnen MPC, De Vries JL (eds) (2006) Critical infrastructures at risk: securing the European electric power system (vel 9). Springer, Dordrecht
4. Linden EV (ed) (2007) Focus on terrorism, 9th edn. Nova Publishers, New York
5. Rasmussen J, Batstone R (1989) Why do complex organisational systems fail? World Bank environmental working paper, No. 20
6. Keating CB, Katina PF (2015) Editorial: foundational perspectives for the emerging complex system governance field. Int J Syst Syst Eng 6(1/2):1–14
7. European Southern Observatory (2008) Science with the VLTI. European Southern Observatory. Retrieved from http://www.eso.org/sci/facilities/paranal/telescopes/vlti/science.html
8. Masters H (2010) Transcript of the accretion of galaxies and stars. Prezi. Retrieved from https://prezi.com/fg0fw5pyu8nk/the-accretion-of-galaxies-and-stars/
9. Ashby WR (1962) Principles of the self-organizing system. In: von Foerster H, Zopf G (eds) Principles of self-organization. Pergamon Press, New York, NY, pp 255–278
10. ASCE (2009) Guiding principles for the nation's critical infrastructure. American Society of Civil Engineers, Reston
11. Price JWH (1998) Simplified risk assessment. Eng Manag J 10(1):19–23
12. Gibson JE, Scherer WT, Gibson WF (2007) How to do systems analysis. Wiley, Hoboken, NJ
13. INCOSE (2011) Systems engineering handbook: a guide for system life cycle processes and activities, (3.2 ed). In: Cecilia H (ed) INCOSE, San Diego, CA
14. Weiss JW, Anderson D (2003) CIOs and IT professionals as change agents, risk and stakeholder managers: a field study. In: Proceedings of the 36th annual Hawaii international conference on system sciences, 2003. https://doi.org/10.1109/HICSS.2003.1174639
15. Holton GA (2004) Defining risk. Financ Anal J 60(6):19–25
16. Knight FH (1921) Risk, uncertainty, and profit. Hart, Schaffner & Marx; Houghton Mifflin Co, Boston, MA
17. Song C (2005) A methodological framework for vulnerability assessment for critical infrastructure systems, hierarchical holographic vulnerability assessment (HHVA), Thesis. ETH Zürich, Zürich. Retrieved from https://www1.ethz.ch/lsa/education/arb/old/archive/da_song_05
18. Katina PF, Pinto CA, Bradley JM, Hester PT (2014) Interdependency-induced risk with applications to healthcare. Int J Crit Infrastruct Prot 7(1):12–26. https://doi.org/10.1016/j.ijcip.2014.01.005
19. Vamanu BI, Gheorghe AV, Katina PF (2016) Critical infrastructures: risk and vulnerability assessment in transportation of dangerous goods—transportation by road and rail, vol 31. Springer, Cham, Switzerland
20. Einarsson S, Rausand M (1998) An approach to vulnerability analysis of complex industrial systems. Risk Anal 18(5):535–546
21. Holmgren A, Molin S, Thedéen T (2001) Vulnerability of complex infrastructure; power system and supporting digital communication system. Presented at the 5th international conference on technology, policy, and innovation, LEMMA Publishers, Utrecht, the Netherlands
22. Turner BL, Kasperson RE, Matson PA, McCarthy JJ, Corell RW, Christensen L, Eckley N, Kasperson JX, Luers A, Martello ML, Polsky C, Pulsipher A, Schiller A (2003)

A framework for vulnerability analysis in sustainability science. Proc Natl Acad Sci 100(14): 8074–8079. https://doi.org/10.1073/pnas.1231335100
23. Tokgoz BE, Gheorghe AV (2013) Resilience quantification and its application to a residential building subject to hurricane winds. Int J Disaster Risk Sci 4(3):105–114. https://doi.org/10.1007/s13753-013-0012-z
24. Sussman JM (2005) Perspectives on intelligent transportation systems. Springer, New York, NY
25. Katina PF (2015) Systems theory-based construct for identifying metasystem pathologies for complex system governance (Ph.D.). Old Dominion University, Virginia, United States
26. Guckenheimer J, Ottino JM (2008) Foundations for complex systems research in the physical sciences and engineering, p 21. Northwestern University: National Science Foundation, Evanston, IL
27. Hammond D (2002) Exploring the genealogy of systems thinking. Sys Res Behav Sci 19(5): 429–439. https://doi.org/10.1002/sres.499
28. Laszlo E (1996) The systems view of the world: a holistic vision for our time. Hampton Press, Cresskill, NJ
29. Bertalanffy L (1968) General system theory: foundations, developments, applications. George Braziller, New York, NY
30. Ackoff R (1999) Re-creating the corporation: a design of organizations for the 21st century. Oxford University Press, Oxford
31. Arbesman S (2016) Overcomplicated: technology at the limits of comprehension. Current, New York
32. Bateson G (1972) Steps to an ecology of mind. Jason Aronson Inc, New York, NY
33. Bostrom N (2014) Superintelligence: paths, dangers, strategies. Oxford University Press, Oxford
34. Capra F (1996) The web of life: a new scientific understanding of living systems. Anchor Books, New York, NY
35. Casti J (2012) X-Events: complexity overload and the collapse of everything. William Morrow, New York, NY
36. Checkland PB (1999) Systems thinking, systems practice. Wiley, New York, NY
37. Churchman CW (1971) The design of inquiring systems. Basic Books, New York, NY
38. Flood RL, Carson ER (1993) Dealing with complexity: an introduction to the theory and application of systems science. Plenum Press, New York
39. François CO (ed) (2004) International encyclopedia of systems and cybernetics. Walter de Gruyter, München, Germany
40. Klir GJ (ed) (1972) Trends in general systems theory, 1st edn. Wiley, New York, NY
41. Martin J (2006) The meaning of the 21st century: a vital blueprint for ensuring our future. Riverhead Books, New York, NY
42. Simon HA (1973) The organization of complex systems. In: Pattee HH (ed) Hierarchy theory: the challenges of complex systems. George Braziller, New York, NY, pp 1–27
43. Skyttner L (2005) General systems theory: problems, perspectives, practice, 2nd edn. World Scientific Publishing Co., Pte. Ltd., Singapore
44. Taleb NN (2010) The black swan: the impact of the highly improbable. Random House Trade Paperbacks Edition, New York, NY
45. von Bertalanffy L (1972) The history and status of general systems theory. Acad Manag J 15(4):407–426. https://doi.org/10.2307/255139
46. Warfield JN (1999) Twenty laws of complexity: science applicable in organizations. Sys Res Behav Sci 16(1):3–40
47. Adams KM, Hester PT, Bradley JM, Meyers TJ, Keating CB (2014) Systems theory as the foundation for understanding systems. Sys Eng 17(1):112–123. https://doi.org/10.1002/sys.21255
48. Gaines BR (1977) Progress in general systems research. In: Klir GJ (ed) Applied general systems research: recent development and trends. Plenum Press, New York, NY, pp 3–28

49. Monod J (1974) On chance and necessity. In: Ayala FJ, Dobzhansky T (eds) Studies in the philosophy of biology. Macmillan Press, London, UK, pp 357–375
50. Clemson B (1984) Cybernetics: a new management tool. Abacus Press, Tunbridge Wells, Kent, UK
51. Katina PF (2015) Emerging systems theory–based pathologies for governance of complex systems. Int J Sys Sys Eng 6(1/2):144–159
52. Stichweh R (2011) Systems theory. In: Badie B, Berg-Schlosser D, Morlino L (eds) International encyclopedia of political science, vol 8. SAGE, New York, NY, pp 2579–2588
53. Strijbos S (2010) Systems thinking. In: Frodeman R, Klein JT, Mitcham C (eds) The Oxford handbook of interdisciplinarity. Oxford University Press, USA, New York, NY, pp 453–470
54. Weinberg GM (1975) An introduction to general systems thinking. Wiley, New York, NY
55. Whitney K, Bradley JM, Baugh DE, Chesterman CW (2015) Systems theory as a foundation for governance of complex systems. Int J Sys Sys Eng 6(1–2):15–32. https://doi.org/10.1504/IJSSE.2015.068805
56. Jackson MC (2003) Systems thinking: creative holism for managers. Wiley, Chichester, UK
57. Katina PF, Calida BY (2017) Complex system analysis for engineering of systemic failures. In: Hopkins M (ed) Systems engineering: concepts, tools and applications. Nova Science Publishers, New York, NY, pp 105–132
58. Jackson MC (1991) Systems methodology for the management sciences. Plenum Press, New York, NY
59. Crownover MWB (2005) Complex system contextual framework (CSCF): a grounded-theory construction for the articulation of system context in addressing complex systems problems. Dissertation, Old Dominion University, Norfolk, VA, United States
60. Katina PF, Keating CB, Gheorghe AV, Masera M (2017) Complex system governance for critical cyber-physical systems. Int J Crit Infrastruct 13(2/3):168–183. https://doi.org/10.1504/IJCIS.2017.088230
61. Katina PF, Keating CB, Gheorghe AV (2016) Cyber-physical systems: complex system governance as an integrating construct. In: Yang H, Kong Z, & Sarder MD (eds) Proceedings of the 2016 industrial and systems engineering research conference. IISE, Anaheim, CA
62. US Congress (2001) Uniting and strengthening America by providing appropriate tools required to intercept and obstruct terrorism (USA PATRIOT ACT) Act of 2001 (No. 147) (p. 115 Stat. 271–402). Washington, DC: 107th Congress. Retrieved from http://www.gpo.gov/fdsys/pkg/PLAW-107publ56/content-detail.html
63. Alur R (2015) Principles of cyber-physical systems. MIT Press, Cambridge, MA
64. Khaitan SK, McCalley JD (2015) Design techniques and applications of cyberphysical systems: a survey. IEEE Syst J 9(2):350–365
65. Adepu S, Mathur A (2016) Generalized attacker and attack models for cyber physical systems, pp 283–292. IEEE, Atlanta, GA, USA. https://doi.org/10.1109/COMPSAC.2016.122
66. European Parliament (2016) Ethical aspects of cyber-physical systems: scientific foresight study (No. PE 563.501). European Parliament, Brussels. Retrieved from http://www.europarl.europa.eu/thinktank/en/document.html?reference=EPRS_STU(2016)563501
67. Kwon C, Liu W, Hwang I (2013) Security analysis for cyber-physical systems against stealthy deception attacks. In: 2013 American control conference, pp. 3344–3349. Washington, DC, USA. https://doi.org/10.1109/ACC.2013.6580348
68. Conti JP (2010) The day the samba stopped. Eng Technol 5(4):46–47
69. Richards G (2008) Hackers vs slackers. Eng Technol 3(19):40–43
70. North American Electric Reliability Council (2003) SQL Slammer worm: lessons learned for consideration by the electricity sector. North American Electric Reliability Council, Princeton, NJ
71. Farwell JP, Rohozinski R (2011) Stuxnet and the future of Cyber War. Survival 53(1): 23–40. https://doi.org/10.1080/00396338.2011.555586
72. Slay J, Miller M (2007) Lessons learned from the Maroochy Water Breach. In: Goetz E, Shenoi S (eds) Critical infrastructure protection, vol 253. Springer, US, Boston, MA, pp 73–82

73. Pasqualetti F, Dörfler F, Bullo F (2013) Attack detection and identification in cyber-physical systems. IEEE Trans Autom Control 58(11):2715–2729
74. Masys AJ (ed) (2015) Disaster management: enabling resilience. Springer, New York, NY
75. Johnson C (2016) Why we cannot (yet) ensure the cyber-security of safety-critical systems. In: SCSC 24th annual symposium, Brighton, UK. Retrieved from http://scsc.org.uk/e378
76. Ali S, Qaisar SB, Saeed H, Khan MF, Naeem M, Anpalagan A (2015) Network challenges for cyber physical systems with tiny wireless devices: a case study on reliable pipeline condition monitoring. Sensors 15(4):7172–7205. https://doi.org/10.3390/s150407172
77. Baugh D (2015) Environmental scanning implications in the governance of complex systems. Int J Sys Sys Eng 6(1–2):127–143
78. Katina PF, Calida BY (2017) Complex system governance: implications and research directions [White paper]. Submitted to the Committee on a decadal survey of social and behavioral sciences for applications to National Security, The national academies of sciences, engineering, and medicine. Retrieved from Available: http://sites.nationalacademies.org/dbasse/bbcss/dbasse_178412
79. Keating CB, Ireland V (2016) Editorial: complex systems governance - issues and applications. Int J Sys Sys Eng, 7(1/2/3):1–21
80. Keating CB, Katina PF (2016) Complex system governance development: a first generation methodology. Int J Sys Sys Eng, 7(1/2/3): 43–74. https://doi.org/10.1504/IJSSE.2016.076127
81. Keating CB, Katina PF, Bradley JM (2014) Complex system governance: concept, challenges, and emerging research. Int J Sys Sys Eng 5(3):263–288
82. Keating CB, Katina PF, Bradley JM (2016) Complex system governance: failure mode effects and criticality analysis application. In: Yang H, Kong Z, Sarder M (eds) Presented at the Proceedings of the 2016 industrial and systems engineering research conference, ISERC, Anaheim, CA
83. Warfield JN (1976) Societal systems: planning, policy and complexity. Wiley, New York, NY
84. Hieronymi A (2013) Understanding systems science: a visual and integrative approach. Sys Res Behav Sci 30(5):580–595
85. Burrell G, Morgan G (1979) Sociological paradigms and organisational analysis. Ashgate Publishing, Burlington, VT
86. Katina PF, Keating CB, Jaradat RM (2014) System requirements engineering in complex situations. Requirements Eng 19(1):45–62
87. Katina PF (2016) Metasystem pathologies (M-Path) method: phases and procedures. J Manag Dev 35(10):1287–1301
88. Mobus GE, Kalton MC (2015) Principles of systems science. Springer, New York, NY
89. Wiener N (1948) Cybernetics: or control and communication in the animal and the machine. MIT Press, Cambridge, MA
90. Beer S (1979) The heart of the enterprise. Wiley, New York, NY
91. Beer S (1984) The viable system model: its provenance, development, methodology and pathology. J Oper Res Soc 35(1):7–25
92. Beer S (1985) Diagnosing the system for organizations. Oxford University Press, Oxford, UK
93. Espejo R, Harnden R (eds) (1989) The viable systems model: interpretations and applications of Stafford Beers' VSM. Wiley, Chichester
94. Keating CB, Morin M (2001) An approach for systems analysis of patient care operations. J Nurs Adm 31(7–8):355–363
95. Keating CB, Bradley JM (2015) Complex system governance reference model. Int J Sys Sys Eng 6(1–2):33–52
96. Carter B (2015) A metasystem perspective and implications for governance. Int J Sys Sys Eng 6(1/2):90–100

97. Djavanshir GR, Khorramshahgol R, Novitzki J (2009) Critical characteristics of metasystems: toward defining metasystems' governance mechanism. IT Prof 11(3):46–49. https://doi.org/10.1109/MITP.2009.45
98. Palmer KD (2000) Meta-systems engineering. In: Proceedings of the tenth annual international symposium of the international council on systems engineering. INCOSE, Las Vegas. Retrieved from http://www.archonic.net/MSE04.PDF
99. Mason RO, Mitroff II (1981) Challenging strategic planning assumptions: theory, cases, and techniques. Wiley, New York, NY
100. Schneider V, Kenis P (1996) Verteilte kontrolle: institutionelle steuerung in modernen gesellschaften. In: Kenis P, Schneider V (eds) Organisation und netzwerk: institutionelle steuerung in wirtschaft und politik (pp. 9–43). Frankfurt: Germany: Frankfurt/Main
101. Schneider V, Bauer JM (2007) Governance: prospects of complexity theory in revisiting systems theory. In: Annual meeting of the midwest political science association, pp 1–36, Chicago, IL. Retrieved from https://www.msu.edu/~bauerj/complexity/schneider.pdf
102. Katina PF, Bradley JM (2016) Towards a systems theory-based curriculum for complex systems governance. ASEE Conferences, New Orleans, LA. https://doi.org/10.18260/p.27069
103. Bradley JM, Katina PF, Keating CB (2016) Complex system governance for acquisition. In: Green J, Snider K (eds) Proceedings of the thirteenth annual acquisition research symposium, pp 196–214. NPS, Monterey, California
104. Keating CB, Bradley JM, Katina PF, Jaradat RM (2017) A systems theoretic-based framework to discover pathologies in Acquisition System Governance. In: Proceedings of the fourteenth annual acquisition research symposium, pp 352–376, Monterey, California: NPS
105. Pyne JC, Keating CB, Katina PF (2016) Enhancing utility manager's capability for dealing with complex issues. Proc Water Environ Fed 2016(8):4207–4232
106. Davidz HL (2017) Systems engineering pathology: leveraging science to characterize dysfunction. In: Annual INOSE international workshop, INCOSE, Los Angeles. Retrieved from www.incose.org/IW2017
107. Katina PF (2016) Systems theory as a foundation for discovery of pathologies for complex system problem formulation. In: Masys AJ (ed) Applications of systems thinking and soft operations research in managing complexity. Springer, Geneva, Switzerland, pp 227–267
108. Keating CB, Katina PF (2012) Prevalence of pathologies in systems of systems. Int J Sys Sys Eng 3(3/4):243–267
109. Troncale L (2013) Systems processes and pathologies: creating an integrated framework for systems science. INCOSE Int Symposium 23(1):1330–1353. https://doi.org/10.1002/j.2334-5837.2013.tb03091.x
110. Fiol CM, Lyles MA (1985) Organizational learning. Acad Manag Rev 10(4):803–813. https://doi.org/10.2307/258048
111. Ansell C, Gash A (2008) Collaborative governance in theory and practice. J Public Adm Res Theor 18(4):543–571
112. Dunsire A (1990) Holistic governance. Public Policy Adm 5(1):4–19
113. Kooiman J (2003) Governing as governance. SAGE Publications Ltd, London, UK
114. Willke H (2007) Smart governance: governing the global knowledge society. Campus Verlag GmbH, Frankfurt, Germany
115. The World Bank (2017) World development report 2017: governance and the law (No. 112303). The World Bank, Washington, DC. Retrieved from http://elibrary.worldbank.org/doi/book/10.1596/978-1-4648-0950-7
116. Jessop B (2003) Governance and metagovernance: on reflexivity, requisite variety, and requisite irony. In: Bang HP (ed) Governance, as social and political communication. Manchester University Press, Manchester, England, pp 142–172
117. Rhodes RAW (2007) Understanding governance: ten years on. Organ Stud 28(8):1243–1264. https://doi.org/10.1177/0170840607076586

118. Brennan NM, Solomon J (2008) Corporate governance, accountability and mechanisms of accountability: an overview. Acc, Auditing Accountability J 21(7):885–906. https://doi.org/10.1108/09513570810907401
119. Lynn L, Heinrich C, Hill C (2000) Studying governance and public management: challenges and prospects. J Public Adm Res Theor 10(2):233–261
120. Bovaird T (2005) Public governance: balancing stakeholder power in a network society. Int Rev Admin Sci 71(2):217–228
121. Keohane R, Nye J (eds) (2000) Governance in a globalizing world. Brookings Institution, Washington, DC
122. Keohane R, Nye J (1989) Power and interdependence. Harper Collins, New York, NY
123. Krahmann E (2003) Conceptualizing security governance. Cooperation Conflict 38(1):5–26. https://doi.org/10.1177/0010836703038001001
124. Kooiman J (2000) Societal governance: levels, models and orders of social-political interaction. In: Pierre J (ed) Debating governance: authority, steering and democracy. Oxford University Press, Oxford, UK, pp 138–166
125. Biermann F, Betsill MM, Gupta J, Kanie N, Lebel L, Liverman D et al (2009) Earth system governance: people, places and the planet. Science and implementation plan of the earth system governance project (No. Earth System Governance Report 1, IHDP Report 20). The Earth System Governance Project, Bonn, IHDP

To Feel Secure or to Be Secure, That Is the Question

Erik Hollnagel

Abstract The title of this volume is "Security by design". This can be read as a statement, as an expression of confidence that it is possible to be secure by design. In that case the obvious next step is to think about possible ways of ensuring security by design, starting, for instance, with the solutions that have been developed to solve the seemingly analogous problems of safety or prevention by design. But what if the title had been "Security by design?" or in other words a question rather than a statement? In that case the title would not be taken as an expression of confidence but rather as an admission of uncertainty or even of impotence, a possible doubt of whether security can be ensured by design. The obvious next step would then be to take a closer look at the two central concepts, namely "security" and "design". The current chapter will do just that.

1 The Meaning of Security

A discussion of the meaning of "security" cannot take place without also discussing the meaning of "safety". Both terms have been used for a very long time and have in many cases been treated as synonyms even though they are not. One way of illustrating the difference between them is to consider, for instance, the onslaught of a severe natural disaster, such as a hurricane or a tsunami. Here safety concerns focus on the immediate consequences of the disaster, on the destruction of the physical environment that may endanger human life and property. In other words, safety is related to whether the people who are exposed to the event will be safe—which in the traditional interpretation means free from harm—while it happens. In contrast to that, security concerns focus on the longer term effects such as the economic or financial security of people that have been subjected to the event, i.e., on how such consequences may affect the ability to continue an existence or

E. Hollnagel (✉)
Teglvænget 5, 2990 Nivå, Denmark
e-mail: hollnagel.erik@gmail.com

A. J. Masys (ed.), *Security by Design*, Advanced Sciences and Technologies for Security Applications, https://doi.org/10.1007/978-3-319-78021-4_8

establish a new one. People who lose their house and possessions lose economic security; they become less able to withstand future problems and perhaps even unable to sustain their own lives.

The same distinction also applies to what can happen at work. Here the safety concerns focus on the acute harm to people—either those who carry out the work or innocent bystanders—that may come from accidents at work as well as on the longer term effects on mental or physical health and well-being. By the same token, security concerns focus on whether, e.g., the job or employment is secure or whether people risk suddenly to lose their livelihood hence their ability to provide for themselves and their family, because they are fired, because the company defaults in one way or the other, or for some other reason.

Despite such superficial similarities, security and safety are clearly different terms and their specific denotations and connotations should be kept separate when they are used. In this book, the scope of security concerns (as expressed by the call for chapters) includes mass migration, terrorist, pandemics, cyberattacks, climate related disasters, energy security, and the Internet of Things. Although any of these areas, and possibly several others as well, may justifiably be seen as given rise to security concerns, it would nevertheless be reasonable to propose that some security concerns, e.g., those associated with natural phenomena or complex societal phenomena such as financial upheavals, belong to a category by themselves. The reason is simply that it seems impossible to do much to guard against them through design—at least if the purpose of design is intentionally to prevent or avoid them. Indeed, to design against something, or to achieve something by design, it is necessary that the systems (that are being designed) are both controllable and understandable. Neither natural disasters nor financial disruptions, among other types of large-scale emergencies, meet these criteria.

1.1 The Lack of Security

Insufficient or deficient security means that the system in question is vulnerable and/or that it is susceptible to threat scenarios. Although the concept of security has been used for thousands of years, e.g., in relation to economic security or the ability to sustain an existence, the term has acquired a new significance through the last 40 years or so because novel types of security concerns have arisen for individuals, organisations, and society. One specific security issue is the vulnerability of information technology and computing systems. This has become an issue because we have made ourselves and our societies dependent on the accurate and reliable functioning of a technology beset with problems that have been characterised as either "wicked" [1], "complex" [2], or "non-trivial" [3]. In other words, we have developed and continue to develop and deploy systems and technologies that we do not completely understand and control, but which we are unwilling—and probably also unable—to give up. The incomplete understanding and control of a system means, of course, that it is vulnerable should anything adversely affect that functioning. Another prevailing issue of security is the ways in which ordinary citizens have come under threat from physical acts of terror and violence that are motivated by political, religious, or socio-economic reasons or problems.

The vulnerability of contemporary Western societies to deliberately harmful actions was precisely characterised already many years ago:

> "Malicious intervention has one crucial difference – so obvious that it is often overlooked. If natural disasters happen to strike a point of weakness, that is an unfortunate coincidence; but malicious actions deliberately seek out and exploit vulnerabilities so as to maximize damage and limit possible responses. Thus identifiable vulnerabilities can invite attack tailored to take advantage of them. If that attack in turn is foreseen, one can try to forestall it by reducing the vulnerabilities that it might exploit. Such reductions will in turn create their own, perhaps different, vulnerabilities – which may be lesser or greater than the original ones – thereby inviting new forms of attack, and so on. This iterative, coevolutionary process reduces total vulnerability to attack only if it carefully anticipates the new vulnerabilities created by responses to earlier ones". [4, p. 14.]

As this description points out, the critical problem is that the systems we build are imperfect, meaning that they include weak points or vulnerabilities. They are imperfect because we are unable completely to describe and analyse them. By the same token, the improvements or reparations that are made in response to a discovered vulnerability will also be imperfect, leading to a *circulus vitiosus* which we so far have been unable to manage, let alone to break free of.

1.2 The Meaning of Safety

The general understanding of safety is well represented by the definition provided by the American National Standards Institute, according to which safety is "the freedom from unacceptable risk". Other definitions mention hazards instead of risk, but by and large the definitions agree that safety is characterised by the *absence* of something that goes wrong—which in the extreme case is interpreted to mean that nothing can go wrong as in the ideal of 'zero accident' or 'zero harm' [5]. Safety thinking is also dominated by the *causality credo* [6], which can be expressed as follows: (1) things that go right and things that go wrong both have their causes, but the causes are different; (2) it is possible to find the causes for adverse outcomes given that enough evidence is collected and once the causes have been found they can be eliminated, encapsulated, or otherwise neutralised; (3) since all adverse outcomes have a cause (or causes) and since all causes can be found, it follows that all accidents can be prevented. The *causality credo* furthermore implies that the events (causes, risks, or hazards) that may be a threat to safety are 'passive' in the sense that they just happen—they are the "unfortunate coincidences" that Lovins & Lovins (op. cit.) referred to. Parts or components in simple systems may fail, break or malfunction or someone may make a mistake or even an "error". In more complicated, intractable, or non-trivial systems there may be latent conditions or hidden weaknesses, such as sneak paths [7, 8] that suddenly come into play and lead to manifest failures. Safety thinking universally assumes that whatever goes wrong does so by itself, and not because it is willed or because someone intends it to go wrong or fail.

1.3 The Difference Between Security and Safety

On the level of manifestations or phenomenology, security and safety may both be defined as a condition without harm, loss, or injury, where nothing untoward or undesired happens. But the essential difference is that for security something happens because it is intended to happen, because someone—an individual or an organisation—has wanted it to happen and has taken steps to ensure that it does. The harm that occurs is the outcome of deliberate actions. On the level of causes, security refers to events that are intentional, whereas safety refers to events that are stochastic or probabilistic by nature. Security is active in the sense that there is an active agent, or agents, that try to do harm while safety is passive in the sense that things just happen. Safety is a condition where people (as individuals, organisations, or societies) are protected from danger, risk or injury; similarly, a system is said to be safe if it is unlikely to cause or lead to danger, risk or injury. Security refers to a condition where people (as individuals, organisations, or societies) are protected against the consequences of planned, malicious and criminal acts or against the acts themselves. The protection covers everything that is of value to an organisation/ individual and the specific intention of the perpetrator is to destroy or harm that which is valued.

Security can therefore in principle be achieved by delaying, preventing, and otherwise protecting against external or internal hazards, failures, and malfunctions as well as actions by criminals and other individuals that threaten, hinder or destroy an organization's "steady state," and deprive it of its intended purpose for being. The question is whether, using this generic definition of security, it is possible to specify the elements of a security program and therefore to ensure security by design.

2 The Limitations of Design

The meaning of design is to plan how something—a mechanism, a production unit, a service, or a system—should work where the plan describes what should happen in minute detail. Design is "a commitment to a plan which is meant to be carried out" [1]. Humans have for millennia been justifiably proud of their ability to design very complicated systems, from the Antikythera Mechanism to the Large Hadron Collider. The manifest ability to design successful mechanical or technological systems—"hard" systems—has become the basis for an undue optimism about how design potentially can provide a solution to all kinds of problems, including those found in "soft" systems and in the "wicked" systems referred to above. This adherence to "solutionism" has been decried from Wiener [9] to Morozov [10], although with limited effect.

The foundation for the design of technological systems is a concise description of a subset of the physical world. The foundation usually comprises a model of the system that is going to be built, a description of how it should work, and a set of assumptions about the environment in which the system is going to exist. In other words, the system as it is imagined to work in a world as it is imagined to be. "Design takes place in the world of imagination, where one invents and manipulates ideas instead of the real thing—in order to prepare the real intervention" [1]. Designing something, whether it is the functioning or service provided by a complex technological or socio-technical system, or the safety or security of the same, thus depends on the ability to imagine both how the system functions, the "inner mechanisms" so to speak, and what may happen in the environment in which it is expected or assumed to function.

2.1 Requisite Imagination

Cybernetics and control theory are two disciplines concerned with how a system or a process can be controlled, so that it will behave or perform as required. The essence of the ability to control a system is expressed by the Law of Requisite Variety [11], which basically says that the variety of a controller should match the variety of the system to be controlled. In a more straightforward fashion this can be interpreted as saying that a regulator of a system must be a model of that system [12]. The variety of the system comes from its functional characteristics as well as the events in the environment, especially the disturbances that may affect it. If we think of the variety as all the different conditions that can possibly exist or all the possible situations that may occur, it stands to reason that the controller must have at least the same variety or richness as the system to be controlled. If not, it will sooner or later be confronted with a situation and a set of conditions for which there is no ready, or even conceivable, response.

Westrum [13] has proposed a concept of *requisite imagination* in analogy with the concept of requisite variety, meaning the ability to speculate constructively about the possible ways in which something will work and especially the possible ways in which something can go wrong. "Requisite imagination is the ability to imagine key aspects of the future we are planning … Most importantly, it involves anticipating what might go wrong, and how to test for problems when the design is developed. Requisite imagination often indicates the direction from which trouble is likely to arrive … It provides a means for the designer to explore those factors that can affect design outcomes in future contexts" [14]. The concept of requisite imagination was later used as the basis for a distinction between three classes of unexpected events called regular threats, irregular threats, and unexampled events, respectively [15].

- Regular threats are events that occur so often that it is cost-effective to develop a standard response to them. The least scary are the predictable threats from inside the system. More disturbing are the predictable external threats that potentially may disrupt the system as a whole.
- Irregular threats are imaginable but usually unexpected. They can occur in so many different ways that it is not only impractical but also too costly to prepare a response to them. The number of similar low-probability but devastating events will in most cases be so large that one cannot prepare responses for all of them. Irregular threats represent a set of recognisable problems of which a subset has solutions that are possible in principle, but not in practice.
- Finally, unexampled events are those that are virtually impossible to imagine and which exceed the responders' collective experience. Since irregular threats and unexampled events are infrequent and unusual, they cannot be treated in the conventional way, e.g., by eliminating them or by designing barriers to block their occurrence. Neither can they easily be described by the linear types of accident models that are *de rigueur* in safety management. Their distinguishing feature seems to be that they emerge out of a situation. Unexampled events require a shift in the mental framework that requires designers as well as responders to go outside of their collective experience envelope, hence leave their zone of comfort.

2.2 Elimination, Prevention, and Protection

Design must be based on some form of understanding or explanation of how a system is assumed to work, of how things happen. Since the beginning of the 20th century an important part of that has been a widely shared understanding of how accidents can happen. According to this description, often referred to as "the anatomy of an accident" [16, 17], the undesirable series of events begins when an unexpected event occurs while the system is in a normal state and performs as it has been designed to (cf., Fig. 1). Unless the unexpected event can be neutralised, quickly or immediately, the system will move from a normal into a disturbed state.

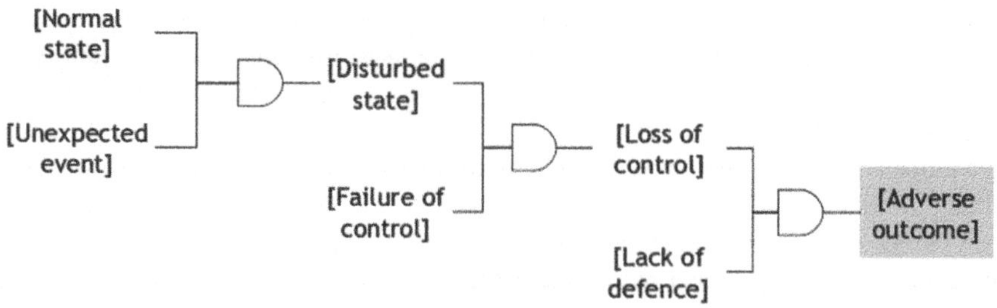

Fig. 1 The anatomy of an accident

In the disturbed state prepared attempts will be made, by people or by automatic systems, to regain control and/or to contain the consequences of the disturbance. If this fails, then the system will enter into a state of loss of control, which most likely will lead to some unwanted occurrences. The system design has hopefully also anticipated this possibility so that specific barriers or defences have been provided. It is only in the case where such barriers are missing or when they fail to work according to their purpose, that the accident actually happens.

The generic description of how an accident unfolds is a useful starting point for thinking about how adverse outcomes either can be prevented or how a system, including the people that directly or indirectly depend on it, can be protected against their consequences. One solution is to make sure that there are as few unexpected events as possible—preferably none at all. This corresponds to the purpose of the classical techniques of risk analysis or hazard analysis, which aim to eliminate known hazards and risks as far as possible. A second solution is to design features or options that make it possible to control and mitigate possible disturbed states, and thereby prevent any further deterioration of system performance. In case the two first solutions are insufficient, a third is to protect against the unwanted outcomes by means of effective defences or barriers. Prevention is, however, as always better than cure.

Although "the anatomy of an accident" nominally is about safety it can of course be used to think about security issues as well. It is straightforward to replace the top state [Adverse outcome] in Fig. 1 with, e.g., [Compromised security] or something similar. The figure then illustrates how security can be compromised and how performance may deteriorate in terms of the gradual transitions from a normal state through a disturbed state to a state of loss of control, all started by an initial unexpected event. The main difference between safety and security is as always that the unexpected event in the case of security is intended rather than random. Harmful intents may furthermore not only be behind the unexpected event but may also affect how well the system is able to control developments and/or protect against the outcomes. In the case of security, multiple threats are likely to be coordinated rather than independent, which makes it questionable whether the experiences from designing for safety can be transferred to designing for security.

The greatest obstacle to designing for security is however that it is fiendishly difficult to imagine where threats may come from and how they will manifest themselves. Adverse outcomes can in principle be prevented, since the number of ways in which something can go wrong or fail is limited. The outcomes are therefore in principle imaginable and can be prevented by means of various forms of barriers or layers of protection, where each solution has known costs, strengths, and weaknesses [18]. The adverse events, the threats that may compromise security, can however not be prevented unless they are imaginable. While it may be possible to design for security in the case of regular threats [15] it is practically impossible to do so for irregular threats and virtually impossible for unexampled events.

Safety concerns typically focus on regular events with occasional forays into the set of irregular events—usually in the aftermath of their occurrence. The focus of security concerns, on the other hand, should be on irregular events and especially on unexampled events. This means that security concerns relate to events that are difficult or impossible to imagine. That unfortunately also means that it is nearly impossible to have security by design: If we cannot think of ways in which something can happen or if there are events that we cannot imagine, then we cannot begin to consider precautions against them, nor possible responses. There is no way of accounting for the deviousness of the human mind, but that is precisely what is behind threats to security.

3 Resilience to the Rescue?

The unfortunate but unavoidable conclusion from the above arguments is that security by design is impossible in principle as well as in practice. While it may be possible to some extent to guard against or prevent unwanted *outcomes* (adverse effects), prevention is not a workable solution for the actions or events that can compromise security. But what if we for a moment abstain from the linear thinking that is inherent in "the anatomy of an accident" and indeed in the general ways we usually understand and explain physical and socio-technical systems alike? Instead of looking at how something may go wrong, fail or be disrupted, we could consider how something functions under various conditions. Instead of focusing on how to *avoid* that an organization's functions are disrupted we could focus on how to ensure that an organization's functions can be *sustained* in as wide a range of conditions as possible. This is precisely what resilience engineering tries to do. According to the current definition [19], the performance of a system is resilient if it can function as required under expected and unexpected conditions alike, where the unexpected conditions are disturbances (threats) but also changes and opportunities. Resilience is thus an expression of how people and organisations cope with everyday situations—large and small—by adjusting their performance to the conditions. Where the current approaches to safety and security focus on possible hazards and threats, on that which may prevent the system from functioning as it should, resilience engineering focuses on what is required to bring about and sustain acceptable performance under a variety of conditions. So instead of looking at possible threats, at what may prevent the functioning, we should look at what is needed to keep the functioning going in as many situations as possible.

Resilience engineering concretely proposes that a system must have four potentials in order to be able to perform in a resilient manner. These are the potential to respond, the potential to monitor, the potential to learn, and the potential to anticipate.

- The potential to respond is obviously fundamental. If an organisation cannot respond it will succumb—and probably sooner rather than later. Conversely, if it can respond it knows what to do when confronted with regular and irregular changes, disturbances, and opportunities. The responses can be in terms of

prepared actions, in terms of adjusting the current mode of functioning, or in terms of inventing or creating new ways of doing things.
- The potential to monitor means that the organisation knows what to look for, and that it is able to keep an eye on that which affects or could affect its performance in the near term—positively or negatively. (In practice this means within the time-frame of ongoing operations whatever they may be.). The monitoring must cover an organization's own performance as well as what happens in the operating environment. Monitoring is essential since without it, every situation that requires a response will be a surprise.
- The potential to learn means that the organisation knows what has happened and that it is able to learn from experience, in particular to learn the right lessons from the right experiences. This includes both single loop learning derived from specific experiences, and the double loop learning that is used to modify the organization's goals or objectives. Learning may also include changes to the values or criteria that co-determine how work is tailored to a situation. Learning is essential since without it an organisation would always respond—and monitor—in the same way.
- Finally, the potential to anticipate means that the organisation knows what to expect, or that it is able to anticipate developments further into the future, such as potential disruptions, novel demands or constraints, new opportunities, or changing operating conditions. Anticipation allows an organisation to consider at least some of the irregular threats and unexampled events before they actually happen, hence to become less vulnerable.

The alternative to design for security can therefore be to design for resilient performance by developing and maintaining an organization's potentials to respond, monitor, learn, and anticipate. In doing so it is essential to acknowledge that the four potentials are interdependent and that they together constitute a whole, an integrated capability. It is consequently inadvisable to consider and develop each potential by itself—although it from a management point of view may seem attractive. It is necessary to consider the potential couplings among the potentials—loose or tight—and the dependencies that often seem to emerge from the dynamics of system performance.

In the traditional thinking prevention and protection refers to structures or to infrastructure. Risks and threats are therefore associated with components or elements and the way they are organised—the architecture—is assumed to be known. Indeed, that is what can be designed. But in the modern world and in today's environment, the functions are more important than structures. It is therefore necessary to think about the organisation in a different way, not as composed of parts (as per the classical system definition) but as being a dynamic configuration where functions are momentarily under control due to compensating changes—as in Weick's [20] concept of dynamic non-events. Instead of thinking about how malicious events can be prevented we should think about ways in which the organization's potentials for resilient performance can be developed and sustained. Security by design should be about how to improve the ways in which the organisation functions, rather than about how to prevent or eliminate threats and disruptions.

References

1. Rittel HWJ (1988) The reasoning of designers. (Arbeitspapier zum international congress on planning and design theory, Boston, August 1987.) Universität Stuttgart: Schriftenreihe des Instituts für Grundlagen der Planung
2. Perrow C (1984) Normal accidents. Basic Books, New York
3. von Foerster H (2002) Understanding systems: conversations on epistemology and ethics (IFSR International series in systems science and systems engineering, vol 17.) Springer, Berlin
4. Lovins AB, Lovins LH (1982) Brittle power. Brick House Publishing Co, Andover, MA
5. Zwetsloot GIJM et al (2013) The case for research into the zero accident vision. Saf Sci 58:41–48
6. Hollnagel E (2014) Safety-I and Safety-II: the past and future of safety management. Ashgate, Farnham, UK
7. Burati DL, Godoy SG (1982) Sneak analysis application guidelines. RADC-TR-82-179. Rome Air Development Center
8. Hahn AH (1991) Applying sneak analysis to the identification of human errors of commission. Reliability Engineering & System Safety 33:289–300
9. Wiener N (1964) God & Golem, Inc. A comment on certain points where cybernetics impinges on religion. The MIT Press, Cambridge, Massachusetts
10. Morozov E (2013) The perils of perfection. The New York Times, 2 March 2013
11. Ashby WR (1956) An introduction to cybernetics. Methuen & Co, London
12. Conant RC, Ashby WR (1970) Every good regulator of a system must be a model of that system. Int J Syst Sci 1(2):89–97
13. Westrum R (1993) Cultures with requisite imagination. In: Wise JA, Hopkin VD, Stager P (eds) Verification ad validation of complex systems: human factors issues. Springer Verlag, Berlin, pp 401–416
14. Adamski AJ, Westrum R (2003) Requisite imagination. The fine art of anticipating what might go wrong. In: E Hollnagel (ed), Handbook of cognitive task design. Lawrence Erlbaum, New Jersey
15. Westrum R (2006) A typology of Resilience Situations. In: Hollnagel E, Woods DD, Leveson N (eds), Resilience engineering. concepts and precepts. Ashgate, Aldershot, UK
16. Green AE (1988) Human factors in industrial risk assessment—some early work. In: Goodstein LP, Andersen HB, Olsen SE (eds) Task, errors and mental models. Taylor & Francis, London
17. Hollnagel E (1995) Latent failure conditions and safety barrier integrity. Invited presentation at Joint OECD/NEA-IAEA symposium on human factors and organisation in NPP maintenance outages: impact on safety. Stockholm, Sweden, 19–22 June
18. Hollnagel E (2004) Barriers and accident prevention. Ashgate, Aldershot, UK
19. Hollnagel E (2017) Safety-II in practice: developing the resilience potentials. Routledge, Abingdon, UK
20. Weick KE (1987) Organizational culture as a source of high reliability. Calif Manag Rev 29(2):112–128

Clustering Heterogeneous Semi-structured Social Science Datasets for Security Applications

D. B. Skillicorn and C. Leuprecht

Abstract Social scientists have begun to collect large datasets that are heterogeneous and semi-structured, but the ability to analyze such data has lagged behind its collection. We design a process to map such datasets to a numerical form, apply singular value decomposition clustering, and explore the impact of individual attributes or fields by overlaying visualizations of the clusters. This provides a new path for understanding such datasets, which we illustrate with three real-world examples: the Global Terrorism Database, which records details of every terrorist attack since 1970; a Chicago police dataset, which records details of every drug-related incident over a period of approximately a month; and a dataset describing members of a Hezbollah crime/terror network in the U.S.

Keywords Clustering · Hashing · Terrorism · Crime · Global terrorism database Chicago policing · Hezbollah

1 Introduction

Social science has conventionally worked with relatively small datasets and with primarily qualitative analysis techniques. However, the cost of collecting data is decreasing while the number of social-science datasets of substantial size—thousands of records and hundreds of attributes—keeps growing. Such datasets typically contain highly heterogeneous information in a semi-structured form: defined

D. B. Skillicorn
School of Computing, Queen's University, Kingston, Canada
e-mail: skill@cs.queensu.ca

C. Leuprecht (✉)
Political Science, Royal Military College of Canada, Kingston, Canada
e-mail: leuprecht-c@rmc.ca

C. Leuprecht
Flinders University of South Australia, Adelaide, SA, Australia

A. J. Masys (ed.), *Security by Design*, Advanced Sciences and Technologies for Security Applications, https://doi.org/10.1007/978-3-319-78021-4_9

records, attributes and fields, but free-form information within them. Excel spreadsheets are often the format of choice.

Social scientists who want to analyze such data face two major difficulties. First, the contents of the fields can be extremely heterogeneous: dates or time intervals, explicitly numerical data (e.g. ages), coded numeric data (e.g. 0 or 1 for presence or absence), categorical data (e.g. country), and short textual fields (e.g. brief descriptions). Standard statistical machinery, such as significance testing, is difficult or impossible to apply to such data. Second, the data is large enough that all but the most simple analysis requires tools and techniques in which most social scientists are not trained. As a result there are a number of large datasets, to which knowledge-discovery techniques could be applied, that are actually (under-)used merely for querying.

This chapter contributes to the development of techniques that can be used to convert such large, heterogeneous datasets into a usable numerical form, the application of clustering analysis to the resulting data, and forms of analysis by overlaying that make it possible to interpret the meaning of clusters in terms of the domain where the data originates. We illustrate with three real-world datasets: the Global Terrorism Database, which records characteristics of every major terrorist attack since 1970, is representative of a national security dataset that purports to capture data across the world; a dataset of police incidents involving drugs in Chicago over a period of slightly more than a month that is representative of patterns across a large metropolitan area; and a dataset of demographics and actions by the members of two connected Hezbollah crime/terrorism groups that is representative of patterns of activity by a particular group. For all three, the visible structures would be difficult to find by directly analyzing the datasets, that is, in the way social scientists habitually approach such dataset

2 Technical Approach

We begin with a dataset that is semi-structured, that is, it has defined records and attributes, but for which the values in each location could be of several types. Often entries are missing, with the corresponding locations left empty. This may mean that the corresponding value does not apply, or that it was not collected, and the size of the dataset usually precludes manual checking. We use the following analysis pipeline:

1. Hash each entry to produce a true numeric value. The intent of this step is to replace the original values, in whatever form they come, with numeric values that correspond to them in some useful way. For consistency, we apply the same algorithm for all types of input attributes: we treat each entry as a string, convert each character to a floating-point value and compute the mean of these values. Thus the length of the input representation does not have much impact.
   The effect of this hashing function is to emphasize similarity and difference, rather than magnitude. It does not necessarily preserve the ordering of ordinal

(or numeric) values. Hence it will not represent an important ordinal attribute with a large range well. The technique relies on the presence of a large number of attributes, many of which are partially correlated with others, which is typical of most real-world datasets. However, there will certainly be some datasets for which a more clever hashing technique will be required. Note also that the 'direction' of magnitude is not necessarily preserved—large magnitudes may be hashed to small numbers, and vice versa.

Some preprocessing is also needed for attributes with only a few values, which are often coded in brief form. For example, performance is improved by replacing binary coded fields (0 and 1) by longer versions such as 00000000 and 11111111. In practice, there is often a significant amount of preprocessing to remove stray characters, and inconsistencies of coding and entry, especially when a dataset has been developed over a period of time.

The result of this step is an entirely numeric dataset in which differences in magnitudes reflect differences in the underlying original data fields.

2. Normalize each column of the dataset. The magnitudes of the hashed entries for each attribute are unpredictable. This step makes the range of values for each attribute roughly the same, and so makes it possible to compare one attribute's value with another mathematically.

   We convert each column to z-scores, by subtracting the column mean from each entry, and then dividing each entry by the column standard deviation. This has the effect of mapping the values in each column so that approximately two-thirds of them lie between −1 and +1.

3. Cluster the records using singular value decomposition [3]. If the numeric dataset has n records and m attributes, then each record has a natural representation as a point in m-dimensional space. Singular value decomposition can be regarded as a transformation of this space in a way that reveals its greatest variation.

   Formally, if A is the $n \times m$ numeric data matrix, then its singular value decomposition (SVD) is

$$A = USV'$$

   where U is $n \times$ m, V is $m \times m$, the dash indicates transposition, S is a diagonal matrix whose entries are non-increasing, and both U and V are orthogonal matrices. This decomposition can be truncated at any $k \leq m$ so that

$$A \approx U_{n \times k} S_{k \times k} V'_{k \times m}$$

   If $k = 2,3$ then each of the n records can be represented as a point in 2- or 3-dimensional space, and directly visualized.

   Truncation reduces the effect of any one attribute on the geometric embedding, as long as many attributes are broadly correlated with many other attributes. This property of the SVD means that inconsistencies introduced by the hashing tend not to distort the results as much as might, at first glance, be expected.

4. Interpret the effect of each attribute on the clustering using overlaying. Each axis of the transformed space is a linear combination of the original attributes, but this is almost always unhelpful when there are many attributes. The plot of a dataset in 3-dimensional space may show the clusters present in the data, but it is not obvious what attribute variation is responsible for the visible variation in the clustering.
   To help with interpretation, we overlay the points of the clustering with colors corresponding to the magnitude of the value of each attribute in turn, producing a new visualization for each one. If an attribute is not related to the visible clusters, then there will be no particular association between each point's position and its color. If there is a relation, then the association between position and color may reveal it.
   This phase cannot be automated, and relies on the pattern-recognition skills of a human analyst.

The beginning and end of this analysis pipeline rely on action by the analyst: cleaning up the data, and interpreting the plots. However, the central part, the SVD and clustering, can be fully automated.

3 Datasets

We use three datasets to illustrate the power of the technique. The first is the Global Terrorism Database [5], which contains records describing every terrorist attack between 1970 and 2011 (in the version we use). For each attack, data concerning timing, geographical location, form of attack, form of target, demographics of attackers, motives and claims, weapons used, casualties, and costs are included when appropriate. The dataset contains 104,687 records and 117 attributes. It has been studied by many groups: the paper by Godwin et al. [2] uses a form of visualization; the paper by Enders et al. [1] applies a kind of calibration to adjust for known issues (largely fixed in recent releases) and to examine temporal patterns of frequencies. The paper by Shafiq et al. [7] attempts prediction from the GTD data, but uses only a few (and the easiest to code) attributes. Another example of analysis of a terrorism dataset is the paper by Guo et al. [4] which merges incident information with aggregated locations to produce maps of terrorism intensity.

The second dataset we use is drawn from a collection of Chicago police reports (https://data.cityofchicago.org/). From this we select only reports of drug incidents. Data concerning time, location, descriptions, and coordinates are included. The dataset contains 35,479 records and 13 attributes.

The third dataset is a collection of records describing 181 members of two connected Hezbollah networks in the U.S. [6]. For each member, 59 attributes were collected. These range from demographics such as place of birth and marital history, to activities associated with the group's criminal, terrorist, and smuggling activities.

4 Results

4.1 Global Terrorism Database

The most striking property of Fig. 1, which represents all terrorist attacks over more than 40 years, is that there are clusters (rather than one big blob or points scattered across the entire space). There are 8 main clusters (labelled with the letters A–H), and these clusters are themselves organized into a "double hinge" structure. We now show how to discover what these clusters represent.

In Fig. 2 we see the same plot, with the variation with respect to some of the attributes shown as color variations. For example, if we look at the overlay derived from the "Country" attribute (Fig. 2a), we see that its variation does not match or agree with the clustering, and so this attribute is not contributing much to the global variation. Figure 2b shows the variation associated with an attribute about the outcomes for hostages in a kidnapping. Other attributes associated with kidnapping and hostage-taking, associated ransom attributes, and duration of incident attributes, show this same variation. Figure 2c shows the variation associated with the number of dead and wounded among the terrorists. Figure 2d shows the variation associated with target type and number of targets. These attributes are also strongly correlated with terrorist nationalities, so we can conclude that targeting and terrorist nationality are correlated. This may reflect the geography of opportunity, but may capture something deeper.

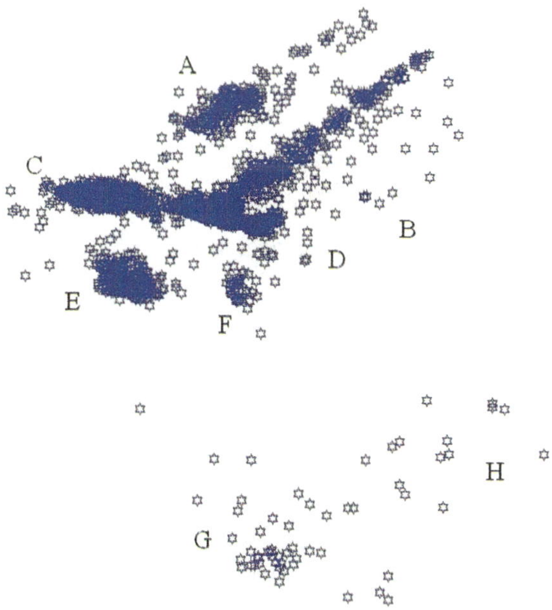

Fig. 1 Clusters in the GTD data

(a) Country attribute (b) Kidnapping

(c) Number of terrorists killed (d) Nationalities and target types

Fig. 2 Overlays of interesting attributes for the GTD clustering

We can now assign meanings to the clusters in Fig. 1. Clusters A and B are hostage taking, while C–H are more conventional attacks. Clusters A, C, E, and G are differentiated from B, D, F, and H by properties of the terrorist groups responsible, primarily whether they were degraded by the attack. Clusters A, B, C, D are differentiated from E and F and also from G and H by the type of target and number of targets—the top layer are single-target events, the second layer (E and F) are two-target events, and the third layer are three-target events (In fact, part of the diffuse structure of G and H is caused by the presence of a few four-target attacks.).

The clustering and subsequent overlaying allow useful information to be gleaned from a dataset that is otherwise hard to work with because of its size and large apparent variation.

4.2 Chicago Drug Incidents

Figure 3 shows the clustering of drug incident records. Again, the clustering is stronger than might have been expected from such disparate records.

Figure 4 shows this clustering overlaid with several attribute values. Figure 4a shows the overlay of the beat in which each incident occurred. It is clear that this attribute has little to do with the macroscopic clustering, but that it (and related geography) explain some of the microscopic structure within each cluster. The next four attributes: whether or not the incident was domestic (Fig. 4b), whether or not an arrest resulted (Fig. 4c), the FBI code associated with the incident (Fig. 4d), and the primary description (Fig. 4e) show that the macroscopic clustering has a Y shape. Domestic versus non-domestic variation goes from lower-right to upper-left; arrests vary from lower-left to upper-right, and both FBI codes and primary code for the incident vary from left to right, with greater variation in the FBI incident.

codes. Figure 4f shows that time of day does not correlate with macroscopic clustering but is reflected in the internal structure of every cluster in a way exactly orthogonal to the beat attribute. This hints that more clusters might be detectable in a higher-dimensional clustering.

Fig. 3 Clusters in the Chicago narcotics dataset

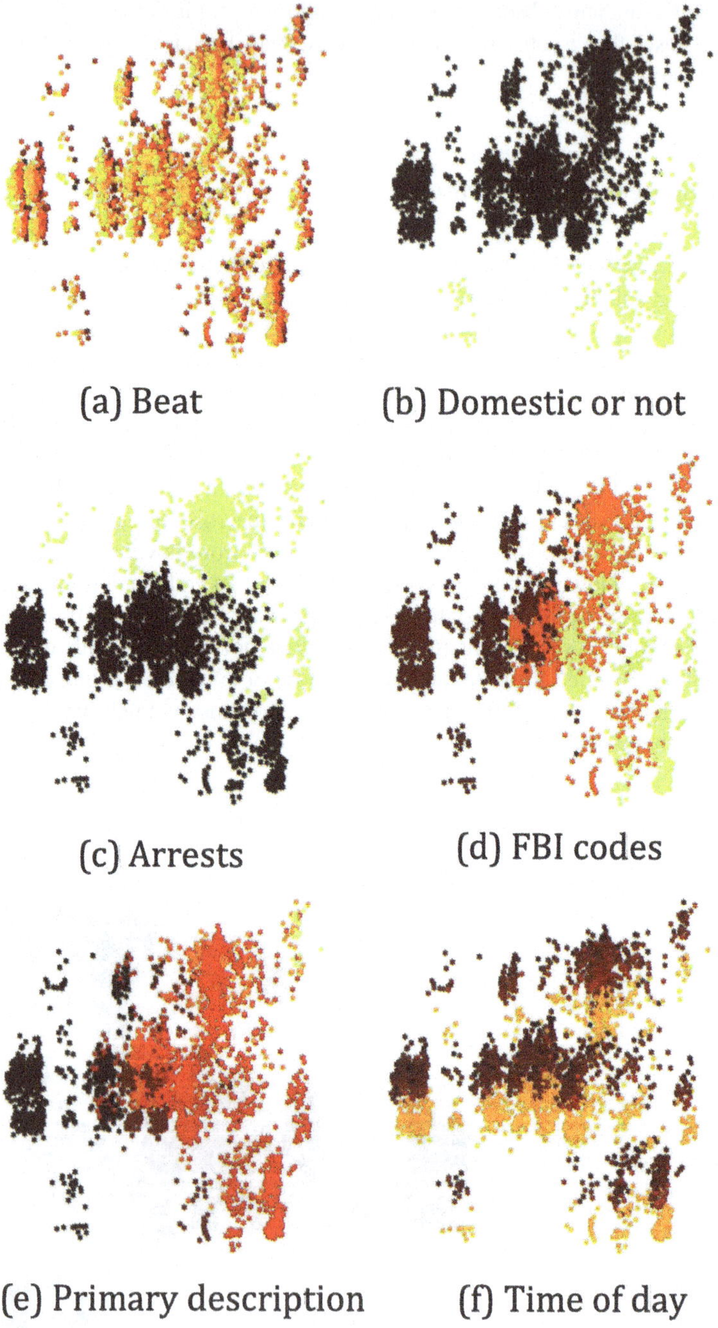

Fig. 4 Overlays of interesting attributes for the Chicago dataset

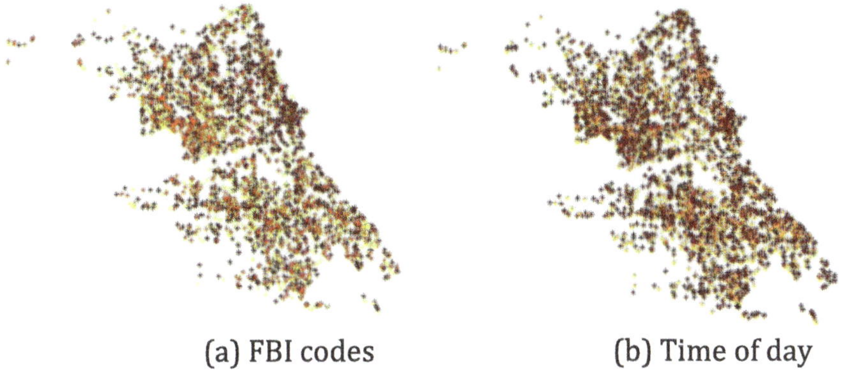

Fig. 5 Attributes related to geographical position

The Chicago data also has geographical coordinates associated with each incident, so that the relationships of attributes to location can be explored. Unsurprisingly, attributes such as beat closely correlate with location. But there are interesting relationships between the kind of incident (using the FBI crime codes and the time of day). These are shown in Fig. 5.

Figure 5a shows that different crimes occur in different areas of the city, which is not surprising, but is extremely useful information for a law enforcement agency. Figure 5b shows that there is also a relationship between location and the time at which crimes happen, again information useful to police.

4.3 Hezbollah Group

Figure 6 shows the clustering of individuals in the Hezbollah dataset. Again, there is significant structure.

Figure 7a–d show common vertical variation relating the kinds of criminal acts carried out by some members, and their demographics related to origin and apparent occupation. Figure 7e, f show common horizontal variation related to marriages. One major activity of this group was creating sham marriages as a way to get residency in the U.S. and several members were married multiple times in attempts to make this happen. Unsurprisingly this is correlated with education, since presumably those with better education had other paths to residency status. The three outlying point in the upper right are members of the group who died, so this variation is associated with dates and locations of their deaths.

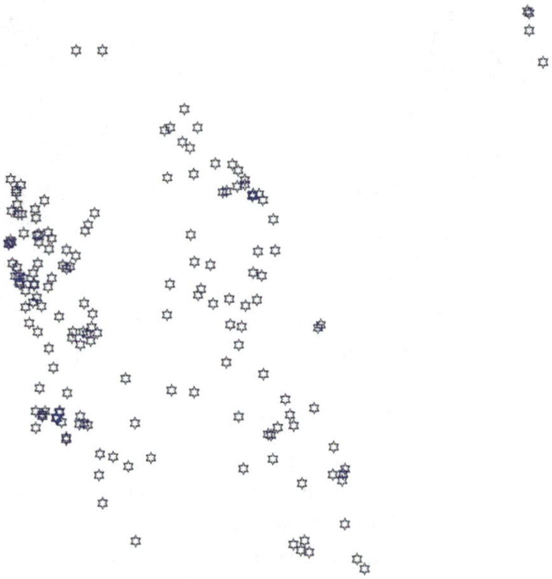

Fig. 6 Clusters for the Hezbollah dataset

Fig. 7 **a–f** Hezbollah dataset

5 Discussion

Large, semi-structured, wildly heterogeneous datasets are common in the social sciences. These properties make analytics difficult, and many are treated simply as databases to be queried: "How many cases like this are present?". However, there is much useful data implicit in such datasets. We have shown that hashing, despite its simplistic assumptions, can convert such data to more readily analyzable form, especially when it is combined with a clustering technique that is insensitive to moderate inconsistencies. The resulting clusterings, as we have shown in the examples, reveal much more implicit structure emerging from the data than might be expected. The use of overlaying makes it possible to infer what aspects of the original data explain the visible variation and clusters. This opens up a large cache of data, currently mostly ignored, to sophisticated analysis.

References

1. Enders W, Sandler T, Gaibulloev K (2011) Domestic versus transnational terrorism: data, decomposition, and dynamics. J Peace Res 48(3):319–337
2. Godwin A, Chang R, Kosara R, Ribarsky W (2008). Visual analysis of entity relationships in global terrorism database. In: Defense and Security 2008, Proceedings of SPIE Vol 6893, 2008
3. Golub GH, van Loan CF (1996) Matrix computations, 3rd edn. Johns Hopkins University Press, Baltimore
4. Guo D, Liao K, Morgan M (2007) Visualizing patterns in a global terrorism incident database. Environ Plan 34:767–784
5. LaFree G (2010) The global terrorism database: accomplishments and challenges. Perspect Terrorism 4(1)
6. Leuprecht C, Walther O, Skillicorn DB, Ryde-Collins H (2016) Hezbollah's global tentacles: the party of god's convergence with transnational organized crime. Terrorism and Political Violence 29(5):902–921
7. Shafiq S, Haider Butt W, Qamar U (2014) Attack type prediction using hybrid classifier. In: Advanced data mining and applications, vol 8933. Springer Lecture Notes in Computer Science, pp 488–498

5 Discussion

Critical Infrastructure Protection and Resilience—Integrating Interdependencies

Frédéric Petit, Duane Verner, Julia Phillips and Lawrence Paul Lewis

Abstract Throughout the world, there is strong recognition that critical infrastructure security and resilience need to be improved. Enhancing the protection and resilience of critical infrastructure has emerged as an urgent goal—a goal made more challenging by the complexity of these systems and their inherent interdependencies. Consideration of critical infrastructure interdependencies, as well as their integration into risk management and business continuity processes, is an important element of *U.S Presidential Policy Directive 21 on Critical Infrastructure Security and Resilience* (PPD–21), the *Canada-United States Action Plan for Critical Infrastructure*, and the *European Program for Critical Infrastructure Protection.* Several international standards also require the consideration of interdependencies between organizations and the effect on their risk management and business continuity practices. All of these documents, from the strategic level through the operational standards, involve the consideration of interdependencies that can exist among infrastructure, how they are managed, and how they affect business continuity, security, and resilience management. A "system of systems" approach can help establish the appropriate scope of an interdependency analysis, as well as the specific assets and/or subsystems for which resilience-related information should be collected.

Keywords Critical infrastructure · Protection · Resilience · Interdependencies

F. Petit (✉) · D. Verner · J. Phillips · L. P. Lewis
Decision and Infrastructure Sciences Division, Argonne National Laboratory,
9700 S. Cass Avenue, Bldg. 221, Argonne, Lemont, IL 60439, USA
e-mail: fpetit@anl.gov

D. Verner
e-mail: dverner@anl.gov

J. Phillips
e-mail: julia.phillips@theperducogroup.com

L. P. Lewis
e-mail: plewis@anl.gov

A. J. Masys (ed.), *Security by Design*, Advanced Sciences and Technologies for Security Applications, https://doi.org/10.1007/978-3-319-78021-4_10

1 Introduction

Throughout the world, there is strong recognition that critical infrastructure security and resilience need to be improved. The world faces significant challenges in preparing for, responding to, and recovering from disasters. Of particular concern are the impacts that natural hazards and manmade threats, including cyber-attacks, could have on critical infrastructure systems. Enhancing the protection and resilience of critical infrastructure has emerged as an urgent goal—a goal made more challenging by the complexity of these systems and their inherent interdependencies. Consideration of critical infrastructure interdependencies, and their integration into risk management and business continuity processes, is an important element of *U.S. Presidential Policy Directive 21 on Critical Infrastructure Security and Resilience* (PPD-21), the *Canada-United States Action Plan for Critical Infrastructure*, and the *European Program for Critical Infrastructure Protection*. Several international standards also require the consideration of interdependencies between organizations and the effect on their risk management and business continuity practices.

A "system of systems" approach can help to establish the appropriate scope of an interdependency analysis, as well as the specific assets and/or subsystems for which resilience-related information should be collected. Using this approach, analyses would consider the high-level context (e.g., a geographic region or an industry sector) and the associated states of these systems, ultimately represented by the most critical assets, to inform the scope and focus of a resilience assessment, including the most critical assets from which to collect interdependency data. The development of a comprehensive and interactive risk and resilience assessment of critical infrastructure considering interdependencies requires integrating multiple areas of expertise (e.g., engineering, social sciences, business continuity, and emergency management) in a combination of top-down and bottom-up approaches.

This chapter presents the classes (i.e., physical, cyber, logical, and geographic) and dimensions (i.e., operating environment, coupling and response behavior, type of failure, infrastructure characteristics, and state of operation) of critical infrastructure operations, and how these influence all components of risk (i.e., threats/hazards, vulnerability, resilience, and consequences). An assessment framework combining top-down and bottom-up data collection and analysis methods to fully consider regional infrastructure interdependencies is discussed and illustrated with an example from the Energy Sector. In particular, this chapter proposes how a combination of failure and hazard analysis techniques could be applied to analyses of all types of systems, subsystems, or an integrated set of systems in order to anticipate and plan for enhancing the protection and resilience of critical infrastructure before a disruptive event.

2 Policies

Improving critical infrastructure protection and resilience has emerged in recent years as an urgent priority. Extreme events have revealed that the consequences of disasters and disruptions extend well beyond the individual facilities or systems directly affected. A better understanding of the complex interactions among critical infrastructure is essential in preparing for, responding to, and recovering from disasters, and supports stakeholders in making better-informed decisions on issues of business continuity, security, and risk management.

Examples of the far-reaching impacts of disasters highlighted by infrastructure interdependencies abound. The terrorist attacks of September 11, 2001, resulted in immediate impacts to electric power, natural gas, communications, emergency services, transportation, and water infrastructure, but spread quickly to include disruptions to air travel, financial markets, and government services. Examination of these impacts following the events highlighted the geographic and temporal nature of cascading and escalating consequences [19]. In August 2003, the disruption in the Ohio electric grid resulted in a blackout affecting the U.S. Midwest and Northeast, as well as Ontario, Canada. The blackout lasted four days, and eight major urban areas in the Great Lakes region were impacted. Within the Energy Sector alone, the disruption in electricity caused petroleum refineries to cease operations and resulted in shutdowns and decreased demand in natural gas pipelines. Beyond the Energy Sector, major metropolitan areas were without water; cell phone use was disrupted due to heavy volume; air travel ceased at affected airports; police directed traffic at major intersections; emergency responders rescued individuals trapped in elevators and subway tunnels; and hospitals powered by generators received more patients, many of whom relied on home ventilators or suffered heat stress; ATMs lost network connections; chemical manufacturing plants were shut down; and grocery stores, restaurants, and individuals at home were without refrigeration and lost food, creating a strain on sanitation systems (Infrastructure Assurance Center [IAC] [32]). After-action reports following Hurricane Katrina in 2005 cited a need for a comprehensive understanding of infrastructure interdependencies in the geographic area affected and the potential national impacts of decisions made, as well as the need for accurate information about the facilities affected, and coordinated information sharing among stakeholders [56, 61]. Infrastructure disruptions in the New York City area during Hurricane Sandy in 2012—including impacts to the electric power system, fuel distribution system, healthcare sector, and transit systems—highlighted the need for local government decision makers to better identify and prioritize critical assets and facilities, and better understand interactions among systems [25].

These catastrophic events illustrate that disruptions at an individual facility often do not end at that facility boundary and are not confined within political or geographic borders. Consider the consequences of the disruption of European air transport after Iceland's Eyjafjallajokull volcano eruption in April 2010 [41] or those related to the failure of nuclear reactors in Japan after the earthquake and tsunami in March 2011 [42]. It is impossible to protect against these types of acute events, but critical infrastructure can be more resilient by promoting adequate measures.

As a result, policies set at the local, state, and national level—as well as policies and standards developed among groups of countries—reflect the importance of understanding critical infrastructure interdependencies. These documents form the basis of current approaches to analyzing and prioritizing infrastructure interdependencies toward a goal of improved security and resilience.

2.1 North America

Following the 2003 blackout, the United States and Canadian governments created a task force to analyze the causes and make recommendations to prevent or mitigate future disruptions (U.S.–Canada Power System Outage Task Force [60]). Plans and strategies developed between the United States and Canada since the event have reiterated the importance of considering interdependencies in risk management processes.

The *Canada-United States Action Plan for Critical Infrastructure*, published in 2010, presents a cross-border approach to strengthening the resiliency of critical infrastructure. One of the three objectives of this action plan is to develop a collaborative risk management approach, which requires identifying and addressing key infrastructure (DHS and Public Safety Canada [18]). The *Joint United States-Canada Electric Grid Security and Resilience Strategy*, developed in December 2016 by the United States and Canadian governments, states the need to integrate infrastructure interdependency considerations to ensure the security and resilience of the electric grid. One of the main goals of this strategy is specifically to "understand and mitigate vulnerabilities from interdependencies with other critical infrastructure" [28].

In Canada, two documents define the risk management approach for strengthening the protection and resilience of critical infrastructure: the *National Strategy for Critical Infrastructure* (National Strategy) and the 2014–2017 *Action Plan for Critical Infrastructure* (Action Plan). The National Strategy defines the concepts and principles for developing partnerships between governments and critical infrastructure, advancing all-hazards risk management, and defining measures to improve information sharing and protection [27]. The concept of interdependencies is specifically highlighted in the National Strategy as an important characteristic of critical infrastructure systems that must be considered in all strategic objectives from the risk management approaches through the sharing and protection of information among key stakeholders [27]. The Action Plan defines the steps required for addressing the objectives defined in the National Strategy. In particular, the Action Plan identifies the action to implement the Regional Resilience Assessment Program across Canada. The goal of this program, similar to the Regional Resiliency Assessment Program (RRAP) in the United States, is to identify and analyze the resilience and interdependencies of critical infrastructure sectors using an all-hazards approach [53].

In the United States, two main documents, issued in 2013, guide the strategy for enhancing the protection and resilience of critical infrastructure: *Presidential Directive* 21 (PPD-21) and the *National Infrastructure Protection Plan* (NIPP). PPD-21 establishes the roles and responsibilities of the Secretary of the U.S. Department of Homeland Security (DHS) to identify and analyze key interdependencies among critical infrastructure sectors. This directive specifically calls for operational and strategic analysis to identify infrastructure interdependencies and to incorporate them into risk assessment and management procedures [58]. The 2013 edition of the NIPP follows PPD-21 and reinforces the need to understand and address risks from cross-sector interdependencies to enhance infrastructure security and resilience [15]. These two documents complement the *National Strategy for Global Supply Chain Security*, issued in 2012. This strategy establishes goals of promoting efficient and secure movement of goods and fostering a resilient supply chain to reduce vulnerability to disruption, prepare for and withstand evolving threats, and recover quickly from disruption [57]. Closely related is the *Critical Foreign Dependencies Initiative* (CFDI), operated within DHS, which has identified and prioritized foreign dependencies of national significance since 2008. The CFDI list includes a "comprehensive inventory of infrastructure located outside the United States that if disrupted or destroyed would lead to loss of life in the United States or critically affect the Nation's economy or national security" [13].

These U.S. national strategic plans are completed by government agencies strategic plans that operationalize the consideration of critical infrastructure interdependencies in risk management approaches. DHS has developed several of these plans.

The DHS Office of Infrastructure Protection *Strategic Plan* 2012–2016 establishes the goals for the DHS Office of Infrastructure Protection to improve risk management activities through better understanding of national and regional infrastructure interdependencies, and improve information sharing among infrastructure stakeholders to enhance understanding of regional and cross-sector interdependencies. The plan also acknowledges the cross-sector nature of infrastructure dependencies and includes objectives aimed at improving coordination with stakeholders and modeling of infrastructure interactions [14]. In 2015, the DHS *National Critical Infrastructure Security and Resilience Research and Development Plan* was developed to guide prioritization of research and development efforts within DHS. Among the tenets articulated for the *Critical Infrastructure Security and Resilience Plan* (CISR), is that "the increasingly interconnected and interdependent nature of critical infrastructure must be better understood and addressed in an integrated and holistic manner, including the identification of systemic as well as localized risks." The plan highlights the need to consider the dynamic linkages among the interdependent infrastructure systems to support risk management, to understand consequences of decisions, and improve collaboration [16].

DHS has developed specific programs, such as the RRAP, to address the objectives of its strategic plans. The RRAP is an interagency and cooperative assessment of specific critical infrastructure within a designated geographic area [17]. RRAP projects are initiated to respond to stakeholders' requirements. Some RRAPs specifically analyze the interconnections existing among critical infrastructure systems and produce maps and diagrams that describe important regional infrastructure relationships, assess the strength of infrastructure connections, and propose resilience enhancement options and operational alternatives to reinforce the robustness of critical infrastructure.

DHS is not the only U.S. agency that has developed plans and programs to integrate the consideration of infrastructure interdependencies in all-hazards risk management processes. The U.S. Department of Energy (DOE) is also active in this area. In the two installments of the *Quadrennial Energy Review*, DOE specifically recognizes the increase in interdependencies among lifeline networks (i.e., energy, communications, transportation systems, and water and wastewater systems) and the importance of considering the interactions in the interconnected global energy market to enhance the resilience and protection of energy systems [20].

In addition to national-level government directives, plans, and policies, numerous private sector industry standards require the consideration of dependencies and interdependencies in their risk management and business continuity practices. Such standards often include identifying internal and external dependencies and interdependencies and critical supply chains. Some of these standards constitute the foundation of the DHS PS-Prep^{TM} program that provides a systemic approach to business continuity and recovery that allow organizations to unify their preparedness activities (Federal Emergency Management Agency [FEMA] [24]):

- National Fire Protection Association (NFPA) 1600 *Standard on Disaster/ Emergency Management and Business Continuity Programs*. This standard recommends that emergency planners "identify interdependencies across functions, processes, and applications to determine the potential for compounding impact in the event of an interruption or disruption" [43].
- American National Standards Institute/American Society for Industrial Security (ANSI/ASI) *SPC.*1-2009 *Standard on Organizational Resilience*. This standard stipulates that "the risk assessment and impact analysis should consider its dependencies on others and others' dependencies on the organization, including critical infrastructure and supply chain dependencies and obligations" [1].
- International Organization for Standards (ISO) 22301 *Societal Security—Business Continuity Management Systems—Requirements* 06-15-2012. This standard specifies that "business impact analysis shall identify dependencies and supporting resources for these activities, including suppliers, outsource partners and other relevant interested parties" [33].

Emphasis on the need to consider interdependencies between facilities and systems in risk management processes is not only included in U.S. or Canadian policies, it also constitutes an important element of worldwide joint strategies and policies.

2.2 Europe

Several countries, including the United Kingdom, France, and Germany, have initiated work to enhance the protection and resilience of critical infrastructure. Concerns about critical infrastructure protection began in the late twentieth century in relation to tensions resulting from the Cold War. After the fall of the Eastern bloc, these concerns faded, only to resurface at the dawn of 2000 because of computer problems anticipated for the new millennium. This need to analyze and protect critical infrastructure has grown with respect to fear of terrorism arising in the past 15 years, from the terrorist attacks of September 11, 2001, in the United States, to the upsurge of events in Europe in 2016 and 2017. Similar to North America, the consideration of resilience options is more recent following the multiplicity of catastrophic natural events during the last decade and the concern over climate change. Consideration of infrastructure interdependencies is at an early stage for most countries, but the Netherlands is an exception. The Dutch work began in 2002 with the implementation of the *Vital Besherming Infrastructuur* program [37]. This program was divided into several stages. The first step was to take an inventory ("quick scan") by identifying critical infrastructure present in the Netherlands and assessing its readiness (vulnerability and protective measures) for various contingency approaches [37]. To perform this inventory, the Dutch sought to identify which infrastructure provides a product or service that is vital to the functioning of the country. Subsequently, they used questionnaires sent to managers of critical infrastructure to determine their interdependencies and their readiness in relation for different hazards.

At a supranational level, the European Union (EU) developed specific programs for the protection of critical infrastructure. The work of the European Commission (Commission) began in June 2004 following a European Council's request to develop a comprehensive strategy for strengthening the protection of critical infrastructure. That same year, the Commission adopted a communication entitled *Critical Infrastructure Protection in the Fight Against Terrorism* [21]. To meet this demand, the Commission proposed a *European Program for Critical Infrastructure Protection* (EPCIP) and a warning system for critical infrastructure, the *Critical Infrastructure Warning Information Network* (CIWIN) [21]. This proposal was accepted in December 2004 [21]. The objective of EPCIP was to ensure adequate and uniform levels of safety for critical infrastructure, to minimize disruption, and provide rapid reaction capabilities [8]. CIWIN is a network that would allow the exchange of best practices by providing a means of transmitting alerts and threat information [8].

In 2006, a proposal for a Council Directive laying the foundation for work on the classification of European Critical Infrastructure (ECI) and the need to improve their protection was put forth. The Council Directive defined a procedure for the identification and classification of ECI and a common approach to assess the need to improve their protection [10, 11]. The Council Directive recommended an all-hazards approach for the EPCIP, combined with a priority given to the risk of terrorism. Critical infrastructure protection remains the responsibility of each country and critical infrastructure owners and operators.

Following the Council Directive, the Commission launched a pilot project involving preparatory actions to strengthen the fight against terrorism. This project included a program for improving the protection of critical infrastructure to support the EPCIP. The objective of this program was to define and develop measures to protect critical infrastructure by the development of improved methodologies [21]. More broadly, concerns about critical infrastructure at the European level were made part of the *European Security Strategy* adopted by the European Council in 2003 [9]. This strategy aimed to achieve comprehensive security at the European level to promote a systemic treatment of security. A willingness exists to consider the interdependencies among critical infrastructure to make the EU more resilient. This is evident in the Horizon 2020 research program's (H2020) funding of projects promoting collaboration among government agencies, academia, and the private sector [22]. For example, the Horizon 2020 program area focusing on Security includes the objective "to foster secure European societies in a context of unprecedented transformations and growing global interdependencies and threats, while strengthening the European culture of freedom and justice" [23].

2.3 Oceania

The laws and standards of Australia and New Zealand, especially their risk management standard (AS/NZS 4360), are the basis for the development of risk management frameworks around the world. Like the United States, Australia is a pioneer in the field of risk analysis and management for critical infrastructure [6, 7]. The main Australian program on critical infrastructure is the *Critical Infrastructure Program for Modelling and Analysis* (CIPMA), which is a government project initiated by the Australian Department of Justice (and its Attorney-General), in partnership with Geoscience Australia and the Commonwealth Scientific and Industrial Research Organization (CSIRO). This program aims to analyze and model the cascading failures of critical infrastructure assets. Since the publication of its *Critical Infrastructure Resilience Strategy* in 2010, the Australian Government has changed its approach to critical infrastructure protection and resilience to adopt a voluntary (non-regulatory) approach, and to promote the continuity of critical infrastructure operations, including the consideration of infrastructure interdependencies [2].

Implementation of the CIPMA approach requires the participation of many industrial partners, because they own and operate approximately 90% of critical infrastructure [3]. This approach is inductive and focuses on large-scale geographic analysis of risk scenarios [12]. CIPMA studies are based on the implementation of an information-sharing network, the Trusted Information Sharing Network (TISN). This network fosters trust between stakeholders and promotes exchange of sensitive data from a safety and security point of view. This approach uses geographic information systems (GISs) for focusing on physical and geographic interdependencies. Studies are based on cascading and escalating failures between critical infrastructure and the analysis of hazards, vulnerabilities, and consequences. However, this approach mainly focuses on only three types of infrastructure networks: energy, finance, and telecommunications. Another Australian program, the *Computer Network Vulnerability Assessment Program* (CNVA), led by the Australian Government's Computer Emergency Readiness Team, is specifically dedicated to the study of vulnerabilities and failures related to the Information Technology and Communications Sectors [59]. This program considers the vulnerability of critical information and communications infrastructure and specific failures that could result. In addition, it addresses cyber vulnerabilities in terms of computer security (e.g., viruses, worms).

2.4 Intergovernmental Organizations

Numerous other critical infrastructure programs exist in other countries throughout the world; however, most are in developmental stages and present similar characteristics to the programs presented previously. International organizations also develop supranational programs. The EU and the North Atlantic Treaty Organization (NATO) are two of these international organizations for which the protection of critical infrastructure represents a challenge [39]. NATO addressed this issue in 2001, when it conducted the review of the readiness of its members in terms of planning and infrastructure mapping [38]. The activities conducted by NATO in this area fall within the broader framework of the *Action Plan for Civil Emergency Planning* on the protection of civilians dealing with chemical, biological, radiological, and nuclear (CBRN) agents, focused mainly on CBRN terrorism [38].

NATO's approach differs from that of the EU, because the objective is not to regulate the protection of critical infrastructure assets. NATO programs promote further standards for the preparation of states; they seek to enhance interoperability through consequence management and to improve resilience of critical infrastructure [38]. In May 2017, 34 NATO and EU nations, as well as private sector partners, held a conference to discuss interdependency and resilience. Findings were published in June 2017 and focused on building strong collaboration, identifying cross-border dependencies and foreign ownership of critical infrastructure, and leveraging tools to mitigate vulnerabilities [55].

Other intergovernmental organizations, such as the World Economic Forum (WEF) and the Organisation for Economic Co-operation and Development (OECD), identify the importance of assessing infrastructure interdependencies. OECD specifically highlights the importance of understanding the characteristics of complex systems for anticipating catastrophic events that could result in global shocks [44]. The WEF's 2013 *Global Risk Report* also highlighted the need to consider interdependencies and prioritize appropriate response measures [65].

These two organizations are especially interested in anticipating events that may require policy interventions and identifying where those interventions should or could occur for maximum efficiency [44, 65].

Even if programs and research on critical infrastructure protection and resilience in development around the world are at different stages of development for the consideration of infrastructure interdependencies, they all present some common elements as discussed in Sect. 2.5.

2.5 Common Elements

The analysis of critical infrastructure interdependencies can be conducted through various approaches. While some countries like Australia, Canada, or the United States are currently developing more operational approaches, other groups, like the European countries and intergovernmental agencies, seem to favor more strategic approaches. This can be explained by the fact that European countries are at the initial stages of their programs in this specific area. Moreover, it seems logical for intergovernmental organizations to focus on large-scale strategic approaches, such as multinational organizations and protection and resilience of critical infrastructure over large geographic areas. The mission of these organizations is to try to organize the integration of national programs.

Whether strategic or operational, the programs addressing the protection and resilience of critical infrastructure have many elements in common. Most of the approaches were developed initially to protect critical infrastructure against cyber dysfunction and, post-2001, against terrorism. All programs and methodologies developed tend to evolve to emphasize on an all-hazards approach, taking into account both natural and anthropogenic hazards. Most of these approaches are inductive and are based on the development of failure scenarios. However, most failure scenario work has tended to focus more on security issues and response to malicious acts. The majority of approaches focus on the relationship between critical infrastructure by analyzing physical and geographical dependencies. In recent years, more programs have started to focus on cyber threats and cyber dependencies, while logical dependencies are rarely taken into consideration.

Enhancing the protection and resilience of critical infrastructure requires consideration of interdependencies at the facility and regional levels. Such an all-encompassing overview accounts for potential vulnerabilities that may exist outside the traditional facility "boundaries." The loss of or degradation to one of the organizations or assets that supplies resources to a critical infrastructure can cause direct and potentially severe impacts. It is also important to consider a facility in its environment and acknowledge the fact that bidirectional relationships between infrastructures exist.

In the last 10 years, critical infrastructure protection programs have rapidly evolved, moving from a purely physical security approach to a more comprehensive consideration of the missions and functioning of critical infrastructure. It is still important to protect the critical infrastructure, but it becomes imperative to ensure their resilience by developing their capability to anticipate, mitigate, respond to, and recover after a disruptive event. This change in paradigm is based on the realization that it is impossible to protect all critical infrastructure from all hazards. Research and risk management for critical infrastructure continue to evolve, but numerous actions and steps still need to be developed for reinforcing the functioning of critical infrastructure and the nations it supports. Events such as Hurricane Sandy, which was the most destructive hurricane in 2012, reminded us of the work that remains to be done for understanding critical infrastructure interdependencies and functioning. Even if significant efforts have been made to better understand and analyze infrastructure interdependencies, there is still a limited understanding of the critical infrastructure system operations and the possible resulting cascading failures. As defined by Flynn [25], critical infrastructure are often managed and regulated in isolation from one another, which hampers private and public sector understandings of the challenges arising from interdependencies. Organizations and jurisdictions need to move beyond putting together disaster response plans that focus only on managing immediate life safety issues to developing recovery plans that allow for the rapid restoration of essential and normal infrastructure functions [25]. This evolution of critical infrastructure system management requires developing new tools to characterize infrastructure interdependencies and identify the systems, assets, and functions that are truly critical for regional resilience.

3 Characterization and Analysis of Critical Infrastructure Interdependencies

Rinaldi et al. [54] defined the concepts of infrastructure dependencies and interdependencies (Table 1).

Over the years, several researchers have created their own taxonomies to categorize and classify infrastructure interdependencies (Table 2).

Despite this array of taxonomic structures, the framework set forth by Rinaldi et al. [54] remains the most useful and comprehensive, as each of the elements identified in subsequent taxonomies can be found in the classes and dimensions established in their taxonomy (Tables 3 and 4).

The interdependency dimensions influence the scope and complexity of analysis; however, it is also necessary to consider the interactions of critical infrastructure with their environment. They can be characterized into three categories (Table 5).

These three categories complement the concepts of classes and dimensions by characterizing the critical infrastructure's interactions with its environment (socio-technical and ecological), and thereby elucidate how failures can cascade and escalate across critical infrastructure systems. A richer understanding of infrastructure interactions also allows for the integration of the characteristics

Table 1 Definitions of infrastructure dependency and interdependency [54]

	Definition
Dependency	Linkage or connection between two infrastructure, by which the state of one infrastructure influences or is reliant upon the state of the other
Interdependency	Bidirectional relationship between two infrastructure in which the state of each infrastructure influences or is reliant upon the state of the other

Table 2 Example of infrastructure interdependency taxonomies [45]

Authors	Dependency and interdependency types
Rinaldi et al. [54]	Physical, Cyber, Geographic, and Logical
Zimmermann [68]	Functional and Spatial
Wallace et al. [64]	Input, Mutual, Exclusive, Shared, Co-located
Pederson et al. [46]	Physical, Informational, Geospatial, Policy/Procedural, and Societal
Visarraga [63]	Physical, Logical, and Functional
Zhang and Peeta [67]	Functional, Physical, Budgetary, and Market and Economic

Table 3 Interdependencies classes [54]

Classes	Definition
Physical	Operations depend on material output(s) of other infrastructure through a functional and structural linkage between the inputs and outputs of two assets
Cyber	Operations depend on information and data transmitted through the information infrastructure via electronic or informational links
Geographic	Operations depend on the local environment, where an event can trigger changes in the state of operations in multiple infrastructure
Logical	Operations depend on the state of other infrastructure via connections other than physical, cyber, or geographical. Logical dependency is attributable to human decisions and actions and is not the result of physical or cyber processes

Table 4 Interdependencies dimensions [54]

Dimensions	Definition
Operating Environment	Characterize how external factors can influence infrastructure operations and connections
Coupling and Response Behavior	Characterize how an infrastructure would react to infrastructure connection disruptions or changes
Type of Failure	Characterize the propagation of consequences resulting from a disruption
Infrastructure Characteristics	Characterize the infrastructure organization and operation
State of Operation	Characterize the critical infrastructure state of operations

Table 5 Interdependencies categories [48]

Category	Definition
Upstream	Products or services provided to one infrastructure by another external infrastructure that are necessary to support its operations and functions
Internal	Interactions among internal operations, functions, and missions of the infrastructure
Downstream	Consequences to a critical infrastructure's consumers or recipients from the degradation of the resources provided by a critical infrastructure

(i.e., categories, classes, and dimensions) of critical infrastructure interdependencies in risk management approaches. Integrating the concepts of infrastructure interdependencies into an infrastructure's risk management process enables an understanding of how they influence all components of risk (i.e., threats, vulnerabilities, resilience, and consequences).

4 Analysis of Interdependencies

Infrastructure interdependency analysis can be analytically complicated, time-consuming, and costly, which in turn can limit the ability of stakeholders to understand and use this information to make risk-informed decisions. In order to manage these complexities, it is necessary to use a process that helps prioritize risk assessment efforts through a "systems approach." Such approaches are based on the assumption that a critical asset can be considered part of a broader system of infrastructure. Higher-level constructs (e.g., a community or a region) include multiple systems. As such, a community or a region operates as a "system of systems." Viewed within this framework, high-level systems analysis can help identify the most critical lower-level systems. This information can, in turn, help determine where to conduct more detailed site assessments focused on the most critical asset-level components [4].

Over the past decade, several modeling and simulation approaches have been developed to better understand infrastructure interdependencies. Some of the most notable work in measuring interdependencies has been done through looking at different consequences from a specific infrastructure; for example, outages and restoration times from electric power compared to restoration times for other types of critical infrastructure such as water, natural gas, and transportation. Generally, all methodologies used for risk assessment and system engineering can also be used for analyzing critical infrastructure interdependencies. Even if most methods addressing infrastructure interdependencies are ad hoc approaches that have been developed for specific uses based on the data available and the goal of the analysis, five main categories of methodologies are primary used [19, 45, 66]:

- Agent-Based Approaches
- Economic Theory-Based Approaches
- Empirical-Based Approaches
- Network-Based Approaches
- System Dynamics-Based Approaches

Table 6 presents the general description of these different approaches.

Each of the modeling and simulation approaches for analyzing critical infrastructure interdependencies presents advantages and limitations (Tables 7 and 8) and have been developed for specific types of critical infrastructure or for specific conditions.

Many analysis approaches focus on physical and cyber connections between critical infrastructure assets. Logical connections seem to be ignored, while geographic dependencies are less frequently considered.

Assessing infrastructure interdependencies requires a scalable approach that combines different analysis approaches to address stakeholder requirements and considers relevant critical infrastructure. The fundamental steps of interdependency analysis can be distilled into a framework that aligns a four-step continuum of maturity [48]:

- *Initial analysis* conducts research open-source information and performs limited analysis of infrastructure interdependencies.
- *Intermediate analysis* combines originally separated infrastructure system models and performs more detailed analysis of infrastructure interdependencies.
- *Advanced analysis* requires the creation of new data collection mechanisms and the integration of assessment approaches and considers several characteristics of critical infrastructure interdependencies.
- *Optimal analysis* anticipates and characterizes, in real time, how all interdependency dimensions influence the protection and resilience of critical infrastructure systems.

Each step in the continuum varies in terms of data required, the type of analysis conducted, and the resulting products [48]. Together, the four steps define a roadmap that can help stakeholders understand and plan for required inputs and desired outputs.

Achievement of the optimal analysis stage requires the development of a "system of systems" approach [4]. This approach includes conducting assessments coordinated both vertically, within a given sector, and horizontally, across sectors [30], and combining top-down and bottom-up data collection; this method also makes it possible for the analysis to fully consider regional infrastructure interdependencies (Fig. 1).

Given the nature of dynamic and uncertain threats, there is a critical need for an integrated approach to optimize resilience and protection of critical infrastructure. A top-down approach provides a simultaneous analysis of an entire system, including study of the systems dependencies and interdependencies. It then enables decision makers to determine protection and resilience measures that can be

Table 6 Descriptions of main interdependencies analysis methodologies

Approach	Description
Agent-Based	Analysis of complex adaptive systems by using a bottom-up approach to simulate the interactions of system components (i.e., autonomous agents) and assess their effects on the system
Economic Theory-Based	Analysis of infrastructure interdependencies as intermediate goods exchanged between a supplier and a user. There are two main economic theory-based approaches: input-output based methods and computable general equilibrium-based methods
Empirical-Based	Analysis of interdependencies based on observation and experience by using historical data in combination with expert judgment
Network-Based	Describe infrastructure systems as networks where infrastructure assets are represented as nodes and the physical connections are represented as arcs. The two main network-based approaches are topology-based methods and flow-based methods
System Dynamics-Based	Analysis of the dynamic behavior of complex systems using a top-down approach to model a system's dynamic and evolutionary behavior by using stock and flow exchanges and causal loops

Table 7 Advantages of main interdependencies analysis methodologies

Approach	Advantages
Agent-Based	• Allows for modeling of all levels of critical infrastructure systems and simulating decision makers' behaviors • Represents real-world interactions
Economic Theory-Based	• Allows for analysis of macroeconomic-level or industry-level interdependencies • Captures nonlinear interactions among infrastructure
Empirical-Based	• Allows for the definition of interdependencies by identifying frequent and significant failure patterns • Identifies the vulnerability of critical infrastructure and potential mitigation measures
Network-Based	• Allows for the identification of infrastructure components and their functionalities, and suggestions on robustness improvements • Captures the flow characteristics of interdependent infrastructure and their operational mechanisms
System Dynamics-Based	• Allows for the modeling of how the operating environment influences the system functioning • Captures the effects of policy and technical factors to reflect infrastructure system evolution

implemented at the system level. A bottom-up approach is more appropriate to determine protection and resilience procedures at the facility level [26]. Combining top–down and bottom–up approaches is a comprehensive method that can be used to support decision-making based on accepted engineering principles [31, 35, 36].

Although the research addressing critical infrastructure interdependencies started more than 15 years ago, modeling, simulation, and visualization tools still perform

Table 8 Limitations of main interdependencies analysis methodologies

Approach	Limitation
Agent-Based	• Highly dependent on assumptions made for developing simulation models • Difficulty in calibrating simulation parameters due to the difficulty in modeling human behavior and possible data gaps
Economic Theory-Based	• Input-output models are not able to analyze interdependencies at the component level and lack the ability to model spatial relationships among infrastructure systems • Only consider normal economic operations and can be difficult to calibrate due to possible lack of data
Empirical	• Heavy reliance on historical data and expert judgments • Potential underreporting of frequent interdependency failures • Lack of standardization across data sources • Difficulty in predicting new disasters or different hazards
Network-Based	• Does not provide enough information about flow performance of real infrastructure systems • Methods are not scalable, and computation costs can be relatively high
System Dynamics-Based	• Requires a large amount of data and associated calibration. • Difficulty in analyzing component-level dynamics • Difficulty in validating simulation models due to previous limitations

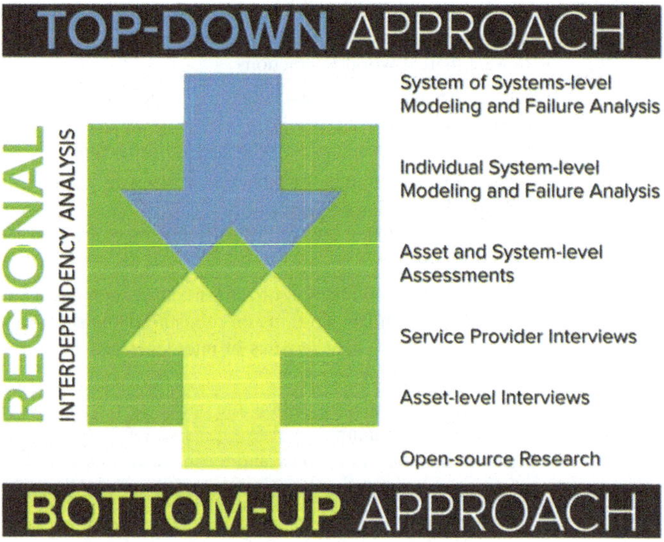

Fig. 1 Regional interdependency analysis overview

at an intermediate analysis level. Developing a comprehensive "system of systems" approach to move toward an advanced infrastructure analysis capability presents several challenges [5]:

- Lack of understanding of infrastructure system operations and of the interactions existing between infrastructure systems.
- Lack of understanding of how infrastructure interdependencies affect the resilience of a region.
- Difficulty in combining simulation models, originally developed in silos and specific to certain critical infrastructure sectors.
- Difficulty in obtaining the data for running the simulation models. Some data are proprietary and some are lacking due to the lack of knowledge.

Furthermore, managing risk associated with infrastructure interdependencies requires not only an understanding of infrastructure failures but also an ability to prioritize protection and mitigation efforts. Without a prioritization process, infrastructure assessment, protection, and mitigation programs are typically guided by expert judgment, and they often do not consider system-level reliability, redundancy, and overall resilience. Although understanding how to prioritize high-consequence failure points for assessments and, ultimately, for protection is essential, the complexity of infrastructure systems can quickly overwhelm decision makers [62]. Protecting critical infrastructure, especially in complex urban areas, should focus on identifying and prioritizing potential failure points that would have the most severe consequences. Optimization algorithms can be used to identify the most consequential failures affecting complex interdependent infrastructure systems regardless of the cause of disruption. Infrastructure system owners and operators and government agencies can then use results from optimization modeling to identify priority assets for in-depth security and resilience assessments, and to inform investment decisions related to critical infrastructure protection and mitigation [62]. Enhancing the resilience of critical infrastructure also requires consideration of the effect of infrastructure interdependencies on response and recovery processes. Modeling simulation can inform planning and help define priority for restoration to shorten resource outages.

Section 5 presents an example of an integrated modeling approach to simulate cascading failures across infrastructure systems.

5 Example

Argonne National Laboratory has developed an assessment framework for anticipating cascading and escalating failures [51]. This assessment framework is composed of three modules:

- Failure Analysis Module
- Infrastructure Interdependency Simulation Module
- Visualization Module

The Failure Analysis Module includes an automated process that models the interaction among hazards, infrastructure assets, and descriptions of asset fragility for defining the initial conditions resulting from a given hazard. The Failure Analysis Module integrates an optimization algorithm to identify what infrastructure assets would generate the most consequential failures.

The Infrastructure Interdependency Simulation Module constitutes the heart of the assessment framework. It uses a data centric modeling and simulation (DCMS) manager to integrate infrastructure simulation models by adopting a generalized integration process that facilitates the exchange of data among infrastructure models with different granularity and network topology complexity. The DCMS manager facilitates the synchronization between multiple disparate models without introducing major modifications to the original stand-alone simulation models [34].

Finally, the Visualization Module utilizes the output of the two previous modules (i.e., Failure Analysis and Infrastructure Interdependency Simulation) to represent the infrastructure service outage areas. The initial application focused on integrating two existing top-down network-based models (i.e., *EPfast* and *NGfast*) to simulate cascading failure among energy transmission systems.

EPfast is an electric power outage area estimation tool for simulating the behavior of large power systems following power disruptions caused by the loss of power system components [50]. The model explores the possibility of uncontrolled islanding caused by cascading failures and estimates the extent (geographic size) and depth (amount load shed) of the power outage.

NGfast uses progressive forward pipeline ownership identification and flow quantification processes to track lost flow volumes caused by a pipeline break or curtailment in natural gas supply. Impacts are measured in terms of extent of gas volume disrupted, states affected, utilities affected, number and type of customers affected, and amount of natural-gas-based capacity affected [49].

The synchronization of *EPfast* and *NGfast* is done through the Infrastructure Interdependency Simulation Module (Fig. 2).

The DCMS manager remains the same, whatever the type of infrastructure simulation model to be integrated. Integration of additional simulation models only requires the development of their corresponding translator models that will transform the output of the new simulation model to inputs for other simulation models, constituting the Infrastructure Interdependency Simulation Module. The identification and characterization of all connecting points between the electric power and natural gas infrastructure systems is the prerequisite for the development of the two translators realizing the integration of *EPfast* and *NGfast*. Figure 3 provides a representative schematic of the connections between the electric power and natural gas networks at the transmission level.

The translator model from *EPfast* to *NGfast* specifically characterizes the connections between the electric power substations and the natural gas processing plants. The translator model from *NGfast* to *EPfast* specifically characterizes the connections between the natural gas city gate stations and the electric power plants. Portante et al. [51] used a mapping algorithm to map existing energy datasets to associate cross-infrastructure nodes that are nearest to one another based on either *spatial distance* (i.e., geographic proximity based on longitude and latitude) or *string distance* (i.e., substation names compared with power plant names).

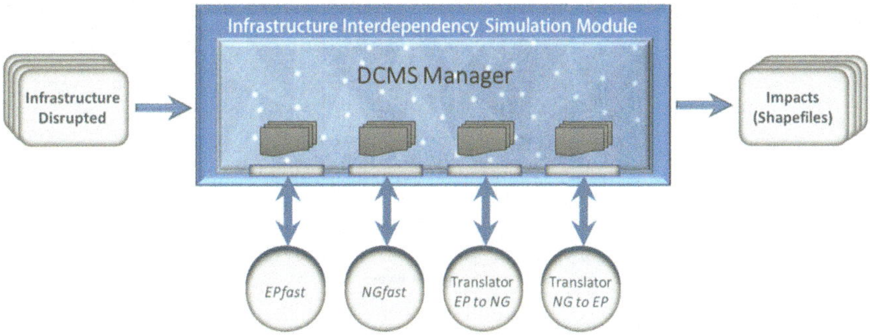

Fig. 2 Overview of the infrastructure interdependency simulation module [51]

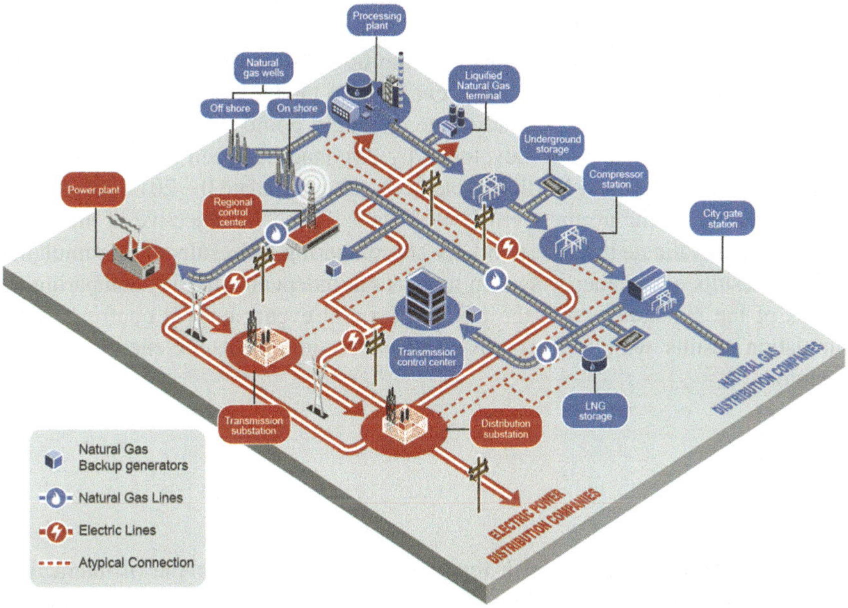

Fig. 3 Interdependencies between electric power and natural gas systems at the transmission level

The connections between electric and natural gas systems can be refined by using infrastructure dependency curves [47]. A dependency curve is a bottom-up approach that characterizes upstream physical dependencies by representing the impact of the loss of resource supply over time. Figure 4 shows an example of the possible operations degradation of a natural-gas-fueled power plant due to a disruption of a natural gas supply.

With no backup available, the loss of natural gas supply results in the almost immediate shutdown of the natural-gas-fueled power plant. Once natural gas supply

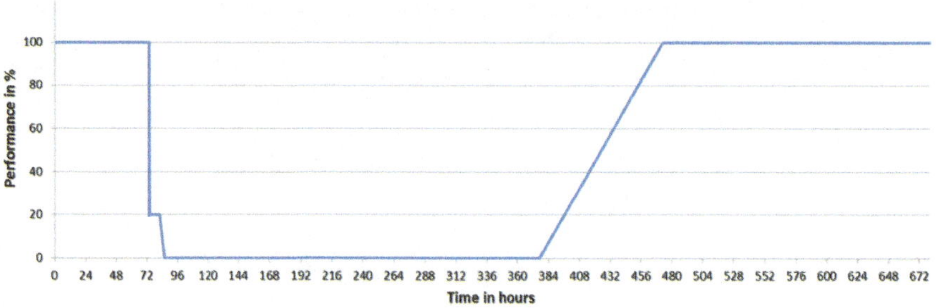

Fig. 4 Natural gas dependency curve

is restored, the power plant is able to ramp back up to normal full operations, accounting for time needed to conduct safety inspections and black start the operations.

The integration of top-down and bottom-up tools with the DCMS manager allows for modeling of the sequence of events from the initiation of the disturbance, to the arrival at the final new steady state operating point.

The assessment framework developed by Argonne has been applied to several case studies in Florida and North Dakota [51] and to simulate the 2014 polar vortex event that affected the northeastern United States [52]. The objective of these case studies was to validate the framework by [1] comparing the results of the simulation with the results of the energy system models ran independently, [2] comparing the results of the simulation with the effects of a real event, and [3] confronting the simulation results with the expertise and expert judgments of energy systems' operators.

5.1 *Florida*

Florida is a terminal state, which represents the simplest example of interactions between natural gas and electric power transportation systems. Cascading failures resulting from a disruption of energy transportation systems do not propagate outside the state. Furthermore, the natural gas system is relatively simple, with only two major high-pressure transmission pipelines serving the state.[1]

The simulation scenario postulates a complete break on a major interstate transmission pipeline supplying natural gas to the state, resulting in a 100% reduction in the flow of gas through the pipeline. The natural-gas-fueled power plants supplied by the pipeline disrupted would immediately cease operation, leading to varying load curtailment intensity ranging from 10 to 100%. The failure

[1]A third pipeline is currently under development.

would escalate due to the de-energization of natural gas processing plants located in load curtailment areas. However, the effects of the shutdown of these natural gas processing plants would be relatively minimal to gas customers in Florida [51].

5.2 North Dakota

Simulation of energy infrastructure interdependencies is more complex in North Dakota, which is an originating state. There are several major production fields in North Dakota. Oil and natural gas are produced and processed in natural gas processing plants, and the refined products are transported through pipelines that traverse the state to deliver natural gas to several downstream states as far away as Illinois.

The simulation scenario postulates the disruption of major substations in western North Dakota close to major natural gas processing plants. The disruption causes an extended power outage over a broad geographic region that encompasses the natural gas processing plants. This would generate a 28% reduction in the natural gas supply to the Southeast [51].

5.3 Northeastern Polar Vortex

In January of 2014, the Midwest, South Central, and East Coast regions of North America experienced extreme cold weather conditions resulting in temperatures 20–30 °F below average. These temperatures resulted in record-high electric and natural gas demands (North American Electric Reliability Council [NERC] 2014). A significant amount of natural-gas-fired generation became unavailable because of curtailments of natural gas in the electric sector. Despite the marked increase in electric demand and the substantial reduction in on-line gas-fired capacity, electric grid operators were mostly able to maintain the electric grid operations by communicating through the NERC Energy Emergency Alert process and implementing mitigation strategy [40].

To reproduce the effect of the polar vortex, [52] developed an approach to identify the pipelines affected and the amount of gas volume diverted from the electric sector. Results of the simulations produced results that closely resemble the actual impacts observed, in both spatial (geographic extent) and quantitative (amount of megawatts lost per service territory) terms. The study demonstrated that the assessment framework and the infrastructure interdependency simulation module has the capacity to simulate and anticipate cascading failures within reasonable error margins [52].

The Assessment Framework and the Infrastructure Interdependency Simulation modules are currently used in several RRAP projects.

6 Future Developments

A comprehensive assessment of all classes of critical infrastructure interdependencies considering all their dimensions requires improving many existing capabilities to include the way data are collected and existing modeling and simulation analysis approaches. This will lead to the development of new models to provide a more holistic understanding of infrastructure interdependencies.

Improvement in data-collection capabilities should build upon existing tools and techniques but will also require the development of new collection mechanisms. Analytic capabilities should evolve toward an integration of methodologies and tools to combine existing mathematical and engineering models, and to develop uncertainty algorithms to extend and predict impacts of disruptions in dependencies. The improvement of analysis capabilities should be guided by the data collected and the types of products required by stakeholders. Moving toward an advancement of risk management that includes interdependencies also requires the enhancement and interactivity of existing visualization capabilities. Development of GIS visualization capabilities is vital for the analysis of critical infrastructure interdependencies, and especially in visualizing cascading and escalating failures at the regional level.

The ultimate goal of analyzing critical infrastructure interdependencies is to better understand and anticipate the consequences of an incident and to support incident management activities. This requires characterizing infrastructure interdependencies and operationalizing their consideration in risk management procedures. Currently, most projects consider physical and cyber interdependencies among lifeline networks (i.e., energy, communications, transportation, and water). Future projects should better analyze logical interdependencies by incorporating human decisions and behaviors into determining the extent of impacts through interconnected systems of critical infrastructure. In particular, the enhancement of risk and resilience management processes requires an understanding of the decisions that critical infrastructure owners and operators make during a crisis, a translation of those decisions into specific actions that change the performance of the infrastructure system, and an analysis of how these actions affect connected infrastructure assets and the overall regional resilience.

7 Conclusion

Critical infrastructure interdependencies are complex elements to consider. They influence all components of risk; they can constitute a threat or hazard, affect the resilience and performance of critical infrastructure, and lead to the propagation of cascading and escalating failures. It is therefore essential to integrate the characterization of interdependencies into risk and resilience assessment methodologies. A data-driven capability that operationalizes the analysis of interdependencies would not only provide an unprecedented level of situational awareness, it would also enable decision makers to anticipate disruptions, which would have a significant impact on regional resilience. Even though the research addressing critical infrastructure interdependencies started more than 15 years ago, modeling, simulation, and visualization tools still perform at an intermediate analysis level. Developing a comprehensive "system of systems" assessment capability requires combining top-down and bottom-up approaches in an adaptive and flexible assessment framework to better understand infrastructure operations, and to anticipate and prepare for potential cascading and escalating failures.

References

1. ASIS (2009) The organizational resilience standard [ASIS SPC. 1-2009]. Available via https://www.ndsu.edu/fileadmin/emgt/ASIS_SPC.1-2009_Item_No._1842.pdf. Accessed 26 Oct 2017
2. Australian Government (2010) Critical infrastructure resilience strategy. Commonwealth of Australia. Available via https://www.tisn.gov.au/Documents/Australian+Government+s+Critical+Infrastructure+Resilience+Strategy.pdf. Accessed 26 Oct 2017
3. Australian Government (2012) Critical infrastructure program for modelling and analysis (CIPMA). Attorney-General's Department. Available via https://www.ag.gov.au/NationalSecurity/InfrastructureResilience/Pages/CriticalInfrastructureProgramforModellingandAnalysisCIPMA.aspx. Accessed 26 Oct 2017
4. Carlson LG, Bassett W, Buehring M, Collins S, Folga B, Haffenden F, Petit J, Phillips D, Verner RW (2012) Resilience theory and applications. ANL/DIS-12-1. Argonne National Laboratory, Argonne, IL. Available via http://www.ipd.anl.gov/anlpubs/2012/02/72218.pdf. Accessed 26 Oct 2017
5. Clifford M, Macal C (2016) Advancing infrastructure dependency and interdependency modeling: a summary report from the technical exchange. Argonne National Laboratory. Available via https://anl.app.box.com/s/3t7mnesdajzl708xy9xo4vczj2qv1wom. Accessed 26 Oct 2017
6. Cobb A (1997) Australia's vulnerability to information attacks. Australian Strategic and Defence Studies Centre, Australia
7. Cobb A (1999) CI attack: an investigation of the vulnerability of an OECD country. In: Bosch JMJ, Luiijf HAM, Mollema AM (eds) NL ARMS—Netherlands annual review of military studies 1999: information operations. Tilburg University Press, Tilburg, Netherlands, pp 201–222
8. Commission of the European Communities (2005) Green paper on a European programme for critical infrastructure protection. COM (2005) 576 final. Bruxelles, Belgique. Available via http://eur-lex.europa.eu/legal-content/EN/TXT/?uri=LEGISSUM:l33260. Accessed 26 Oct 2017
9. Council of the European Union (2003) A secure Europe in a better world, European security strategy. Brussels, 12 December. Available via https://www.consilium.europa.eu/uedocs/cmsUpload/78367.pdf. Accessed 26 Oct 2017
10. Council of the European Union (2008) Council directive 2008/114/EC of 8 December 2008 on the identification and designation of European critical infrastructures and the assessment of the need to improve their protection. Official Journal of the European Union. Available via http://eur-lex.europa.eu/LexUriServ/LexUriServ.do?uri=OJ:L:2008:345:0075:0082:EN:PDF. Accessed 26 Oct 2017

11. Council of the European Union (2010) Draft internal security strategy for the European Union: 'towards a European security model', Brussels, 23 February 2010. Available via http://register.consilium.europa.eu/pdf/en/10/st05/st05842-re02.en10.pdf. Accessed 26 Oct 2017
12. CSIRO (2012) Protecting Australia's critical infrastructure with CIPMA. Commonwealth Scientific and Industrial Research Organisation. Site Internet du Commonwealth Scientific and Industrial Research Organisation. Available via http://www.csiro.au/partnerships/CIPMA.html. Accessed 26 Oct 2017
13. DHS (2009) National infrastructure protection plan 2009. Available via https://www.dhs.gov/xlibrary/assets/NIPP_Plan.pdf. Accessed 26 Oct 2017
14. DHS (2012) Office of infrastructure protection strategic plan: 2012–2016. Available via http://www.dhs.gov/sites/default/files/publications/IP-Strategic-Plan-FINAL-508.pdf. Accessed 26 Oct 2017
15. DHS (2013) National infrastructure protection plan 2013: partnering for critical infrastructure security and resilience. Available via https://www.dhs.gov/sites/default/files/publications/national-infrastructure-protection-plan-2013-508.pdf. Accessed 26 Oct 2017
16. DHS (2015) National critical infrastructure security and resilience research and development plan. Available via http://www.dhs.gov/sites/default/files/publications/NationalCISRRDPlan_Nov2015.pdf. Accessed 26 Oct 2017
17. DHS (2017) Regional resiliency assessment program. Available via https://www.dhs.gov/regional-resiliency-assessment-program. Accessed 26 Oct 2017
18. DHS and Public Safety Canada (2010) Canada-United States action plan for critical infrastructure. Available via https://www.dhs.gov/xlibrary/assets/ip_canada_us_action_plan.pdf. Accessed 26 Oct 2017
19. DOE (2001) Critical infrastructure interdependencies, impact of the September 11 terrorist attacks on the world trade center—a case study. Office of Critical Infrastructure Protection, 28 p
20. DOE (2017) Quadrennial energy review. Available via https://energy.gov/epsa/initiatives/quadrennial-energy-review-qer. Accessed 26 Oct 2017
21. Europa (2012) European programme for critical infrastructure protection. Available via http://europa.eu/legislation_summaries/justice_freedom_security/fight_against_terrorism/l33260_en.htm. Accessed 26 Oct 2017
22. European Commission (2017a) Horizon 2020—the EU framework programme for research and innovation. Available via http://ec.europa.eu/programmes/horizon2020/. Accessed 26 Oct 2017
23. European Commission (2017b) Horizon 2020—the EU framework programme for research and innovation—security. Available via http://ec.europa.eu/programmes/horizon2020/en/area/security. Accessed 26 Oct 2017
24. FEMA (2017) PS-Prep™ program resources. Available via https://www.fema.gov/program-resources. Accessed 26 Oct 2017
25. Flynn S (2015) Bolstering critical infrastructure resilience after superstorm Sandy: lessons for New York and the nation. Georges J. Kostas Research Institute for Homeland Security, Center for Resilience Studies, Northeastern University. Available via https://repository.library.northeastern.edu/files/neu:m0419677k. Accessed 26 Oct 2017
26. Gopalakrisshnan K, Peeta S (2010) Sustainable and resilient critical infrastructure systems—simulation modeling, and intelligent engineering. Springer, Berlin, Heidelberg
27. Governments of Canada (2009) National strategy for critical infrastructure. Available via https://www.publicsafety.gc.ca/cnt/rsrcs/pblctns/srtg-crtcl-nfrstrctr/index-en.aspx, Accessed 26 Oct 2017
28. Governments of the United States and Canada (2016) Joint United States–Canada electric grid security and resilience strategy. Available via https://www.nrcan.gc.ca/sites/www.nrcan.gc.ca/files/energy/pdf/JOINT%20GRID%20SECURITY%20AND%20RESILIENCE-Strategy_en.pdf. Accessed 26 Oct 2017

29. Hasan S, Foliente G (2015) Modeling infrastructure system interdependencies and socioeconomic impacts of failure in extreme events: emerging R&D challenges. J Int Soc Prev and Mitig Nat Hazards, May 31, 2015. Available via https://link.springer.com/article/10.1007/s11069-015-1814-7. Accessed 26 Oct 2017
30. Hokstad PI, Utne B, Vatn J (2012) Risk and interdependencies in critical infrastructures—a guideline for analysis. Springer-Verlag, Berlin, Heidelberg
31. Hollnagel E (2014) Safety-I and safety-II—the past and future of safety management. CRC Press, Taylor & Francis Group, Boca Raton, London, New York
32. IAC (2003) Infrastructure interdependencies associated with the August 14, 2003, electric power blackout. Argonne National Laboratory, 21 p
33. ISO (2012) ISO 22301:2012—societal security—business continuity management systems—requirements. Available via https://www.iso.org/standard/50038.html. Accessed 26 Oct 2017
34. Joshi R (2011) Data-centric architecture: a model for the era of big data, the world of software development, Dr. Dobb's. Available via http://www.drdobbs.com/architecture-and-design/data-centric-architecture-a-model-for-th/229301018. Accessed 26 Oct 2017
35. Leveson NG (2011) Engineering a safer wafer world – systems thinking applied to safety, engineering systems. The MIT Press, Cambridge, MA. Available via https://mitpress.mit.edu/books/engineering-safer-world. Accessed 26 Oct 2017
36. Linkov IE, Anklam ZA, Collier D, DiMase OR (2014) Risk-based standards: integrating top–down and bottom–up approaches. Environ Syst Decisions 4(1): 134–137. Available via http://link.springer.com/article/10.1007/s10669-014-9488-3. Accessed 26 Oct 2017
37. Luiijf E, Burger H, Klaver M (2003) Critical infrastructure protection in the Netherlands: a quick-scan. In: Gattiker UE (ed) EICAR Conference Best Paper Proceedings. Danemark, Copenhagen, p 19
38. NATO (2007) 162 CDS 07 E rev 1—the protection of critical infrastructures. North Atlantic Treaty Organization
39. NATO (2012) Protecting critical infrastructure: a challenge for NATO and its partners. North Atlantic Treaty Organization. Available via http://www.nato.int/cps/en/natolive/news_88054.htm. Accessed 26 Oct 2017
40. NERC (2014) Polar vortex review. Available via http://www.nerc.com/pa/rrm/January2014PolarVortexReview/Polar_Vortex_Review_29_Sept_2014_Final.pdf. Accessed 26 Oct 2017
41. New York Times (2010) Iceland volcano eruption of 2010 (Eyjafjallajokull volcano). The New York Times, Times topics, Tuesday, April 20, 2010. Available via http://topics.nytimes.com/top/news/international/countriesandterritories/iceland/eyjafjallajokull/index.html. Accessed 26 Oct 2017
42. New York Times (2011) Powerful quake and tsunami devastate northern Japan, Friday, March 11, 2011. Available via http://www.nytimes.com/2011/03/12/world/asia/12japan.html?_r=1&pagewanted=all. Accessed 26 Oct 2017
43. NFPA (2013) NFPA® 1600—standard on disaster/emergency management and business continuity programs, 2013th edn. NFPA, Quincy, MA, USA
44. OECD (2011) Future global shocks. Available via http://www.oecd.org/governance/48329024.pdf. Accessed 26 Oct 2017
45. Ouyang M (2014) Review on modeling and simulation of interdependent critical infrastructure systems. Elsevier Reliab Eng Syst Saf 121:43–60
46. Pederson P, Dudenhoeffer D,. Hartley S, Permann M (2016) Critical infrastructure interdependency modeling: a survey of U.S. and international research. INL/EXT-06-11464. Idaho National Laboratory, Idaho Falls, USA. Available via http://cip.management.dal.ca/publications/Critical%20Infrastructure%20Interdependency%20Modeling.pdf. Accessed 26 Oct 2017
47. Petit F, Wallace K Phillips J (2014) Interactive dependencies curves for resilience management. J Bus Continuity Emergency Plan 8(2):141–155. Henry Stewart Publications London, United Kingdom

48. Petit F, Verner D, Brannegan D, Buehring W, Dickinson D, Guziel K, Haffenden R, phillips, Peerenboom J (2015) Analysis of critical infrastructure dependencies and interdependencies. ANL/GSS-15/4. Argonne National Laboratory, Argonne, IL, USA. Available via https://www.osti.gov/scitech/biblio/1184636-analysis-critical-infrastructure-dependencies-interdependencies. Accessed 26 Oct 2017
49. Portante EC, Craig BA, Folga SM (2007) NGfast: a simulation model for rapid assessment of impacts of natural gas pipeline breaks and flow reductions at U.S. state borders and import points. In: Henderson SG,. Biller B, HsiehM-H, Shortle J, Tew JD, Barton RR (eds), Proceedings of the 2007 winter simulation conference. IEEE, pp 1118–1126. Available via http://ieeexplore.ieee.org/xpls/abs_all.jsp?arnumber=4419711&tag=1. Accessed 26 Oct 2017
50. Portante EC, Craig BA, Talaber Malone L, Kavicky J. Folga SM (2011) EPFast: a model for simulating uncontrolled islanding in large power systems. In: Jain S, Creasey RR, Himmelspach J, White KP, Fu M (eds), Proceedings of the 2011 winter simulation conference. IEEE, pp 1758–1769. Available via http://ieeexplore.ieee.org/stamp/stamp.jsp?arnumber=06147891. Accessed 26 Oct 2017
51. Portante EC, Kavicky JA, Craig BA, Talaber LE, Folga SM (2017a) Modeling electric power and natural gas systems interdependencies. ASCE J Infrastructure Syst 23(4), December 2017. Available via http://ascelibrary.org/doi/abs/10.1061/%28ASCE%29IS.1943-555X.0000395. Accessed 26 Oct 2017
52. Portante EC, Kavicky JA,. Craig BA, Talaber LE, Folga SM (2017b) Simulation of the January 2014 polar vortex and its impacts on interdependent electrical natural gas infrastructure. In: Winter simulation conference 2017. Las Vegas, NV, December 3–6, 2017
53. Public Safety Canada (2014) 2014–2017 action plan for critical infrastructure. Available via https://www.publicsafety.gc.ca/cnt/rsrcs/pblctns/pln-crtcl-nfrstrctr-2014-17/index-en.aspx. Accessed 26 Oct 2017
54. Rinaldi SM, Peerenboom JP, Kelly TK (2001) Identifying, understanding, and analyzing critical infrastructure interdependencies. IEEE Control Syst Magazine, December 2001. Available via http://www.ce.cmu.edu/~hsm/im2004/readings/CII-Rinaldi.pdf. Accessed 26 Oct 2017
55. Supreme Allied Command Transformation and City of Norfolk (2017) Building resilience: collaborative proposals to help nations and partners. June 2017. Available via https://www.norfolk.gov/DocumentCenter/View/30383. Accessed 26 Oct 2017
56. The White House (2006) The federal response to hurricane Katrina, lessons learned. Available via https://georgewbush-whitehouse.archives.gov/reports/katrina-lessons-learned. Accessed 26 Oct 2017
57. The White House (2012) National strategy for global supply chain security. Available via https://obamawhitehouse.archives.gov/sites/default/files/national_strategy_for_global_supply_chain_security.pdf. Accessed 26 Oct 2017
58. The White House (2013) Presidential policy directive 21: critical infrastructure security and resilience. Available via https://www.dhs.gov/sites/default/files/publications/national-infrastructure-protection-plan-2013-508.pdf. Accessed 26 Oct 2017
59. TISN (2012) TISN for critical infrastructure resilience, trusted information sharing network. Available via https://www.tisn.gov.au/Pages/default.aspx. Accessed 26 Oct 2017
60. U.S.-Canada Power System Outage Task Force (2004) Final report on the August 14, 2003, blackout in the United States and Canada: causes and recommendations. Available via https://energy.gov/sites/prod/files/oeprod/DocumentsandMedia/BlackoutFinal-Web.pdf. Accessed 26 Oct 2017
61. U.S. House of Representatives (2006) A failure of initiative – final report of the select bipartisan committee to investigate the preparation for and response to hurricane Katrina. Available via https://www.uscg.mil/history/katrina/docs/USHouseOfRepKatrina2006MainR1eport.pdf. Accessed 26 Oct 2017
62. Verner D, Kim K, Petit F (2017) A prioritization algorithm to identify tipping points in critical infrastructure systems. In: 85th military operations research society symposium. June 19–22. United States Military Academy, West Point, NY

63. Visarraga DB (2011) Understanding complex system: infrastructure impacts. Los Alamos National Laboratory. LA-UR-11-01344. Available via http://permalink.lanl.gov/object/tr?what=info:lanl-repo/lareport/LA-UR-11-01344. Accessed 26 Oct 2017
64. Wallace WA, Mendonca DM, Lee EE, Mitchell JE, Chow J (2003) Managing disruptions to critical interdependent infrastructures in the context of the 2001 World Trade Center attack. In: Monday JL (ed) Beyond September 11th: an account of post-disaster research, special publication #39. University of Colorado, Natural Hazards Research and Applications Information Center, pp 165–98
65. WEF (2013) Global risks 2013 eighth edition. Available via http://reports.weforum.org/global-risks-2013/. Accessed 26 Oct 2017
66. Xiao N, Sharman R, Rao HR, Upadhyaya S (2008) Infrastructure interdependencies modeling and analysis—a review and synthesis. In: AMCIS 2008 Proceedings. 224. Available via http://aisel.aisnet.org/amcis2008/224/?utm_source=aisel.aisnet.org%2Famcis2008F224&utm_medium=PDF&utm_campaign=PDFCoverPages. Accessed 26 Oct 2017
67. Zhang P, Peeta S (2011) A generalized modeling framework to analyze interdependencies among infrastructure systems. Transp Res Part B Methodological 45(3):553–579
68. Zimmerman R (2001) Social implications of infrastructure network interactions. J Urban Technol 8:97–119

The Application of Warden's Concentric Ring Theory to Applegate's Offensive Cyber Manoeuvre Strategy in Modern Conflicts

David G. Wood

Abstract Cyber has become an essential element for governments, militaries and societies. This importance makes it a target when planning a military campaign. The targeting model developed by Colonel John Warden has demonstrated flexibility in that it can be applied to any entity that can be analyzed as a system. Scott Applegate has accurately described three different strategies that can be employed when waging offensive cyber warfare. The targeting model developed by Colonel John Warden can be applied to each of these cyber maneuver strategies to analyse and attack an opponent's cyber capabilities. There is anecdotal evidence that these different cyber strategies have been employed in recent world conflicts. Real world examples accessed from open source literature have been analysed using Warden's theory, to demonstrate that countries have already started basing their attacks against their opponent's cyber capabilities using this model.

Keywords Cyber strategies · Concentric ring theory

1 Introduction

With the evolution of airpower, the achievement of air superiority became a necessary requirement in warfare. Once achieved, airpower would then support the troops on the ground and warships on and under the sea in order to achieve land and sea superiority. Colonel John Warden took the notion of airpower one step further with the introduction of his concentric ring theory. It was first demonstrated during the Gulf War in 1992. His theory categorized an enemy as a system of concentric rings; with the fielded fighting forces on the outside ring and leadership at the center ring.[1] By attacking as far into the ring system as was possible, Warden demonstrated

[1]Warden III [35, 108].

D. G. Wood (✉)
Royal Canadian Air Force, 101 Colonel By Drive, Ottawa, ON K1A 0K2, Canada
e-mail: David.Wood3@forces.gc.ca

A. J. Masys (ed.), *Security by Design*, Advanced Sciences and Technologies for Security Applications, https://doi.org/10.1007/978-3-319-78021-4_11

that one did not have to fight all of the fielded fighting forces found on the outside ring. One could attack further into the ring system, and end the conflict far sooner than normally anticipated.[2] Warden described the Gulf War as being the "…first true 'inside to outside' war, beginning with the most important central ring in Baghdad and working its way to the outermost ring of fielded forces".[3] The success of Gulf War I was stark evidence of this theory in action, with the ground component of the war being measured in metrics of either hours or days.[4] Warden's concentric ring theory was an effective targeting model when applied to Iraq as a system.

What airpower was to military planners in the 20th Century, cyber power is now taking this role on in the 21st Century. There is anecdotal evidence that in recent conflicts involving Israel, Russia and the United States, that achieving cyber dominance or cyber superiority at the outset of a conflict is a necessary component of their warfighting strategy. Cyber is emerging as an important area of warfare. Notably, prior to commencing a conflict with Georgia, Russia launched cyber attacks in conjunction with their military operation.[5] Bonner noted that the "…2008 Russia-Georgia war marks the only public incidence of cyber power integrated with traditional kinetic military operations."[6]

Applegate applied maneuver theory to cyber warfare, and proposed different forms of maneuver that an attacker could incorporate into their overall strategy. For offensive cyber maneuver,[7] he described three different approaches:

a. Exploitive Maneuver—securing information for advantage at the tactical, operational or strategic level[8];
b. Positional Maneuver—compromising or outright seizing of key nodes in the cyber environment, and then utilizing these nodes for your benefit[9]; and
c. Influencing Maneuver—Applegate describes an influencing maneuver as "…the process of using cyber operations to get inside an enemy's decision cycle or even to force that decision cycle though direct or indirect actions."[10]

Has Applegate defined different strategies that can be conducted when waging cyber warfare? Does his model make sense? Recent cyber conflicts do appear to match the different approaches presented in Applegate's model. In some cases, the attacker wanted to deny the internet to the country under attack. This was

[2]Warden III [34, 69].

[3]Ibid., 78.

[4]Center of Military History [32, 34]

[5]NATO Review Magazine [23].

[6]Bonner III [6: 103].

[7]Applegate [2: 7].

[8]Ibid., 7–8.

[9]Ibid., 8.

[10]Applegate [2: 9].

demonstrated in 2007 with Russian aggression against Estonia,[11] and in 2008 during the Russian invasion of Georgia.[12] These strategies correspond most closely with Applegate's definition of Influencing Maneuver.[13]

The United States is not following this strategy, however, in its war against the Islamic State. In this case, the Americans are allowing their opponent access to the internet. This example most closely correlates with Applegate's definition of Exploitive Maneuver, whereby an attacker captures "...information resources in order to gain a strategic, operational or tactical competitive advantage."[14]

In the final type of example, cyber was used in a manner similar to that noted by McGuffin and Mitchell, that being as a supporting capability to a larger military operation.[15] Applegate referred to this as being Positional Maneuver, whereby "... key physical or logical nodes in the information environment..."[16] are captured or compromised. Applegate used as an example the 2007 Israeli attack on a Syrian reactor, which was preceded by an apparent electronic and cyber attack.[17] In keeping with McGuffin and Mitchell's thinking that cyber would be supporting a larger operation,[18] Applegate notes that using "...positional maneuver prior to the initiation of actual kinetic combat operations set them up for success and illustrates the potential decisive nature of this form of cyber maneuver,".[19]

Given the importance of cyber for individuals, militaries and societies, and the paradigm shift going on between the timing for achieving air superiority versus the timing for achieving cyber superiority; the question must now be asked—how would cyber warfare be fought in a modern conflict? Analysis of recent cyber conflicts will show that certain countries appear to have already developed detailed cyber strategies, and they are employing these strategies. This paper will argue that Applegate has accurately described three different strategies that can be employed when waging offensive cyber warfare, and that the targeting model developed by Colonel John Warden can be used to analyse and attack an opponent's cyber capabilities. Real world examples accessed from open source media will be analysed, to demonstrate that countries have already started basing their attacks against their opponent's cyber capabilities using this model.

[11]Herzog [16: 51].

[12]Clarke and Knake [10, 19].

[13]Applegate [2, 9].

[14]Ibid., 7.

[15]McGuffin and Mitchell [21: 411].

[16]Applegate [2: 8].

[17]Ibid.

[18]McGuffin and Mitchell [21: 411].

[19]Applegate [2: 9].

2 Warden Applied to Cyber

2.1 Early Cyber Warfare

One of the first examples of cyber being considered as part of a larger military plan was an idea presented to General Norman Schwarzkopf during the First Gulf War (1990–1991), just prior to Warden's targeting plan being implemented. Clarke and Knake refer to an occasion where Special Operations Command presented a plan that would see a team of special forces soldiers attacking an Iraqi radar installation just prior to the commencement of hostilities. They would be accompanied by computer experts, who would upload software into the radar network, causing "… computers on the network all over the country to crash and be unable to reboot."[20] General Schwarzkopf did not have confidence in either the plan nor the personnel proposing it, and as a result he rejected it.[21] At the time, he felt that should "…you want to make sure their air defense radars and missiles don't work, blow them up first. That way they stay dead."[22]

By 2003, before the commencement of hostilities for the Second Gulf War, the American military's perception of cyber as a weapon had evolved. Clarke and Knake refer to the United States military hacking into the secure military computer network used by the Iraqi military.[23] The Americans were overt in their activities this time, however.

They sent emails to Iraqi officers on Iraq's "closed loop" network,[24] just prior to the start of the war. United States Central Command told the officers that the American's quarrel was with Saddam Hussein and his sons, and that they had no quarrel with any Iraqi soldier who parked his tanks and vehicles and abandoned them.[25] This plan was successful. When the Americans moved into Iraq, "… many units had neatly parked their tanks in rows outside their bases, thus allowing U.S. aircraft to neatly blow them up. Some Iraqi army commanders sent their troops on leave in the hours before the war."[26]

---

[20] Clarke and Knake [10, 9].

[21] Ibid.

[22] Ibid.

[23] Ibid.

[24] Clarke and Knake [10, 9].

[25] Ibid., 10.

[26] Ibid.

2.2 *Russia Versus Estonia*

In April 2007, the Government of Estonia arranged to move a statue called the "Bronze Soldier"[27] This action sparked a riot by the minority Russian community within Estonia.[28] Of greater significance, however was that between 27 April to 18 May 2007, powerful cyber attacks were launched—specifically distributed denial-of-service (DDoS) attacks—"…targeting the country's infrastructure (that) shut down the websites of all government ministries, two major banks, and several political parties. At one point, hackers even disabled the parliamentary email server."[29] The attacks were allegedly wrought by pro-Russian hackers, however Clarke and Knake note that this was not some amateur hacker attack. This was a well-organized, elaborate attack, which involved "…targeting Internet addresses most people would not know, not those of public webpages, but the addresses of servers running parts of the telephone network, the credit-card verification system, and the Internet directory."[30] While Herzog acknowledges that "…we may never know the true extent of Kremlin involvement in the cyber attacks on Estonia, it is clear that Russian officials encouraged the hackers…"[31] Further, he assesses that the Russians "…tolerated and encouraged the cyber attacks, and the Kremlin may have even colluded with the hackers responsible for the strikes."[32]

2.3 *Israel Versus Syria*

Later that year, on September 6, 2007, at a "…North Korean—designed nuclear weapons plant…"[33] under construction in Syria, an Israeli strike package of F-15 Eagles and F-16 Falcons launched an attack that completely destroyed the facility.[34] What is significant about this attack is that despite being equipped with a modern Russian air defence system, the Syrians were unaware of the incoming attack. In particular, what "…appeared on the radar screens was what the Israeli Air Force had put there, an image of nothing."[35] The Syrians immediately went to the Russians to ask what happened.[36] The Russians were distressed, not only because they were not exactly sure what was compromised in their system, but also because

[27]Herzog [16, 50].

[28]Ibid.

[29]Ibid., 51.

[30]Clarke and Knake [10, 15].

[31]Herzog [16, 53].

[32]Ibid., 55.

[33]Clarke and Knake [10, 3].

[34]Ibid., 2.

[35]Ibid., 5.

[36]Ibid.

they were about to sell a similar system to Iran.[37] Exactly how Israel defeated the Syrian/Russian radar system is something known to only a select group. The consensus is that Israel launched a cyber attack to defeat the Syrian/Russian radar system, somehow transmitting "…1' s and 0' s to control what the Syrian air defense radars saw."[38] Applegate described the attack as being "…a combination of both electronic and cyber-attacks which caused all of Syria's air defense radar systems to go offline for the duration of the raid."[39] By launching a cyber attack instead of a kinetic strike against Syrian air defence radars, the Israeli's kept the element of surprise until the last minute, and "…in the age of cyber war, the Israelis ensured that the enemy could not even raise its defenses."[40] Applegate notes that Israeli confidence in launching this form of attack implies that they "…had already gained the necessary level of access into these systems and had pre-positioned themselves to carry out this attack."[41]

Clarke and Knake propose three possibilities as to how the Israelis managed to launch their cyber attack against the Syrian/Russian radar system. The first would involve a stealth drone, that would transmit data packets down to the radar system, telling it to display a blank screen, while simultaneously not reporting any defect to an operator.[42] Clarke and Knake note that the American military "… has a similar cyber-attack system, code-named Senior Suter."[43] The second would involve Israel or an ally somehow inserting a digital back door or "Trojan Horse,"[44] that would respond to pre-arranged codes or signals to display a blank screen.[45] The third would be if an Israeli operative somehow accessed the fibre-optic cable connecting the air defence system, and triggered the Trojan that way.[46]

3 Russia Versus Georgia

The next example of cyber attacks being employed and coordinated as part of a larger campaign occurred during the August, 2008 Russian invasion of Georgia.[47] DDoS style attacks were launched against "…Georgian news and government

[37]Clarke and Knake [10, 6].

[38]Ibid.

[39]Applegate [2: 8–9].

[40]Clarke and Knake [10, 6].

[41]Applegate [2: 9].

[42]Clarke and Knake [10, 6–7].

[43]Ibid., 7.

[44]Ibid.

[45]Ibid., 8.

[46]Ibid.

[47]Shakarian [28, 29, 63].

websites."[48] Shakarian noted that these attacks occurred the day prior to the ground invasion, and concluded that "…the hackers knew about the date of the invasion beforehand."[49] Clarke and Knake noted the "…intensity and sophistication…"[50] of the attack, which (as noted also by Shakarian),[51] was initiated "…just as the ground fighting broke out."[52] As a result of this significant cyber attack, hackers took control of the .ge web domain,[53] Georgia was cut-off from the global banking network (including credit card transactions),[54] and the government had to transfer the Georgian President's webpage to a site in the United States.[55] Chayes states that this war represented the first time where there was a coordinated cyber "…attack synchronized with major combat actions in the other warfighting domains."[56]

3.1 Stuxnet Virus

In 2008, Israel suspected that Iran was planning on developing nuclear weapons using material from its Natanz nuclear facility, and requested specialized American bunker busting bombs and American authority to route an air attack through Iraqi airspace in order to destroy the facility.[57] Talbot notes that the Israeli attack plan was quite advanced, and included 100 fighter aircraft involved in "…a contingency exercise flown over the Mediterranean in 2008…"[58] Sanger reported that President Bush did not agree to this request, instead, he "…told the Israelis that he had authorized new covert action intended to sabotage Iran's suspected effort to develop nuclear weapons…"[59] This covert action appears to have been using the Stuxnet virus to attack hardware within the Natanz facility.

The Stuxnet virus was first detected in June 2010,[60] however Chen notes that evidence has been reported by Microsoft that would date the code to as early as January 2009.[61] Lindsay states that the Stuxnet virus was a "…US-Israeli…"[62]

[48]Ibid.
[49]Ibid.
[50]Clarke and Knake [10, 19].
[51]Shakarian [28, 29, 63].
[52]Clarke and Knake [10, 19].
[53]Ibid.
[54]Ibid., 20.
[55]Ibid., 19.
[56]Chayes [8, 477].
[57]Sanger [26].
[58]Talbot [31, 74].
[59]Sanger [26].
[60]Lindsay [20, 365].
[61]Chen [9, 2].
[62]Lindsay [20, 366].

initiative; just one element of a "...broader US cyber campaign against Iran code-named "Olympic Games"."[63] What is significant about the Stuxnet virus is that it "...is the first instance of a computer network attack known to cause physical damage across international boundaries."[64] How significant this attack was is up for debate. McGuffin and Mitchell note that the damages resulting from Stuxnet "... were not described as an armed attack by the targeted state."[65] The targeted state was Iran,[66] and the targeted systems were the uranium enrichment centrifuges at Iran's Natanz facility.[67] Chen states that the virus attacked the software (specifically the supervisory control and data acquisition software, or SCADA) that ran the programmable logic controllers (PLCs) which allowed users to enter information and control the centrifuges in question.[68] By interfering with this software, Shakarian notes that Stuxnet would then set a rotational frequency speed close to the maximum potential speed of the centrifuges in question, eventually damaging them.[69] The true impact of this attack is known only to the Iranians, and damage estimates vary in the literature. Chen noted a mysterious decrease of "...15% in production in 2009, around when Stuxnet is believed to have been spreading."[70] Shakarian notes that between 2009 and 2010, Iran "...decommissioned and replaced about 1000 IR-1 centrifuges at the Natanz FEP (6 cascades of 164 centrifuges each)."[71]

In addition to setting back Iran's nuclear program, the Stuxnet attack also prevented a potential Middle East conflict from breaking out. Sanger reported that the Bush Administration was worried at the time that an "...airstrike could ignite a broad Middle East war in which America's 140,000 troops in Iraq would inevitably become involved."[72] This war did not occur, however, due to the damaging cyber attack that was launched.

3.2 United States Versus Islamic State

The Russian campaigns in Estonia and Georgia were examples where Russia sought to achieve cyber superiority over the countries in question, and then deny them access to the internet. The American efforts to defeat the Islamic State,

---

[63] Ibid.
[64] Ibid., 365.
[65] McGuffin and Mitchell [21, 411].
[66] Lindsay [20, 366].
[67] Ibid.
[68] Chen [9, 3].
[69] Shakarian [28, 29, 4].
[70] Chen [9, 3].
[71] Shakarian [28, 29, 5].
[72] Sanger [26].

referred to also as ISIS, ISIL or Daesh, are markedly different. Members of the Islamic State have been very active on social media. In addition to broadcasting prisoner videos, members of the Islamic state would post everything from selfies to trophy photos to threats against other countries. Lesaca estimates that as of 2015, Russia alone was threatened 25 times via the internet, and France was threatened 20 times.[73] Many people were understandably outraged at what was being posted by the Islamic State on social media. An unnamed source from one of the major social media sites acknowledged this very fact in an article in the Guardian, stating "... there are lots of people who want us to take these accounts down or block them."[74] The main reason they do not, however, is because "...the government intelligence and military want us to keep them up, because that's how they track them."[75] One specific example was provided by the Commander Air Combat Command, General Herbert "Hawk" Carlisle, who provided an example where USAF members "... recognized a comment on social media and turned that into an airstrike that resulted in three Joint Direct Attack Munition (JDAM) bombs destroying an Islamic State in Iraq and Syria (ISIS) headquarters building."[76]

In April of 2016, however, the United States openly acknowledged a new front in its cyber campaign. Instead of simply gathering intelligence from social media, the United States overtly announced that it was engaging in cyber operations designed to "...disrupt the ability of the Islamic State to spread its message, attract new adherents, circulate orders from commanders and carry out day-to-day functions, like paying its fighters."[77] Pomerleau quotes Defense Secretary Ashton Carter, who stated the objectives to be "...interrupt ISIL command and control, interrupt its ability to move money around, interrupt its ability to tyrannize and control population, interrupt its ability to recruit externally—all of that it does in a cyber-enabled way...."[78] In addition to disrupting the Islamic State in general, Sanger notes that this strategy also has the benefit of rattling "...the Islamic State's commanders, who have begun to realize that sophisticated hacking efforts are manipulating their data."[79] One tactic, which both rattles commanders and delivers kinetic effects against fighters on the ground is a process whereby American cyber operatives first study Islamic State commanders, and learn their "...online habits."[80] The Americans then "...imitate them or to (sic) alter their messages, with the aim of

[73]Lesaca [19].

[74]Arthur [3].

[75]Ibid.

[76]Hoffman [17].

[77]Sanger [25].

[78]Pomerleau [24].

[79]Sanger [25].

[80]Ibid.

redirecting militants to areas more vulnerable to attack by American drones or local ground forces."[81] In another example, the Americans would target finances, by "… using cyberattacks to interrupt electronic transfers and misdirect payments."[82]

3.3 Russia and the 2016 American Election

On October 7th, 2016, the Department of Homeland Security issued a press release, stating that the United States "…is confident that the Russian Government directed the recent compromises of e-mails from US persons and institutions, including from US political organizations."[83] Specifically, they concluded that the mass leaks of hacked information and emails by Guccifer 2.0, WikiLeaks and DC Leaks were "… consistent with the methods and motivations of Russian-directed efforts…",[84] with the intent of interfering "…with the US election process."[85] Noting that Russia has employed these same tactics across both Eurasia and Europe, with the intent of influencing public opinion[86]; Homeland Security officials conclude that "…only Russia's senior-most officials could have authorized these activities."[87] Meyer notes that the "…hack has especially targeted individuals around Democratic nominee Clinton…,"[88] was ordered by the Russian government leadership, and "…is an attempt to influence the presidential election and advance the broader strategic objectives of the Putin regime."[89]

These are all impressive examples of cyber warfare in action, each with varied intent and purpose. The question must now be asked, however, are any patterns emerging? Can any models or systems be applied or derived from these examples? Are these just random actions, striking at diverse targets, or can some form of framework be applied to what has occurred (whether intended by the aggressors or not)?

3.4 Applegate's Cyber Maneuver Theory

Applegate applied maneuver theory to cyberspace.[90] He noted that the overall goal of cyber maneuver was the same as kinetic forms of maneuver, namely to "…secure

[81]Ibid.

[82]Ibid.

[83]Department of Homeland Security Press Office [12].

[84]Ibid.

[85]Ibid.

[86]Ibid.

[87]Ibid.

[88]Meyer [22].

[89]Ibid.

[90]Applegate [2, 7].

positional advantages in respect to an enemy or competitor state..."[91] While doing so, he noted (as did McGuffin and Mitchell)[92] that cyber maneuver is *different* from kinetic forms of maneuver, with Applegate noting that the manner in which cyber maneuver is executed is "...conducted at machine speeds inside a virtual construct."[93] Cyber effects are somehow different as well, however, in that following known cyber attacks such as Stuxnet, the targeted state did not view the outcome as having been an "...armed attack..."[94] While McGuffin and Mitchell argued that this fact contributed to cyber not warranting domain status,[95] Applegate observed that this was a unique feature of cyber warfare—the duality that anonymity and difficulties with attributing the attack result in no retaliations, in comparison with real world actions that would have resulted in open conflict.[96] Above all else, however, the effect that cyber maneuver generates is to "...influence human and machine behaviour."[97]

The idea of using maneuver to avoid battle (or conduct warfare by different means) also correlates with Delbrück's thinking. He agreed with Turenne,[98] preferring a "...maneuver strategy that is clever and active but avoids combat....,"[99] with the result that "...one could turn completely away from the decision by battle...,"[100] and that a nation could achieve its aims in a different way.[101] Applegate's maneuver theory corresponds to that proposed by Delbrück, and agrees with Warden as well.

Applegate's theory outlines different operational forms of maneuver. They could be employed separately, or together, in a phased approached. In particular, he described the elements of offensive cyber maneuver[102]:

a. Exploitive Maneuver—securing information for advantage at the tactical, operational or strategic level[103];
b. Positional Maneuver—compromising or outright seizing of key nodes in the cyber environment, and then utilizing these nodes for your benefit[104]; and

[91]Ibid.

[92]McGuffin and Mitchell [21, 410].

[93]Applegate [2, 7].

[94]McGuffin and Mitchell [21, 411].

[95]Ibid.

[96]Applegate [2, 14].

[97]Ibid., 4.

[98]Delbrück [11, 335].

[99]Ibid.

[100]Ibid., 387.

[101]Ibid.

[102]Applegate [2, 7].

[103]Ibid., 7–8.

[104]Ibid., 8–9.

c. Influencing Maneuver—Applegate describes an influencing maneuver as "…the process of using cyber operations to get inside an enemy's decision cycle or even to force that decision cycle though direct or indirect actions." For example, somehow compromising an enemy's C2 system, injecting or changing data, and influencing a commander such that he begins to doubt his systems and/or slow down his decision loop."[105]

Has Applegate defined different strategies that can be conducted when waging cyber warfare? Does his model make sense? Looking back at the examples provided earlier in this chapter, it appears that in some cases, the attacker wanted to deny and disrupt the internet use of the country under attack. This was demonstrated in 2007 with Russian aggression against Estonia,[106] and in 2008 during the Russian invasion of Georgia.[107] These strategies correspond most closely with Applegate's definition of Influencing Maneuver, where cyber operations are conducted "…to gain and maintain information superiority and dominance and to maintain freedom of maneuver in cyberspace."[108]

The United States is not following this strategy, however, in its war against the Islamic State. In this case, the strategy adopted by the Americans it to *not* deny access or use of the internet to its opponent. Both the Islamic State and individual fighters are being allowed access to the internet, via their computer systems and personal cell phones. Access to social media sites has been allowed as well (despite pressure otherwise), with the United States taking full advantage of targeting opportunities that arise due to poor operational security on the part of Islamic State fighters.[109] This example most closely correlates with Applegate's definition of Exploitive Maneuver, whereby an attacker captures "…information resources in order to gain a strategic, operational or tactical competitive advantage."[110] Applegate notes that how the information gained from this form of maneuver is applied "…makes it a valid and dangerous form of cyber maneuver."[111]

In the final type of example, cyber was used in a manner similar to that noted by McGuffin and Mitchell, that being as a supporting capability to a larger military operation.[112] Applegate referred to this as being Positional Maneuver, whereby "…key physical or logical nodes in the information environment…"[113] are captured or

[105]Applegate [2, 9].
[106]Herzog [16, 51].
[107]Clarke and Knake [10, 19].
[108]Applegate [2, 9].
[109]Hoffman [17].
[110]Applegate [2, 7].
[111]Ibid.
[112]McGuffin and Mitchell [21, 411].
[113]Applegate [2, 8].

compromised. The example that Applegate himself gives for this type of maneuver is the 2007 Israeli attack on a Syrian reactor, which was preceded by an apparent electronic and cyber attack.[114] In keeping with McGuffin and Mitchell's thinking that cyber would be supporting a larger operation,[115] Applegate notes that using "...positional maneuver prior to the initiation of actual kinetic combat operations set them up for success and illustrates the potential decisive nature of this form of cyber maneuver."[116]

Whether intended or not, each of the examples given at the beginning of this chapter seem to fit one of Applegate's three maneuver models. In some cases, an attacker seeks to obtain cyber superiority, and then deny internet access to its opponent.[117] In other cases, an attacker intentionally allows a defender to continue to access the internet, while reaping the treasure trove of intelligence leaked by an unsuspecting foe.[118] Or, an attacker may use cyber to support a larger operation.[119] To date, Applegate's maneuver strategy seems to quite accurately describe the different cyber strategies that have been conducted during recent campaigns.

3.5 *Warden and Cyber*

Can one apply Warden's targeting theory to each one of Applegate's maneuver strategies? Warden found most complex systems could be broken down into sub-elements. The first step in applying Warden's model to any potential system is systems understanding. As discussed earlier, for most systems, be they biological, industrial or natural, Warden found that each could be broken down into five separate components:

a. "leadership or direction;
b. organic essentials;
c. infrastructure;
d. population; (and)
e. a fifth component that protects the system from outside attack or general degradation."[120]

[114] Ibid.

[115] McGuffin and Mitchell [21, 411].

[116] Applegate [2: 9].

[117] Ibid.

[118] Ibid., 7.

[119] Ibid., 8.

[120] Warden [35].

Can this model be applied to cyber? Arwood et al. noted that the systems and targeting model developed by Warden were "...developed to provide insight into how a complex system (nation-state, drug cartel, terrorist group) would be attacked via its centers of gravity, with emphasis on defeating the organization."[121] In comparing cyber power to air power, Arwood et al. note that the core of "... strategic air power theory was the idea that wars could be won by striking at the heart of the enemy rather than having to grind through a protracted terrain conflict."[122] They note that a unique feature of air power was the compressed "...time and distance..."[123] on the battlefield, and that similar features are shared by cyber; notably that attacks may be launched "...from anywhere in the world..."[124]

When noting the similarities between air power and cyber power, one must also consider the issue of domain superiority. A key tenet in Warden's theory using airpower was that both strategic and operational air superiority was crucial to the success of the overall plan.[125] How would one define cyber superiority? In building his definition for cyber superiority, Bryant used the air domain as a template.[126] He argued that cyber superiority "...will be local and transient."[127] Further, Bryant cited Joint Publication 3-12, Cyberspace Operations, which states cyberspace superiority is the "...degree of dominance in cyberspace by one force that permits the secure, reliable conduct of operations by that force, and its related land, air, maritime, and space forces at a given time and place without prohibitive interference by an adversary."[128]

When trying to apply Warden's theory to Applegate's maneuver strategies, a logical counter argument would be that perhaps the attackers were not thinking of Warden's model when they conceived of their attack. This issue was addressed by Distelzweig in his analysis of Operations Unified Protector and Odyssey Dawn in Libya. There, he concluded that the "...planning and execution of Operations Odyssey Dawn and Unified Protector followed Colonel John A. Warden's theory of warfare, based on the importance of air superiority and attacking the enemy as a system, **even if this was not the intended methodology**."[129]

If one accepts that argument, can Warden's concentric ring theory be applied to Applegate's cyber-maneuver model; namely the three distinct offensive strategies which appear to be emerging?

---

[121] Arwood et al. [4, 34].

[122] Ibid., 32.

[123] Ibid.

[124] Ibid.

[125] Warden III [35, 118].

[126] Bryant [7, 26].

[127] Ibid., 39

[128] Director, Joint Staff [13, GL-4].

[129] Distelzweig [14, 2].

3.6 *Warden and Applegate's Exploitive Maneuver Strategy*[130]

This targeting model would be similar to the cyber strategy adopted by the Americans against ISIS. The enemy would not be denied access to the internet nor cyber capabilities, allowing the attacker to exploit the enemy through the internet. Applying Warden's template, the model[131] would look like this:

a. Fielded military—The United States overtly announced that it was engaging in cyber operations designed to "…disrupt the ability of the Islamic State to spread its message, attract new adherents, circulate orders from commanders and carry out day-to-day functions, like paying its fighters."[132] This strategy directly affects the fielded fighting forces of the Islamic State; by impacting recruiting efforts, and by affecting the confidence that Islamic State fighters have in their State, in particular if an outside power can disrupt their pay system.[133] One tactic which would have a significant impact on morale would be spoofing orders given by their own commanders (whom the Americans have studied and learned their "…online habits…"[134]). While directly affecting the inner leadership ring, this tactic would also affect the troops on the ground as well. Once Islamic State fighters became aware that the Americans were capable of spoofing their commanders online, they would never know if orders being passed were those of their commanders ordering them to battle, or if they were spoofed orders from the Americans, ordering them to a location where death from lethal coalition airpower would be waiting for them. The overall American strategy against the Islamic State's fighting forces was to permit them to stay online, while reaping the benefits of the intelligence that could be gained while concurrently disrupting operations and causing a loss of confidence with respect to their leadership and the orders they were being given.
b. Population—The American cyber strategy with respect to the population appears focused on disrupting "…the ability of the Islamic State to spread its message…"[135] Pomerleau notes that part of their strategy is to "…interrupt its ability to tyrannize and control (the) population…"[136] By interfering with the Islamic State's ability to communicate, influence and terrorize the population, the Americans are influencing the population itself.
c. Infrastructure—According to American Treasury Secretary Jacob Lew, "…Isis needs access to the international financial system for oil equipment, weapons,

---

[130] Applegate [2, 7].

[131] Warden III [35, 108].

[132] Sanger [25].

[133] Ibid.

[134] Sanger [25].

[135] Ibid.

[136] Pomerleau [24].

communications equipment and other imported items which requires them to move funds and that provides opportunities for attack."[137] Along with other tactics, the Americans would target finances, by "...using cyberattacks to interrupt electronic transfers and misdirect payments."[138] Pomerleau notes that the Americans would use cyber against the Islamic State to "...interrupt its ability to move money around..."[139] By attacking financial infrastructure, the Americans concurrently disrupted two important system essentials—oil (sold for money), and money itself. With respect to other possible infrastructure cyber targets, Agence France-Presse cited a 2015 briefing to a Senate panel by the director of the National Security Agency, Admiral Michael Rogers. Specific mention was made of, "...critical infrastructure networks—power grids, transportation, water and air traffic control, for example—where a computer outage could be devastating."[140]

d. System Essentials—American Treasury Secretary Jacob Lew assessed oil and oil revenues to be an Islamic State strength, noting that "...Isis has reaped an estimated $500 m from black market oil and millions more from the people it brutalises (sic) and extorts."[141] This financial strength, however, is also something that could be targeted. In another example, the Americans would target finances, by "...using cyberattacks to interrupt electronic transfers and misdirect payments."[142] The United States would also attack the Islamic State's financiers, oil and money directly. In 2015, American Special Forces conducted a raid that killed senior Islamic State financier Abu Sayyaf, obtaining "...reams of data on how ISIS operates, communicates and earns its money," the official told CNN, referring to some of the communications elements, such as computers, seized in the raid."[143] Exploiting the intelligence gleaned from the computers taken during the Sayyaf raid, along with "...a combination of satellite imagery, electronic intercepts and informers' tips, analysts have tracked Islamic State operatives storing huge amounts of cash in bank vaults, private residences and other hiding places."[144] The result, as of May 2016, has been "...21 strikes on cash storage and distribution sites since October, destroying what the Central Command said was hundreds of millions of dollars."[145] The American cyber strategy with respect to system essentials appears to be to use it in conjunction with other kinetic means to disrupt and destroy essential resources such as oil and money.

[137] Associated Press at the United Nations [5].
[138] Sanger [25].
[139] Pomerleau [24].
[140] Agence France-Presse [1].
[141] Associated Press at the United Nations [5].
[142] Sanger [25].
[143] Starr and Smith-Spark [30].
[144] Schmitt [27].
[145] Schmitt [27].

e. Leadership—The example cited earlier, of American hacking directly into the secure military computer network used by the Iraqi military,[146] is a good example of targeting the center leadership ring in an exploitive cyber maneuver strategy. By sending personal emails directly to Iraqi officers,[147] prior to the commencement of hostilities, the Americans influenced these leaders to park the fighting vehicles under their command and send personnel on leave.[148] The Americans did not have to fight these armoured vehicles nor those troops on the ground."[149] By employing a similar exploitive strategy against the Islamic State, the United States is influencing the leadership a different way. The Americans are disrupting the ability of the State to "…circulate orders from commanders and carry out day-to-day functions…."[150] The Americans are being overt in their actions, however, with the effect of rattling "…the Islamic State's commanders, who have begun to realize that sophisticated hacking efforts are manipulating their data."[151] Finally, by studying Islamic State commanders, and learning their online habits, the Americans intermittently spoof them by sending false orders to their troops.[152] The American cyber strategy with respect to leadership appears to be disrupting the ability of the Islamic State commanders to command effectively, and to affect the confidence they have with respect to the information they are receiving, and the orders they are giving.

3.7 *Warden and Applegate's Influencing Maneuver Strategy*[153]

This targeting model would be similar to the cyber strategy adopted by the Russians against Estonia in 2007 and Georgia in 2008. The attacking state would seek to achieve cyber superiority, and then would attempt to deny access to the internet for the country under attack. Applying Warden's template, the model[154] would look like this:

a. Fielded military—Hollis notes that the first cyber target attacked by Russia in their 2008 campaign against Georgia was a hacking forum.[155] While not

[146] Clarke and Knake [10, 9].
[147] Ibid.
[148] Ibid., 10.
[149] Ibid.
[150] Sanger [25].
[151] Sanger [25].
[152] Ibid.
[153] Applegate [2, 9].
[154] Warden [35, 108].
[155] Hollis [18, 3].

military targets per se, in cyber warfare this group does constitute "fielded military". Hollis states that by attacking the Georgian hacking community first, "…Russian-supported hacker militia pre-emptively (sic) tried to forestall or mitigate a counter-attack (or returning fire) from Georgian hackers."[156] This attack contributed to Russian cyber superiority. Chayes notes that the deluge of DDoS cyber attacks against Estonia in 2007 left their military "…unable to communicate."[157] One year later, during the 2008 war between Russia and Georgia, the Russians again launched DDoS attacks against Georgian military networks[158]; again impacting the military's ability to communicate. The overall Russian cyber strategy against the military appeared to be to shut down its communications, impacting it ability to operate.

b. Population—The cyber attacks by Russia against Estonia had a significant impact upon the Estonian population. Chayes notes that "Estonia was a highly-wired society…,"[159] however the conveniences of having such a wired society also served as a vulnerability. Herzog notes that "…97% of bank transactions occur online; and in 2007, 60% of the country's population used the Internet on a daily basis."[160] The ability of the population to properly function as a society "…was nearly brought to a halt in less than a month…"[161] due to the Russian cyber attacks. When the Russians conducted similar attacks against Georgia in 2008, the result was that the Russians "…sowed panic and confusion among the Georgian civilian population because it was unable to communicate with its government."[162] The Russian cyber strategy against the population appeared to be to disrupt, blind and create fear and confusion.
c. Infrastructure—Clarke and Knake noted the complexity and sophistication of the cyber attacks being waged against Estonian infrastructure in 2007. The specificity of targets made them surmise that this was not some amateur hacker attack. The attack was well organized and elaborate, targeting obscure yet specific webpages such as "….the addresses of servers running parts of the telephone network, the credit-card verification system, and the Internet directory."[163] Herzog notes that "…Estonia relies on the Internet for its critical infrastructure; electronic networks are integral to the functioning of government operations, electric power grids, banking services, and even Tallinn's water supply."[164] Further, the Estonian government has adopted a model known as

---

[156]Hollis [18, 3].
[157]Chayes [8, 476].
[158]Ibid., 477.
[159]Ibid., 476.
[160]Herzog [16, 51].
[161]Chayes [8, 476].
[162]Chayes [8, 477].
[163]Clarke and Knake [10, 15].
[164]Herzog [16, 51].

"…paperless government…,"[165] and is dependent on access to the internet. By attacking the websites that they did, the Russian cyber strategy effectively shut down "…the websites of all government ministries, two major banks, and several political parties. At one point, hackers even disabled the parliamentary email server."[166] The Russian cyber strategy against vital infrastructure appears to have been to shut down key infrastructure needed for the government to govern, and the society to function effectively.

d. System Essentials—by launching cyber attacks against financial websites, Russia directly affected a system essential—money—in a society where "…97% of bank transactions occur online…"[167] Herzog notes that the attacks "…prevented credit card and automatic teller machine transactions from occurring for several days."[168] The Russian cyber strategy against system essentials seems to be to target digital transfers of money, in a highly wired and connected country.

e. Leadership—During the 2007 cyber attacks against Estonia, repeated attacks "…crashed Estonia's Internet system, leaving the government—including the president, parliament, police, and military—unable to communicate."[169] This had a direct impact on the ability of Estonian leadership, at all levels, to lead or govern. By conducting similar attacks against Georgia one year later, with similar effect, the Russians again shut down government and military networks and websites.[170] Chayes notes that in addition to this impacting the Georgian leadership's ability to lead and govern, "…the attackers sowed panic and confusion among the Georgian civilian population because it was unable to communicate with its government."[171] As a result of this significant cyber attack, Russian hackers took control of the (Georgian) .ge web domain,[172] Georgia was cut-off from the global banking network (including credit card transactions),[173] and had to transfer the Georgian President's webpage to a site in the United States.[174] The cyber strategy adopted by the Russians against the Georgian leadership appears to be to deny them access to the internet, and separate the leadership from the people in order to stoke fear and uncertainty.

[165]Ibid.

[166]Herzog [16, 51].

[167]Ibid.

[168]Ibid., 52.

[169]Chayes [8, 476].

[170]Ibid., 477.

[171]Chayes [8, 477].

[172]Clarke and Knake [10, 19].

[173]Ibid., 20.

[174]Ibid., 19.

3.8 Warden and Applegate's Positional Maneuver Strategy[175]

This targeting model would be similar to the cyber strategy adopted by the Israelis against Syria. The attacker would infiltrate the defender's cyber systems, and would compromise them in order to support a kinetic attack. In this case, Warden's template[176] would still be used, however it would be against "...key physical or logical nodes in the information environment which can then be leveraged during follow-on operations."[177] Warden's theory would then be applied against these nodes, which could "...be viewed as centers of gravity in the information environment and gaining logical control of these nodes will give the attacker key advantages and leverage during the escalation of conflict..."[178] As a result, when employing positional maneuver strategy, an attacker would direct efforts against specific nodes and specific vulnerabilities within a defender's cyber systems.

3.9 Summary

Has Applegate accurately defined different offensive cyber maneuver strategies, and can Warden's targeting theory be applied to these different strategies? Is there merit in incorporating cyber into a military campaign? Although a cyber attack was dismissed by General Schwarzkopf as being a reliable means of attacking Iraq's air defence network during the first Gulf War,[179] by the time the Second Gulf War erupted just over a decade later; cyber warfare capabilities, and military leaders' confidence in them, had improved considerably.[180] The first decade of the twenty-first century saw cyber warfare being used increasingly as part of overall campaign strategies. Demonstrating the flexibility of cyber warfare, it was used to effect quite different outcomes in each of the manners in which it was employed. With the STUXNET example, it was used to prevent an escalation to a larger scale kinetic conflict. With the Israeli attack against a Syrian nuclear reactor, cyber warfare was employed in a singular contained military operation. Finally, the examples of Russian employment in Estonia and Georgia, along with the American example against the Islamic State, demonstrated how cyber warfare was being employed in major conflicts.

---

[175] Applegate [2, 8].
[176] Warden III [35, 108].
[177] Applegate [2, 8].
[178] Applegate [2, 8].
[179] Clarke and Knake [10, 9].
[180] Ibid.

Three distinct maneuver strategies emerged, which closely matched Applegate's model of offensive maneuver; these were exploitive, positional and influencing maneuver. The Russian strategies against Estonia and Georgia most closely matched Applegate's definition of influencing maneuver. The American strategy against the Islamic State most closely matched Applegate's definition of exploitive maneuver. Finally, cyber warfare could be employed in a supporting role to a military operation, in a manner similar to that proposed by McGuffin and Mitchell, which Applegate himself called positional maneuver.

Warden's concentric ring theory was then applied to influencing and exploitive maneuver. Analysis of open source information pertaining to the wars against Estonia, Georgia and the Islamic State demonstrated that Warden's targeting model could be applied to offensive cyber maneuver strategies. A generic targeting model, based on Applegate's offensive cyber maneuver strategies and Warden's targeting theory will be discussed next.

4 Generic Cyber Strategy

In the last decade, Russia and the United States have conducted two completely different cyber strategies as part of larger military campaigns. Both appear to have been effective. While these strategies were described anecdotally in the previous chapter, can a generic model be extracted for each strategy? Secondly, how would cyber be employed as part of a larger military campaign? These questions will be studied in this chapter.

4.1 Warden Applied to Applegate's Exploitive Cyber Maneuver

For exploitive cyber maneuver, an attacker would allow an opponent to have access to the internet, and the attacker would be overt with its cyber strategy. The effect desired would be to subvert the confidence of fielded fighting forces and the population. While the exact methods are closely held, the means most likely employed to accomplish this would-be software and password hacking. With respect to Warden's theory, the generic model would be described as follows:

a. Fielded military. With this model, an attacker would conduct cyber warfare against command and control, logistics and administrative networks. The aim would be to disrupt fuel, supply, administration and pay systems in order to affect troop confidence and morale.[181] Using enemy leadership electronic accounts and addresses, an attacker would transmit spoofed orders to enemy fighters to direct them where an attacker would want them to go.[182] The attacker would be overt in its actions, with the result that even legitimate orders from enemy commanders would be suspect. Once the fielded fighting forces became aware that an attacker possessed this capability, they would never know if messages from their superiors were proper commands, or spoofed orders directing them to a location where enemy aircraft were waiting to attack;

b. Population. Here, an attacker would aim to disrupt the ability of enemy leadership to deliver its messages to the population. The effect desired would be to prevent enemy leadership from exerting influence over the population[183];
c. Infrastructure. In this situation, the objective would be to attack infrastructure by conducting cyber warfare against the enemy's financial and banking networks.[184] In addition, electrical systems, water systems, internet, pipelines, transportation systems and air traffic control systems could all be attacked.[185] The effect desired would be to disrupt or shut down infrastructure, and thus paralyze the state under attack;
d. System essentials—A system essential in any state is money. The means of conducting banking and commerce are also quite easy to attack. Here, one would attack the financial and banking systems via the internet[186]; and
e. Leadership. In this situation, an attacker could communicate directly with mid-level leadership,[187] subverting the higher ruling authority. While spoofing leader online identities, email accounts and promulgating false orders was referred to earlier as creating an effect within the fielded fighting force, this tactic would also affect enemy leadership by subverting their authority.[188] By allowing a defender unrestricted use of the internet, an attacker could exploit poor operational security. By doing so, an attacker would determine locations of key leaders, headquarters and fighting units, and then target them with kinetic strikes.[189]

[181]Sanger [25].

[182]Sanger [25].

[183]Ibid.

[184]Pomerleau [24].

[185]Agence France-Presse [1].

[186]Sanger [25].

[187]Clarke and Knake [10, 9].

[188]Sanger [25].

[189]Hoffman [17].

4.2 *Warden Applied to Applegate's Influencing Cyber Maneuver*

For influencing cyber maneuver, an attacker would deny or disrupt an opponent on the internet. At least initially, an attacker's posture would be covert. The effect desired would be to prevent an opponent's ability to access the internet. The means most often employed to accomplish this to date has been distributed denial of service (DDOS) attacks, which involve large numbers of random computers around the world sending information to a website all at the same time. The result is that the website becomes overwhelmed and cannot respond to legitimate requests. With respect to Warden's theory, the generic model would be described as follows:

a. Fielded military. With this model, an attacker would disrupt or deny military command, control and communications systems, hindering the ability of leadership and fielded fighting forces to communicate with each other.[190] An attacker could also disrupt sensitive yet critical military warning and offensive technologies (such as air defence),[191] along with weapons systems, logistics and administrative systems in order to paralyze an opponent's military forces;
b. Population. Here, an attacker would conduct cyber attacks against government[192] and banking websites.[193] By shutting down these websites, the effects desired would be to disrupt and blind both the government and population, and to create fear and confusion;
c. Infrastructure. In this situation, an attacker would attack servers running the telephone network, internet and credit card verification systems.[194] Electrical systems, water systems, internet, pipelines, transportation systems and air traffic control systems could all be attacked.[195] The intent would be to shut down these systems, paralyzing the state;
d. System essentials. As with exploitive cyber maneuver, a system essential in any state is money. Here, an attacker would attack the banking and financial systems, depriving the state and society of being able to access and transfer money online[196]; and
e. Leadership. Here, an attacker would conduct cyber attacks against government websites, preventing leadership from being able to communicate with the population, government services and the military.[197] As with the model for population above, by shutting down these websites, the effects desired would be

[190]Chayes [8, 476].
[191]Clarke and Knake [10, 6].
[192]Chayes [8, 477].
[193]Herzog [16, 51].
[194]Clarke and Knake [10, 19–20].
[195]Agence France-Presse [1].
[196]Herzog [16, 52].
[197]Chayes [8, 476].

to disrupt and blind both government and the population, and to create fear and confusion. The government would simply not be able to communicate its messages to the population.

4.3 Warden Applied to Applegate's Positional Cyber Maneuver

For positional cyber maneuver, an attacker would allow an opponent to access the internet. An attacker's posture would be covert, in order to not alert an opponent as to what they are doing. When employing positional maneuver strategy, an attacker would direct efforts against specific nodes and specific vulnerabilities within a defender's cyber systems.[198] The effect desired would be to access command and control systems, weapons systems, and critical sensors, and allow an attacker to have the equipment perform in a manner desired by the attacker. The means most often employed to accomplish this would be via cyber hacking.

4.4 Comparison with Other Models

The generic models developed to describe Warden's theory applied to Applegate's offensive cyber maneuver strategies will now be compared and contrasted to other applications of Warden's theory to different systems. These comparisons can be found in Table 1.

Similarities between systems have been highlighted in bold in Table 1. Based on these highlighted similarities, it is assessed that there are numerous commonalities between the proposed application of Warden to Applegate's Offensive Cyber Maneuver Strategies, two models proposed by Warden against a state and a drug cartel, and a model proposed by Hazdra against the Al Qaeda terrorist network.

4.5 Cyber Warfare as Part of a Larger Military Campaign

When building the operational design for a campaign, Bonner outlined where cyber would fit into the overall plan. Since World War I, achieving air superiority at the

[198] Applegate [2, 8].

Table 1 Comparison with other models

Warden's rings	Warden		Hazdra	Offensive cyber maneuver	
	Warden's model applied to a state[a, 1]	Warden's model applied to a drug cartel[b, 2]	Hazdra's application of warden to Al Qaeda[c, 3]	Warden's model applied to exploitive cyber Maneuver	Warden's model applied to influencing Maneuver
Leadership	"Government: **communication** security	"**Leader**: **communication** security	**Leadership:** **bin Laden** **Al-Zawahiri** **Communication** Alliances	Impersonate leadership online **Communicate false messages** **Communicate with mid-level leaders and influence** **Exploit poor OPSEC, target leadership kinetically**	**Communication** **government Websites**
System essentials	Energy: Electricity Oil Food **Money**	Coca source plus conversion	"**Money** Weapons False Documents Sanctuaries"	**Money** via banking system	**Money** **Banking** **Money transfer**
Infrastructure	**Roads** **Airfields** Factories	**Roads** **Airways** Sea Lanes	"**Transportation infrastructure** of host states" "Al Qaeda Companies and Businesses" "Al Qaeda Terrorist Training Camps"	Banking Finance Electricity Water Internet **Transportation** **Air Traffic** Control	Electrical Water Internet Pipelines **Transportation** **Air Traffic Control** Telephone Internet Credit Card

(continued)

Table 1 (continued)

Warden's rings	Warden		Hazdra	Offensive cyber maneuver	
	Warden's model applied to a state[a, 1]	Warden's model applied to a drug cartel[b, 2]	Hazdra's application of warden to Al Qaeda[c, 3]	Warden's model applied to exploitive cyber Maneuver	Warden's model applied to influencing Maneuver
Civilian population	**People**	Growers Distributors Processors	**Muslim world**	**Government communications with population**	**Government communications with population**
Fielded military	**Military** Police Firemen"	**Street Soldiers**"	**"Recruits and Al Qaeda's networked terrorist cells**"	**Fielded Military**: Command, control networks Logistics networks Administrative networks Spoof leadership orders	Communication Sensors Weapons systems

[a]Colonel John A. Warden, "Air theory for the twenty-first century"
[b]Ibid
[c]Lieutenant-Colonel Richard J. Hazdra, "Al Qaeda as a System"
[1]Warden III [33, 107]
[2]Ibid.
[3]Hazdra [15, 1–24].

Sources Colonel John A. Warden III, "Air theory for the twenty-first century", 107; Lieutenant-Colonel Richard J. Hazdra, "Al Qaeda as a System", 16; David E. Sanger, "U.S. Cyberattacks Target ISIS in a New Line of Combat;" Mark Pomerleau, "Cyber operations come out of the shadows;" Agence France-Presse, Cyber Attackers Leaving Warning 'Messages': NSA Chief.; Richard A. Clarke and Robert Knake, Cyber War: The Next Threat to National Security and What to Do About It?; Michael Hoffman [17], "US Air Force Targets and Destroys ISIS HQ Building Using Social Media; " Antonia Chayes, "Rethinking Warfare: The Ambiguity of Cyber Attacks;" Stephen Herzog [16] "Revisiting the Estonian cyber attacks: Digital threats and multinational responses; "Scott D. Applegate, "The Principle of Maneuver in Cyber Operations" beginning of a campaign quickly became a necessity.[199] Achieving air superiority at the beginning provided attacking forces with "…the ability to exploit airpower for reconnaissance, mobility, and attack without prohibitive enemy interference."[200] Once air superiority was achieved, it would then support the ground war through air interdiction, which "…destroys or interrupts those elements of an enemy's system of supply or communication for a sufficient time that the degradation will immediately or in due course prove fatal to his continuance of effective operations."[201] Because of this, achievement of air superiority has always been towards the beginning of a campaign's operational design.

Shakarian notes that Russia preceded its military campaign against Georgia by launching a cyber attack.[202] This attack "…targeted Georgian news and government websites…,"[203] with the result that the population was blinded and confused and the government lost its ability to communicate with the people. Similarly, Bonner states that when developing the operational design for a campaign, planners "…should have as their priority the attainment and maintenance of cyber superiority…"[204] Once achieved, cyber would then play a supporting role similar to what airpower plays once air superiority is achieved (in this case, supporting the ground war). Bonner states that once cyber superiority is achieved, it would then support "…kinetic operations with a focus on supporting the air campaign."[205]

4.6 Summary

Applegate has accurately described three different strategies that can be employed when waging offensive cyber warfare. Warden's targeting model can be applied to each of these, and generic targeting models for cyber warfare can be developed. A generic exploitive cyber maneuver strategy would allow an opponent to continue to access the internet, and an attacker would be overt with their intentions. The effect desired would be to subvert the confidence of the fielded fighting forces and population in their government. While the means to accomplish this are clandestine, malicious software and password hacking would be the most likely means of conducting this form of cyber warfare. The fielded fighting forces and population

[199]Bonner [6, 109].
[200]Ibid.
[201]Ibid.
[202]Shakarian [28, 29, 63].
[203]Ibid.
[204]Bonner [6, 109].
[205]Bonner [6, 109].

would be affected by disrupting command, control and logistics networks. With respect to infrastructure and system essentials, the attacker would disrupt financial and banking networks (with money being a system essential). In addition, with respect to infrastructure, electrical, water, internet, pipeline, transport and air traffic control systems could all be targeted, with the intent of paralyzing the state.

For influencing cyber maneuver, the attacker would deny or disrupt an opponent's access to the internet. The attacker's posture would be covert (at least initially), with the effect desired being to prevent an opponent from accessing the internet. Up to this point in time, the means most often employed to accomplish this has been a distributed denial of service attack, which overwhelms a site under attack. For fielded military and the population, an attacker would focus on government and military communications sites, along with banking websites. For infrastructure, the targets would be very similar to exploitive cyber maneuver, and would involve financial and banking networks (with money being a system essential), along with electrical, water, internet, pipeline, transport and air traffic control systems. The intent would be to paralyze the state. For system essentials, the target again is money, with financial and banking websites being targeted. Finally, for leadership, government websites would be attacked, preventing the leadership from being able to lead the people.

For positional cyber maneuver, an attacker would direct efforts against software vulnerabilities within the computer systems of an opponent's command, control, weapons or sensor systems, with the intent of the attacker being able to have these systems do or display what the attacker wants. These devices would then be working for the attacker, not the defender who paid for this equipment.

When comparing Applegate's exploitive and influencing maneuver against how Warden applied his model to a state and a drug cartel; and against how Hazdra applied Warden's model to Al Qaeda, numerous similarities quickly became apparent across all five applications. One can conclude that the application of Warden's theory to each of the cyber maneuver models is consistent with other applications of his model.

Finally, a paradigm shift was noted in campaign operational design. While the achievement of air superiority is still very important, it is now apparent that cyber superiority must be achieved first, and then cyber is used to aid in the achievement of air superiority. This new maxim has been seen in recent military campaigns.

5 Conclusion

Applegate has accurately described three different strategies that can be employed when waging offensive cyber warfare; exploitive, positional and influencing maneuver. The targeting model developed by Colonel John Warden can be used to analyse and attack an opponent's cyber capabilities for each of these offensive cyber maneuver strategies. An analysis of several recent military campaigns and operations demonstrate that countries have already started basing their attacks against their opponent's cyber capabilities using these models, along with incorporating cyber into the operational designs of their overall military campaigns.

Cyber is interwoven into our lives, our militaries and into society, and is an ever-expanding realm of human activity. These essential, interwoven characteristics of cyber are what warrant it being targeted by militaries in a conflict. Attacking cyber capabilities is a natural extension of warfare. Recent examples in Georgia and Estonia have demonstrated that when cyber has been incorporated into the operational design of a larger military campaign, the result is paralysis within the country being attacked.

Although a cyber attack was dismissed by General Schwarzkopf as being a means of attacking Iraq's air defence network during the first Gulf War,[206] by the time the Second Gulf War erupted just over a decade later; cyber warfare capabilities, and military leaders' confidence in them, had improved considerably.[207] The first decade of the twenty-first century saw cyber warfare being used increasingly as part of overall campaign strategies. Demonstrating the flexibility of cyber warfare, it was used to effect quite different outcomes in each of the manners in which it was employed. With the STUXNET example, a clandestine computer code attack against equipment in an Iranian nuclear facility prevented an escalation to a larger scale kinetic conflict. When Israel attacked a Syrian nuclear reactor, cyber warfare was employed in a supporting role in a singular, contained military operation. Finally, the examples of Russian employment in Estonia and Georgia, along with the American example against the Islamic State, demonstrated how cyber warfare was being employed in major conflicts, when incorporated into the operational design of a much larger military campaign.

Three distinct maneuver strategies emerged, which closely matched Applegate's model of offensive maneuver; these were exploitive, positional and influencing maneuver. The Russian strategies against Estonia and Georgia most closely matched Applegate's definition of influencing maneuver. The American strategy against the Islamic State most closely matched Applegate's definition of exploitive maneuver. Finally, cyber warfare could be employed in a supporting role to a military operation, in a manner similar to that proposed by McGuffin and Mitchell, which Applegate himself called positional maneuver.

Warden's concentric ring theory was then applied to influencing, positional and exploitive maneuver. Analysis of open source information pertaining to the wars against Estonia, Georgia and the Islamic State demonstrated that Warden's targeting model could be applied to cyber warfare.

A generic exploitive cyber maneuver strategy would allow an opponent to continue to access the internet, and an attacker would be overt with its intentions. The effect desired would be to subvert the confidence of the fielded fighting forces and population in their government and military leadership. While the means to

[206]Clarke and Knake [10, 9].

[207]Ibid.

accomplish this are clandestine, malicious software and password hacking would be the most likely means of conducting this form of cyber warfare. The fielded fighting forces and population would be affected by disrupting command, control and logistics networks. With respect to infrastructure and system essentials, the attacker would disrupt financial and banking networks (with money being a system essential). In addition, with respect to infrastructure, electrical, water, internet, pipeline, transport and air traffic control systems, these could all be targeted, with the intent of paralyzing the state.

For influencing cyber maneuver, the attacker would deny or disrupt an opponent's access to the internet. The attacker's posture would be covert (at least initially), with the effect desired being to prevent an opponent from accessing the internet. Up to this point in time, the means most often employed to accomplish this has been a distributed denial of service attack, which overwhelms a website. For fielded military and the population, an attacker would focus on government and military communications sites, along with banking websites. For infrastructure, the targets would be very similar to exploitive cyber maneuver, and would involve financial and banking networks (with money being a system essential), along with electrical, water, internet, pipeline, transport and air traffic control systems. The intent would be to paralyze the state. For system essentials, the target again is money, with financial and banking websites being targeted. Finally, for leadership, government websites would be attacked, preventing the leadership from being able to lead the people.

For positional cyber maneuver, an attacker would direct efforts against software vulnerabilities within the computer systems of an opponent's command, control, weapons or sensor systems, with the intent of the attacker being able to have these systems act or display in a manner desired by the attacker. These devices would then be working for the attacker, not the defender who is relying on this equipment.

When analysing exploitive and influencing maneuver against how Warden applied his model to a state and a drug cartel; and against how Hazdra applied Warden's model to Al Qaeda, numerous similarities quickly became apparent across all five applications. One can conclude that the applications of Warden's theory to each of the cyber maneuver models are consistent with other applications of his model.

Finally, a paradigm shift was noted in campaign operational design. While the achievement of air superiority is still very important, it is now apparent that cyber superiority must be achieved first, and then cyber is used to aid in the achievement of air superiority. This new maxim has been seen in recent military campaigns.

Thus, Applegate has accurately described three different strategies that can be employed when waging offensive cyber warfare, and that the targeting model developed by Colonel John Warden can be used to analyse and attack an opponent's cyber capabilities.

Acknowledgements This chapter was an extract from "*Warden's Concentric Ring Theory Applied to Applegate's Offensive Cyber Maneuver Strategy*,"[208] by Major David G. Wood (Royal Military College of Canada, 2017). The author was supervised by Professor Paul Mitchell, Director of Academics and Associate Dean of Arts (CFC) at Canadian Forces College.

References

1. Agence France-Presse (2015) Cyber attackers leaving warning 'Messages': NSA chief. 19 Mar http://www.securityweek.com/cyber-attackers-leaving-warning-messages-nsa-chief. Accessed 20 Nov 2016
2. Applegate SD (2012) The principle of Maneuver in cyber operations. In: 2012 4th international conference on cyber conflict (CYCON). IEEE Communications Society Tallinn, Estonia, pp 1–13
3. Arthur C (2014) Taking down ISIS material from Twitter or YouTube not as clear cut as it seems. The Guardian. 23 June. https://www.theguardian.com/world/2014/jun/23/taking-down-isis-youtube-twitter-google-video. Accessed 5 Oct 2016
4. Arwood S, Mills RF, Raines RA (2010) Operational art and targeting strategy for cyberspace operations. IOSphere (Spring), pp 30–36
5. Associated Press at the United Nations (2015) United Nations adopts plan to attack Islamic State's funding. 17 December. https://www.theguardian.com/world/2015/dec/17/united-nations-plan-islamic-state-funding-terrorist-group-al-qaida. Accessed 9 Oct 2016
6. Bonner EL (2014) Cyber power in 21st century joint warfare. Joint Force Q 74 (3rd Quarter):102–109
7. Bryant WD (2013) Cyberspace superiority: a conceptual model. Air University, Maxwell AFB
8. Chayes A (2015) Rethinking warfare: the ambiguity of cyber attacks. Harvard Nat Secur J 6:474–519. http://harvardnsj.org/wp-content/uploads/2015/06/Chayes.pdf
9. Chen TM (2010) Stuxnet, the real start of cyber warfare? [Editor's Note]. IEEE Netw 24 (6):2–3
10. Clarke RA, Knake R (2010) Cyber war: the next threat to national security and what to do about it?. Harper Collins. Kindle Edition, Toronto
11. Delbrück H (1975) English language edition published 1985. History of the art of war within the framework of political history. Translated by Walter J. Renfroe, Jr. Vol. IV. Greenwood Press, Westport
12. Department of Homeland Security Press Office (2016) Joint statement from the department of homeland security and office of the director of national intelligence on election security. 7 October. https://www.dhs.gov/news/2016/10/07/joint-statement-department-homeland-security-and-office-director-national. Accessed 8 Oct 2016
13. Director, Joint Staff (2013) Joint publication 3-12 (R) cyberspace operations. Joint Chiefs of Staff, Washington
14. Distelzweig K (2014) Operations Odyssey Dawn and unified protector: another win for Warden's Theory. Monograph, School of Advanced Military Studies, Army Command and General Staff College, Fort Leavenworth, Kansas: Army Command and General Staff College
15. Hazdra, Lieutenant-Colonel RJ (2006) Al Qaeda as a system. Research project. U.S. Army War College, Carlisle Barracks, Carlisle, PA, pp 1–24
16. Herzog S (2011) Revisiting the Estonian cyber attacks: digital threats and multinational responses. J. Strateg Secur 4(2):49–60

[208]Wood [36].

17. Hoffman M (2015) US air force targets and destroys ISIS HQ building using social media. DEFENSETECH. 3 June. http://www.defensetech.org/2015/06/03/us-air-force-targets-and-destroys-isis-hq-building-using-social-media/. Accessed 5 Oct 2016
18. Hollis D (2011) Cyberwar case study: Georgia 2008. Small Wars J 11 (6 Jan 2011), 1–10
19. Lesaca J (2015) Fight against ISIS reveals power of social media. The Brookings Institution. 19 November. https://www.brookings.edu/blog/techtank/2015/11/19/fight-against-isis-reveals-power-of-social-media/. Accessed 5 Oct 2016
20. Lindsay JR (2013) Stuxnet and the limits of cyber warfare. Secur Stud 22(3):365–404
21. McGuffin C, Mitchell P (2014) On domains: cyber and the practice of warfare. Int J (The Authors) 69(3):394–412
22. Meyer J (2016) Russia Hack of U.S. politics bigger than disclosed, includes GOP. 8 October. http://www.nbcnews.com/news/us-news/russia-hack-u-s-politics-bigger-disclosed-includes-gop-n661866. Accessed 10 Oct 2016
23. NATO Review Magazine n.d Cyber Timeline. http://www.nato.int/docu/review/2013/cyber/timeline/EN/index.htm. Accessed 14 Aug 2016
24. Pomerleau M (2016) Cyber operations come out of the shadows. Defense systems. 5 May. https://defensesystems.com/articles/2016/05/05/us-cyber-war-isis.aspx. Accessed 5 Oct 2016
25. Sanger DE (2016) U.S. Cyberattacks Target ISIS in a New Line of Combat. The New York Times. 24 April. http://www.nytimes.com/2016/04/25/us/politics/us-directs-cyberweapons-at-isis-for-first-time.html. Accessed 5 Oct 2016
26. Sanger DE (2009) U.S. rejected aid for Israeli raid on Iranian nuclear site. 10 January. http://www.nytimes.com/2009/01/11/washington/11iran.html. Accessed 20 Nov 2016
27. Schmitt E (2016) U.S. says its strikes are hitting more significant ISIS targets. 25 May. http://www.nytimes.com/2016/05/26/us/politics/us-strikes-isis-targets.html?_r=0. Accessed 10 Oct 2016
28. Shakarian P (2011) Stuxnet: cyberwar revolution in military affairs. Small Wars Journal (United States Military Academy) 1–10. smallwarsjournal.com
29. Shakarian P (2011) The 2008 Russian cyber campaign against Georgia. Mil Rev 91(6):63–68
30. Starr B, Smith-Spark L (2015) Abu Sayyaf, key ISIS figure in Syria, killed in U.S. raid. 17 May. http://www.cnn.com/2015/05/16/middleeast/syria-isis-us-raid/. Accessed 9 Oct 2016
31. Talbot BJ (2011) Stuxnet and after. J Int Secur Aff Fall/Winter—Number 21:69–78
32. United States. Center of Military History, United States Army (2010) War in the Persian Gulf —operations desert shield and desert Storm August 1990—March 1991. Center of Military History Publication, Washington
33. Warden III JA (1995) Air theory for the twenty-first century. In: Schneider BR, Grinter LE (ed) Battlefield of the future: 21st century warfare issues (Air University Press) 3:103–124
34. Warden III JA (1992) Employing air power in the twenty-first century. In: Shultz RH, Pfaltzgraff RL Jr (eds) The future of air power in the aftermath of the gulf war. Air University Press, pp 57–82
35. Warden JA (1995) The enemy as a system. Airpower J 9(1). http://www.airpower.maxwell.af.mil/airchronicles/apj/apj95/spr95_files/warden.htm. Accessed 11 Sept 2016
36. Wood, Major DG (2017) Warden's concentric ring theory applied to applegate's offensive cyber Maneuver Strategy. Directed Research Project, Royal Military College of Canada, 2017. https://www.cfc.forces.gc.ca/259/290/301/286/wood.pdf

Fentanyl Crisis: A National Security Matter

Corey Tsang and Anthony J. Masys

Abstract The complexity of the current threat landscape associated with terrorism and criminal networks continues to be a top national and global security agenda item. Amongst the global security challenges, synthetic drugs constitute one of the most significant drug problems worldwide (UNODC in Global smart update: fentanyl and its analogous—50 years on, 2017 [54]). Fentanyl and non-pharmaceutical fentanyls (NPFs) have moved beyond being a concern to emerging as a national and global security matter. In 2015, more than 52,000 Americans (144/day) died from drug overdose. As reported by Chester (Fentanyl: the next wave of the opioid crisis, 2017 [9]), 63% of overdoses involved opioids whereby they are having '…a considerable impact on public health and public safety in communities across the United States'. In Canada, the Fentanyl-related deaths has markedly increased across British Columbia, Alberta, Ontario and Quebec, ranging from a doubling to 20-fold increase from 2009 to 2014 (CCENDU in Deaths involving fentanyl in Canada, 2009–2014, 2015 [4]). This increasing trend is noted internationally with increasing mortality in Australia (Roxburgh et al. in Drug Alcohol Rev 32:269–275, 2013 [48]) and the Nordic countries (Denmark, Finland, Iceland, Norway, Sweden) (Simonsen et al. in Forensic Sci Int 248:172–180, 2015 [50]). Richmond (An introduction to systems thinking. High Performance Systems, Inc. NH: Hanover, 2001 [46]) argues that 'The way we think is outdated. As a result, the way we act creates problems, and then we are ill-equipped to address them because of the way we think'. His argument stems from the '…fact that we have been applying the same set of non-solutions to the crucial social problems such as drugs, poverty and crime for most of the last fifty years without any positive results'. This chapter examines the national and global fentanyl crisis and introduces a holistic perspective of the problem space that leverages social innovation and macro-social analysis (Boivin in Networks and network analysis for defence and security.

C. Tsang
University of Waterloo, Waterloo, Ontario, Canada

A. J. Masys (✉)
University of South Florida, Tampa, FL, USA
e-mail: tmasys@health.usf.edu

A. J. Masys (ed.), *Security by Design*, Advanced Sciences and Technologies for Security Applications, https://doi.org/10.1007/978-3-319-78021-4_12

Springer, Berlin, 2014 [1]; Masys in Exploring the security landscape—non-traditional security challenges. Springer, Berlin, 2016 [37]) to examine the public health and criminal dimensions.

Keywords Fentanyl · Opioids · National security · Synthetic drugs Social innovation · DODAF

1 Introduction

As reported by Gladden et al. [22: 837], in 2015 both the Drug Enforcement Administration (DEA) and Centers for Disease Control and Prevention (CDC), respectively, '…issued nationwide alerts identifying illicitly manufactured fentanyl (IMF) as a threat to public health and safety'. Novel synthetic opioids offer enormous profit potential and as such are being trafficked by organized crime groups nationally and internationally. As argued by the United Nations Office on Drugs and Crime [54], Synthetic drugs constitute one of the most significant drug problems worldwide. In the US alone, more than 52,000 Americans, or approximately 144 people each day died from a drug overdose. In 2015, 9580 drug overdose deaths involved synthetic opioids other than methadone (a statistical category that is dominated by fentanyl), up from 3105 such deaths in 2013, a 209% increase [9]. Overdoses from synthetic opioids (specifically fentanyl) have seen dramatic increases globally. This is both a global and national security crisis. Not only posing a risk to users but clandestine fentanyl reprocessing/production labs pose a significant safety risk to the public and first responders as well. Exposure to fentanyl and fentanyl analogues may result in serious medical distress, including overdose or death due to the nature of its toxicity.

Law enforcement agencies are therefore seeking for more proactive strategies in targeting these criminal network structures more effectively. This starts with a better understanding of the way they operate and adapt over time [13: 39].

This chapter examines the national and global fentanyl crisis and introduces a holistic perspective of the problem space that leverages social innovation and macro-social analysis [1, 34] to design strategies to combat the fentanyl crisis.

2 Landscape of the Opioid Crisis: Fentanyl

According to the latest report from the International Narcotics Control Board (INCB), the US and Canada continues to lead the world in opioid consumption [25: 226]. A standardize unit of "defined daily doses for statistical purposes" (S-DDD) is used to allow comparison between opioids. The US consumes 47 580 S-DDD, followed by Canada at 34 444 S-DDD. The breadth of the crisis is wide, extending into Europe with Germany, Denmark and Austria at 30 796, 22 670, 21

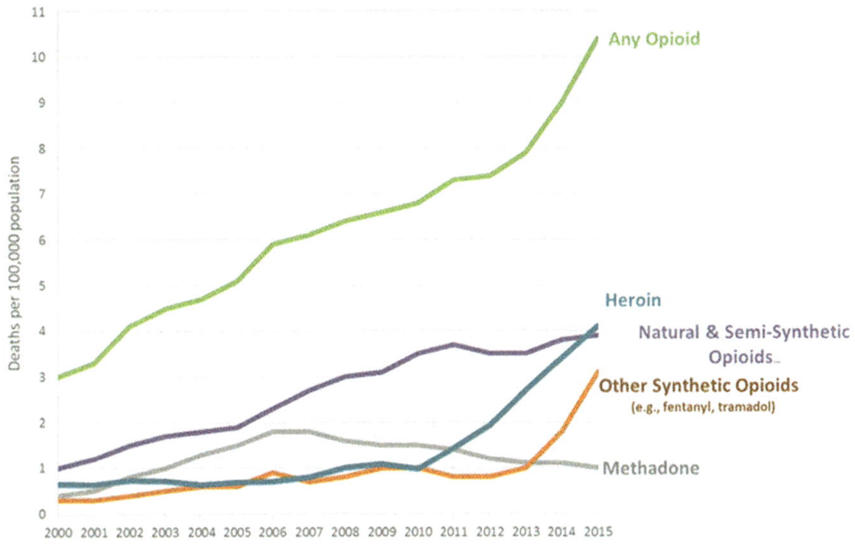

Fig. 1 Overdose deaths involving opioids in the United States from 2000 to 2015, by opioid [7, 8]

448 S-DDD respectively. Australia and New Zealand also have high opioid consumption rates at 16 591 and 12 084 S-DDD, respectively.

Fentanyl is a large player in Canada, with over 35% of overall prescription opioid consumption [18]. The consumption of Fentanyl has more than doubled from 2003 to 2013 and surpassing the United States, despite a 500% increase in its use from 1997 to 2007 [18].

The overdose-related deaths involving opioids has significantly risen in the United States over the period of 2000–2015 (Fig. 1). There is a sharp increase of synthetic opioid (including Fentanyl) related deaths since 2013.

In Canada, the Fentanyl-related deaths has markedly increased in Canada's largest four provinces (British Columbia, Alberta, Ontario and Quebec) ranging from a doubling to 20-fold increase from 2009 to 2014 [4]. This increasing trend is noted internationally with increasing mortality noted in Australia [48] and the Nordic countries (Denmark, Finland, Iceland, Norway, Sweden) [50].

The potential economic burden of prescription opioid overdose, abuse, and dependence was $78.5 billion US dollars in 2013 alone [19]. This is distributed through lost productivity, criminal justice system and the health insurance industry [19].

3 Pharmacology

Fentanyl is a synthetic opioid developed in 1960 [51], see Fig. 2. The opioid medication class was originally derived from the opium poppy plant in the early 19th century [27].

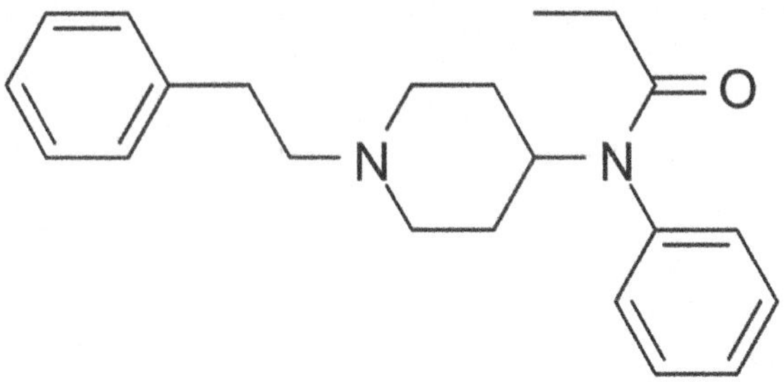

Fig. 2 Fentanyl chemical structure

It has a relative potency of approximately 80–100x more than morphine [43].

Opioids primarily act on the opioid receptors Mu (u), and to a lesser extent, Delta (S) and Kappa (K). Different opioids act on the Mu receptor with varying affinities. These receptors are distributed in the central nervous system (CNS), as well as the heart, lungs, liver, gastrointestinal and reproductive tract [17, 53]. When opioids act on these receptors, they are responsible for its therapeutic and adverse effects. For example, activation of the Mu_1 receptor in the brain causes analgesia, while activation of Mu_2 causes potentially fatal adverse effects, such as respiratory depression and sedation (Trescot et al. 2014).

Fentanyl's chemical properties differ from other opioids, which may contribute to its susceptibility of abuse. For example, fentanyl is more lipophilic than other opioids, making it easier to cross cellular lipid membranes [12]. This property makes the drug suitable for novel drug administration, such as the transdermal patch. Its properties also allow for rapid transportation into the CNS, providing a much faster onset of action [31], contrasting a lesser lipophilic opioid, such as morphine.

4 Pharmacokinetics and Formulations

Fentanyl was originally formulated as an intravenous injection for anesthetic purposes. The need for less-invasive, safer and convenient options led to the development of non-intravenous formulations.

When administered orally, fentanyl is extensively metabolized in the liver to inactive metabolites, rendering the route unsuitable for drug administration [31]. Other routes that would avoid first-pass metabolism by the liver, includes transdermal or transmucosal (i.e. via the oral or nasal mucosa).

The first was a transdermal patch in the 1980 used for cancer pain [51]. The main disadvantage of transdermal delivery is the relatively slow-onset of action (12–24 h). This led to the development of multiple transmucosal options, which have an onset of action between 2 and 15 min, depending on the exact product (e.g. lozenge, sublingual tablet, sublingual spray, buccal tablet, buccal film, nasal spray) [31].

Each product's onset of action is loosely related to how long its duration of action. In general, the faster the onset of action the shorter the duration of action. Two contrasting examples are the transdermal patch with an onset of action of 12–24 h and a duration of 12 h, after patch removal [31]. The sublingual tablet has an onset of action of 15 min but a duration of 60 min [31].

5 Indications and Regulations

Drugs are regulated and approved for use in Canada by Health Canada, the US by the Food and Drug Administration (FDA) and in Europe by the European Medicines Agency (EMA). Each jurisdiction has its specific wording on how the drug is to be used, but in general:

Long-acting formulas:

- *Transdermal patch*: Management of pain severe enough to require daily, continuous, long-term opioid treatment, and:
  - that is opioid-responsive; and
  - for which alternative options are inadequate.
  - Only for use in patients who are already receiving opioid therapy at a total daily dose of at least 60 mg/day morphine equivalents (ME) [26].

Short-acting formulas:

- *Sublingual tablet and buccal soluble film*: Management of breakthrough pain in patients with cancer, 18 years and older, who are already receiving, and who are tolerate to, opioid therapy for their persistent baseline cancer pain [35, 41].
- *Injectable (intravenous or intramuscular)*:
  - For analgesic action short duration during the anesthetic periods, premedication, induction and maintenance, and the immediate postoperative period (recovery room) as the need arises.
  - For use as a narcotic analgesic supplement in general or regional anesthesia.
  - For administration with a neuroleptic such as droperidol injection as an anesthetic premedication, for induction of anesthesia, and as an adjunct in the maintenance of general and regional anesthesia.
  - For use as an anesthetic agent with oxygen in selected high risk patients, such as those undergoing open heart surgery or certain complicated neurological or orthopedic procedures [49].
- *Injectable (epidural)*: for the post-operative management of pain following general surgical procedures and cesarean sections [49].

Other dosage forms exist, but are not available in Canada: nasal spray, sublingual spray, lozenge and iontophoretic transdermal system. Fentanyl can also be extemporaneously compounded into unique dosage forms such as, lollipops, chewable gummy gels and solution for inhalation.

6 Current Prescribing

Before starting the transdermal fentanyl patch, an important safety consideration is a patient's prior opioid exposure, measured in ME. Due to fentanyl's high potency, if a patient has not been sufficiently exposed to previous opioids, they could experience life-threatening adverse effects. The minimum prior to starting transdermal fentanyl is 60 mg of ME per day.

A Canadian longitudinal study examined previous opioid exposure in patients who were newly-prescribed the fentanyl patch [21]. The fentanyl initiation was deemed 'safe' if the previous opioid prescriptions met the 60 mg of morphine equivalent per day. The data was collected through Manitoba's provincial drug program information network on 11,063 patients aged 18 years or older between April 1 2001 and March 13 2013. The study found that 74.1% of new prescriptions were deemed unsafe. In other words, the patients had inadequate past exposure to opioids, prior to beginning fentanyl. This did improve over the study period to 50% by 2012 ($p < 0.001$).

7 Abuse and Mitigation Strategies

In Canada, illicit fentanyl is mainly driven by diversion of pharmaceutical fentanyl (mainly transdermal patches) and importation from abroad, notably China [4].

This is facilitated by certain factors, for instance, previously worn transdermal fentanyl devices, may contain 28–84% of the initial drug [32]. The authors advocated for disposal policies, which was not in existence until regulation was passed in Ontario in 2015 which mandated patients to return their used patches to a pharmacy before receiving new ones.

Prescription monitoring systems are central databases, which monitors the dispensing of narcotics at the community pharmacy level. These systems are active in several provinces in Canada and states in the United States [10] to reduce the effect of prescription forgery and double-doctoring (the use of multiple physicians to obtain multiple narcotic prescriptions).

Designer drugs or non-pharmaceutical fentanyls (NPFs) are not detected by routine urine toxicology tests [52]. Urine tests also have a number of limitations including the potential for tampering and adulteration. The test is also indiscriminate between individual opioids which complicates the interpretation.

Naloxone is an antidote used in suspected or confirmed opioid overdose situations where potentially fatal symptoms, such as respiratory depression or CNS depression is observed [29]. Overdose education and community naloxone distribution (OEND) consists of 'educate laypersons to recognize opioid overdose and instruct them how to administer naloxone to reverse respiratory depression' [29]. This harm reduction strategy is employed in Canada, the United States and in certain areas of Europe (Denmark, Germany, Estonia, Ireland, Italy, England and Norway) [5, 15, 29].

8 Discussion

Organized crime groups are a continuous threat to global and national security by causing harm to the economic and social fabric of society. This complex threat landscape challenges law enforcement and public health agencies who are struggling with important questions: How can we detect these criminal groups and their activities? What are the best strategies to disrupt them effectively? And how do they develop resilience against interventions? [13: 39].

As described in Prekupec et al. [45: 256] 'Novel synthetic opioids (NSOs) include various analogs of fentanyl and newly emerging non-fentanyl compounds. Together with illicitly manufactured fentanyl (IMF), these drugs have caused a recent spike in overdose deaths, whereas deaths from prescription opioids have stabilized. NSOs are used as stand-alone products, as adulterants in heroin, or as constituents of counterfeit prescription medications. During 2015 alone, there were 9580 deaths from synthetic opioids other than methadone. Most of these fatalities were associated with IMF rather than diverted pharmaceutical fentanyl'. This certainly captures the essence of the current global and national fentanyl crisis. UNODC [54: 2] reports that 'a growing interplay between these new drugs and traditional illicit drug markets is being observed. By July 2016, the emergence of NPS had been reported from 102 countries and territories. Trends on the synthetic drug market evolve quickly each year'. The crisis is not solely a law enforcement issue but a public health and safety issue as well. Risk to law enforcement officers and first responders has been well reported. For example, UNODC [54: 6] reports that 'overdose may also result from handling strong synthetic opioids without the precautions that prevent the substance from being inhaled or absorbed through the skin or mucous membranes. Contact with fentanyl or its analogues is so hazardous that both Canada and the USA recorded incidents of hospitalization of law enforcement officers that carried out seizures of such chemicals'.

With that being said, both law enforcement strategies and social innovation strategies emerge as part of the solution space. As described by Moore and Westley [36] 'complex challenges demand complex solutions. By their very nature, these problems are difficult to define'. Social Innovation is the process of developing and deploying effective solutions to challenging systemic social and environmental issues in support of social progress. In so doing, social innovation is aimed at improving human well-being and thereby resonates with the current global and national fentanyl crisis. Social innovation recognizes that 'action, reaction and interactions are dynamical, emergent, unpredictable and sometimes iterative.

Operating under complex conditions requires ongoing adaptation' [42: 251]. Regarding this notion of adaptation, we draw upon [23: 23] who argue that '...our actions are bound to have systemic consequences that we cannot foresee before they occur and often fail to understand afterwards; this is true for us individually, but potentially even more so for policy makers and institutions seeking to provide guidance and management in this highly complex environment'.

9 Law Enforcement

The complexity of the fentanyl crisis is fueled by the significant financial gains for traffickers and counterfeiters to import and sell these substances. Law enforcement challenges have been exacerbated by the '...explosive growth in the market for NPS, fueled by entrepreneurs and organized crime groups who have exploited both the manufacturing capacity in Asian countries and the emergence of globalized trade [45: 256]. With the upsurge of clandestine manufacturing of fentanyl stemming from the availability of the precursors, materials and equipment for manufacturing, the production and distribution of Illegal Manufactured Fentanyl increased to unprecedented levels [22]. For example, in North America, '...the pills and powders containing such substances sold on the illicit market pose a threat to public health because of the variable quantity and potency of the active components, which in extreme cases, such as with carfentanil, may be 10,000 times more potent than morphine' [54: 3]. In Europe as well, Estonia in particular '...stands out as having an endemic problem, with entrenched patterns of fentanyl use, including injection and associated high mortality rates' [37: 626].

O'Connor [40: 5] reports that China has helped fuel a fentanyl crisis in the United States, 'exporting a range of fentanyl products to the United States, including raw fentanyl, fentanyl precursors, fentanyl analogues, fentanyl-laced counterfeit prescription drugs like oxycodone, and pill presses and other machinery necessary for fentanyl production. These products are sent to small-scale drug distributors and criminal organizations across the United States who package and sell the product'. By virtue of this global clandestine supply chain (described in detail in O'Connor [40]), the fentanyl crisis is transnational highlighting that strategies to curb the influx of fentanyl will require cross border collaboration.

10 Criminal Business Model

Understanding the criminal 'business model' is key in supporting disruptive strategies to battle the fentanyl crisis. Prekupec et al [45: 258] argue that '...it is estimated that a single kilogram of NSO can be used to manufacture hundreds of thousands of counterfeit prescription tablets, which can produce millions of dollars in revenue for traffickers. Heroin production cost est. $65,000 per kilogram wholesale, vs. illicit Fentanyl cost est. $3500 per kilogram. 'Drug dealers thus face strong incentives to mix fentanyl with heroin and other street drugs. The drug appears to significantly reduce market prices of illicit opioids (and some other substances), while dramatically increasing risk' [20: 605].

As described in Masys [34], it is not enough to just understand the existence of connections between actors in a criminal network. Context is a key attribute. Crime script analysis is a methodology that supports contextual understanding of criminal networks. According to Levi and Maguire [30], crime scripts are an innovative way to gain a more detailed understanding of complex forms of crime and design prevention measures. Crime scripts map the sequence of actions used by offenders during crime commission. The potential for crime scripts to capture the sequential detail of crime-commission processes is important for understanding complex crimes, such as the operations involved in drug laboratories, human trafficking, financial and cyber crime.

A business purpose is established that evolves into a sequence of events that needs to be executed in order to commit a criminal offence. Although the complexity of this sequence varies across different crime types, every criminal conspiracy needs a division of tasks. In a criminal network the responsibilities following from these tasks are often divided between the actors according to experience, skills or knowledge. By mapping the roles of these actors within the network structure different crimescripts of the criminal activities can be identified [2]. A combination of SNA and scripting is useful for understanding the deeper operational structures within criminal networks and the interdependences between actors within their illegal activities. This level of understanding is necessary for designing strategic prevention measures that have lasting reduction effects on crime. It is about revealing potential 'dark network vulnerabilities'.

In supporting the crime scripting methodology, one can conceive of the criminal network supporting the fentanyl crisis as an enterprise. As described in Masys [34], DoDAF defines a way of representing an enterprise architecture that enables stakeholders to focus on specific areas of interests in the enterprise while retaining sight of the big picture. The DoDAF methodology does this by dividing a problem space into manageable pieces that correspond to stakeholder viewpoints, which are further defined as models (Fig. 3).

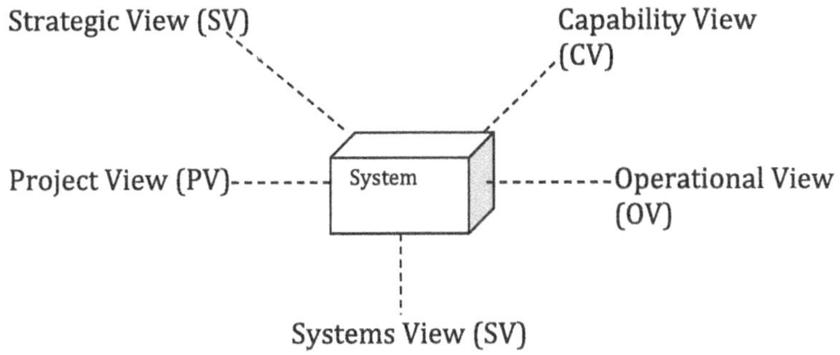

Fig. 3 DODAF views

DoDAF organizes models into views (DODAF [11]: 3.1–3.105)

- **Strategic View**: **not part of the traditional DODAF family of views**, but an essential element in understanding the conceptual goals, vision, mandate of the criminal or terrorist networks.
- **Capability View**: describes capability requirements, delivery timing, and deployed capabilities.
- **Operational View**: describes operational scenarios, activities, and requirements that support capabilities.
- **Project View**: describes relationships between operational and capability requirements and various projects to deliver capabilities.
- **Systems View**: models identify and describe system resource flows, organizational activities performed or supported by system functions.

Through the DoDAF views, interrelationships between systems/service functions, operational activities, operational nodes, data and information objects, technical standards, rules, policies, timelines, and a number of project and development artifacts are realized.

As shown in Table 1, different views afforded by DODAF reveals different perspectives and context associated with the network. In this way the combination provides a more holistic view and thereby facilitates a greater informed targeting analysis.

We begin with a high level crime script 'functional mapping' for fentanyl (Fig. 4) depicting the strategic vision of the criminal organization as the driver.

An OV-1 of the high-level functional model depicted in Fig. 4 may focus on any one or a collection of the functional areas listed. For example, a supply chain OV-1

Table 1 Operational views [11]

DODAF view	Description
OV-1: operational concept	Presents the concepts of operation of a described architecture
OV-2: organizations and resources	Presents resources that are used by organizational performers
OV-3: organizations, activities, and resources	Presents resources that are consumed and produced by activities performed by organizational performers
OV-4: organizational relationships	Presents the composition and relationships among organizational performers
OV-5a: operational activity hierarchy	Presents the hierarchical structure of organizational activities
OV-5b: operational activities	Presents activities performed by organizational performers to consume and produce resources
OV-6a: operational rules	Presents rules that constrain organizational activities
OV-6b: operational state transitions	Presents the states of resources consumed and produced by activities performed by organizational performers
OV-6c: operational activity sequences	Presents sequences of activities performed by organizational performers

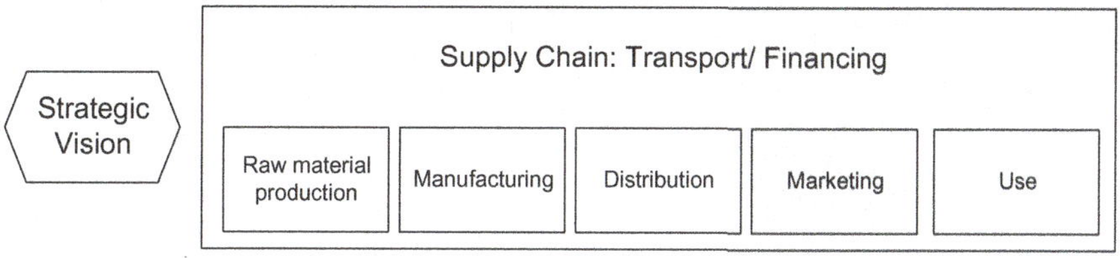

Fig. 4 High level crime script for fentanyl

Fig. 5 OV-1 transborder fentanyl supply chain

that focuses on the raw and precursor material may show a global pathway highlighting the transborder characteristics of the fentanyl crisis (Fig. 5).

Expanding Fig. 5 OV-1 into an OV-4 would show the actors involved in the supply chain. As described in Masys [33], mapping affiliation networks associated with the supply chain would help to better understand how fentanyl operations are strategically, operationally and tactically conducted and how organizational networks become entangled in the crime scripting. Roberts and Everton [47] show the value of analyzing a network for context. For explanatory purposes, Fig. 6a–f shows a terrorist network from 6 perspectives illustrating that all links are not equal and that actors are multiplex.

Through the DoDAF views, interrelationships between systems/service functions, operational activities, operational nodes, data and information objects, timelines, are realized thereby shedding light on the enterprise architecture supporting the fentanyl crisis. Examples of the application of DODAF in support of counter terrorism strategy development is discussed in detail in Masys [34].

By combining the DODAF methodology with social network analysis, what emerges is contextual understanding of the criminal enterprise architecture. As described in Duijn et al. [14: 3], the most central actor isn't necessarily the network member with the most leadership potential. One must identify the actor's role (strategic, operational or tactical) in addition to their skillsets in order to create value for strategic intervention development. What this suggests is that disruption strategies developed through SNA, DODAF and crime scripting are better informed.

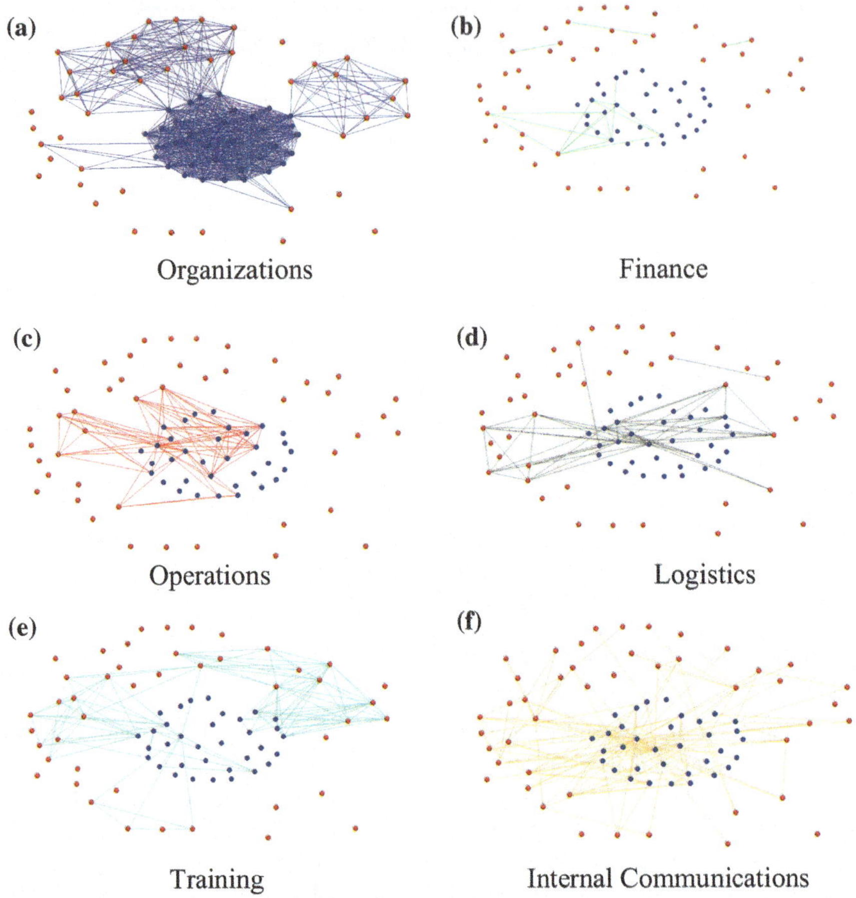

Fig.s 6 a–f Noordin top network analysis ([33] from Roberts and Everton [47])

One can draw a parallel regarding the application of such views to the fentanyl crisis. Chester [9] further expands upon the crime scripting similarities between heroin and illicit fentanyl. He describes '…similarities in production, trafficking, and consumption, it is important that we address concerns regarding heroin and illicit fentanyl together within the broader context of the opioid crisis. The same drug trafficking organization can manufacture and package both heroin and clandestinely produced fentanyl. These organizations likely use the same supply routes and distribution methods for both drugs'. With this in mind, insights from exploitation of heroin intervention strategies may be transferable to fentanyl. This can be explored and validated through a DODAF analysis of both drug crime scripts.

A similar approach to examining the problem space from a public health perspective would be useful. In this sense, mapping the public health enterprise

Table 2 Operational views for law enforcement and public health

DODAF view	Law enforcement	Public health/social intervention
StratV: strategic view	Presents criminal business purpose across the complete supply chain. This identifies the criminal supply chain element mission and vision	Presents public health strategic vision and mission. As well include mission and vision of social intervention strategies
CapV: capability view	Presents the capability requirements that support the criminal supply chain. This will identify gaps and vulnerabilities in the supply chain	Presents the capability gaps and opportunities that reside in the public health intervention strategy
OV-1: operational concept	Presents the concepts of operation associated with the illicit fentanyl supply chain	Presents the concepts of operation associated with the public health view and interventions associated with fentanyl crisis
OV-2: organizations and resources	Presents resources that are used by organizational performers along the complete supply chain	Presents resources that are used by organizational performers in support of public health matters
OV-3: organizations, activities, and resources	Presents resources that are consumed and produced by activities performed by the actors along the crime script associated with fentanyl supply chain	Presents resources that are consumed and produced by activities performed by the actors within the public health and social services domain
OV-4: organizational relationships	Presents the composition and relationships among criminal supply chain	Presents the composition and relationships among public health, social services, law enforcement and victims
OV-5a: operational activity hierarchy	Presents the hierarchical structure of the criminal network identifying associated roles and skill sets	Presents the hierarchical structure of the public health domain identifying roles and responsibilities
OV-5b: operational activities	Presents activities performed by the criminal network via a crime script representation	Presents activities performed by the public health and social services. This will identify opportunities for collaboration and outcome performance measurement

architecture as it pertains to the fentanyl crisis would highlight capability gaps across the domains of People, Processes and Technology. Table 2 highlights the complementary DODAF views applied to law enforcement and Public Health.

The factors that influence the fentanyl crisis are manifold. Complex issues need complex. responses. Law enforcement is only one piece of this complex threat landscape. Although disrupting and dismantling the criminal networks is a key element to the fentanyl crisis, a public health and social services approach requires '…soft organizational capacities, including communication, trust building, diplomacy, networking, political advocacy, and leadership are critical for health systems improvement' [24: 147]. As noted by Dr. Leana Wen, commissioner of health for the city of Baltimore 'It's not just physical pain that we're treating. It's also something else. And when we look at the areas that are the hardest hit by the opioid epidemic as of late, it's also the areas that are the hardest hit by unemployment, unstable housing, where people have unstable futures and where if tomorrow is no better than today maybe opioids are the way to go' [3].

11 Social Innovation

Moore and Westley [36] argue that 'complex challenges demand complex solutions'. Social innovation has recently emerged '…as a response to an acceleration of global crises and so-called 'wicked problems' characterised by multiple and contradictory analyses and diagnoses'. This includes such wicked problems as: climate change; social breakdown; health care; poverty; growing inequality; rising incidences of chronic long-term conditions and [39: 7]. This certainly resonates with the fentanyl crisis and calls forth for the application of social innovation. Social innovation is the process of developing and deploying effective solutions to challenging and often systemic social issues. Phills et al. [44: 36] define social innovation as: 'a novel solution to a social problem that is more effective, efficient, sustainable'. As noted in Murray et al. [38: 10], 'NESTA defines social innovation as: innovation that is explicitly for the social and public good. It is innovation inspired by the desire to meet social needs which can be neglected by traditional forms of private market provision and which have often been poorly served or unresolved by services organised by the state…Social innovation seeks new answers to social problems by: identifying and delivering new services that improve the quality of life of individuals and communities'. The DODAF representations described in Table 2 highlights the gaps and opportunities in an integrative, comprehensive and responsive social intervention (social innovation) strategy. In addressing the fentanyl crisis, social innovation that is scalable and sustainable requires collaborative efforts across key stakeholders from law enforcement and public health. The opioid crisis cannot be solved in one way alone; rather it will take a collaborative effort between numerous disciplines and initiatives.

As described in Kaundert and Masys [28], the process model for social innovation recognizes the requirement for feedback and reflective practices (Fig. 7).

Prekupec et al. [45: 261–262] discuss how '…within the United States, the DEA is expanding its 360 Strategy which leverages partnerships at federal, state, and local levels on 3 different fronts: law enforcement, diversion control, and demand reduction (DEA 2016). With regard to US government expenditures for drug policy, it should be noted that two-thirds of expenditures are currently spent on law enforcement and supply reduction [16]. Recently, there has been a call for a more equitable distribution of resources to fund prevention, treatment, and harm reduction. Multiple strategies, including targeted education interventions for primary prevention, greater access to medication-assisted therapies, and increased

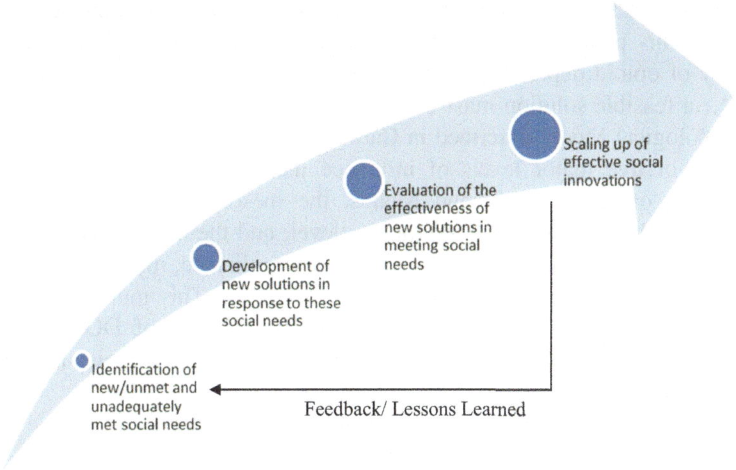

Fig. 7 Social innovation process

Fig. 8 OV-1 depicting the multiple dimensions associated with the fentanyl crisis

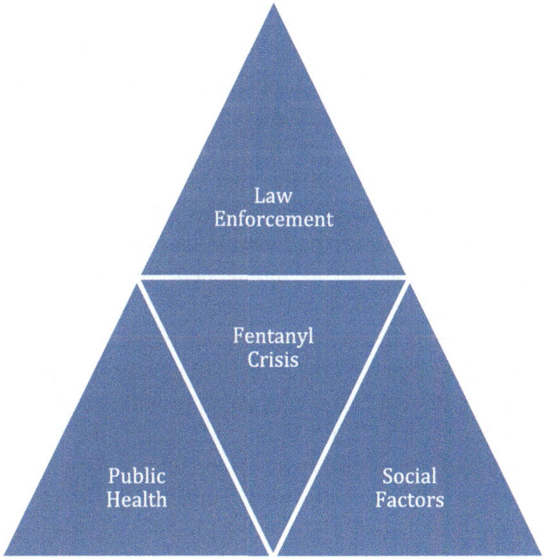

availability of naloxone to prevent overdose deaths, are currently being explored'. This approach maps well to the OV-1 in Fig. 8 that depicts an integrated, comprehensive and responsive approach that is rooted in social innovation.

Carney [6] provides an overview of current strategies associated with the opioid crisis. Citing research, Carney [6] notes that 'abstinence oriented interventions such as detoxification and relapse prevention, maintenance treatment, and harm reduction measures such as needle exchange programs are among the solutions proposed. This platform shows promise as it considers the complex nature and longevity of opioid dependence'. With the complexity associated with the opioid crisis, '...a feasible solution must use an approach more broad in nature' [6]. The Social Ecological Model described in Carney [6] provides a framework that is '... comprised of five major levels of influence including the individual level, the microsystem of close family and friends, the mesosystem or organizations of involvement, the exosystem or community level, and the macrosystem or societal cultural level'. This framework calls forth for a coordinated, multifaceted approach across all levels to influence an individual's recovery. This multiple perspective approach resonates with the enterprise architecture views of DODAF and the intervention strategies of social innovation. It also reflects the complex nature of the current opioid crisis.

The OV-1 depicted in Fig. 8, highlights the interdependent nature of the solution space in addressing the fentanyl crisis.

With this understanding of the multiple dimensions of the problem space and the complex interdependencies that cross them emerges the requirement for developmental evaluation as part of the social innovation strategy. Developmental evaluation, as depicted in Fig. 7, 'supports learning to inform action that makes a difference. This often means changing systems, which involves getting beyond surface learning to deeper understandings of what's happening in a system' [42: 11]. This encompasses double loop learning that goes '...beyond the single loop of identifying the problem and finding a solution to a second loop that involves questioning the assumptions, policies, practices, values and system dynamics that led to the problem in the first place and intervening in ways that involve the modification of underlying system relationships and functioning' [42: 11].

Considering the development and operationalization of intervention strategies, we are reminded of the unintended consequences actions can often have. Goldin and Mariathasan [23: 23] argue that '...our actions are bound to have systemic consequences that we cannot foresee before they occur and often fail to understand afterwards; this is true for us individually, but potentially even more so for policy makers and institutions seeking to provide guidance and management in this highly complex environment'. This is reflected in the causal loop diagram of Fig. 9. Public health and law enforcement interventions often fail because of a lack of attention to social, cultural and historical contexts and engagement with the people they are designed to benefit. The need to tailor delivery strategies to local contexts is a common finding from social science studies of NTD programs [24: 147]. Including the community is thereby part of the solution space.

With that being said, social innovation across the dimensions involves 'experimenting, adapting and developing what is done in response to program feedback, changing conditions, new insights, and emergent challenges....They assume a world of multiple causes, diversity of outcomes, inconsistency of interventions, interactive effects at every level' [42: 41]. This mental model of the complex landscape associated with the fentanyl crisis is essential. The analysis of the fentanyl crisis shows the requirement for a number of views/perspectives of this

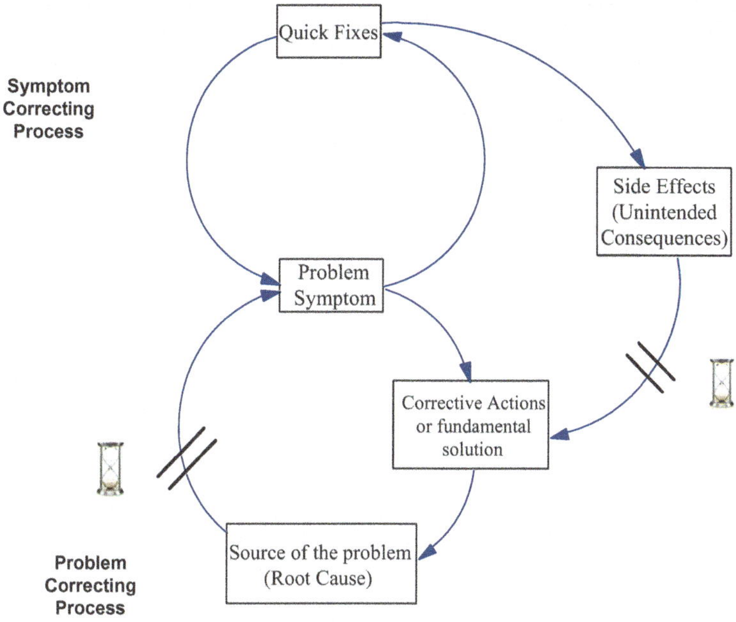

Fig. 9 Fixes that fail

complex threat landscape. DODAF provides such views as a Strat V, Cap V, and OV that can support this multiple perspective of the problem space. From here a metadata analysis can be conducted to support design of disruption strategies. Essentially the development of the Fentanyl Enterprise architecture will help identify strategies to disrupt criminal organizations involved in this public safety crisis as well as address the public health and social factors dimensions.

12 Conclusion

As discussed in Suzuki and El-Haddad [52: 107] 'Despite the concerted efforts to address the opioid crisis in this country, in recent years there has been a dramatic increase in overdoses from synthetic opioids, specifically fentanyl and related analogs'. It is recognized that the factors influencing the fentanyl crisis are manifold. An urgent, collaborative public health and law enforcement response is needed to address the increasing problem of IMF and fentanyl deaths [22: 840]. Herein lays the benefit of a social innovation strategy that works across the solution space of law enforcement, public health and social services. Strategies derived from DODAF representations serve to support social innovation by providing a systems perspective. This coupled with SNA and crime scripting can produce a strategy that is integrated, comprehensive and responsive.

References

1. Boivin R (2014) Macrosocial network analysis: the case of transnational drug trafficking. In: Masys AJ (ed) Networks and network analysis for defence and security. Springer, Berlin
2. Bruinsma G, Bernasco W (2004) Criminal groups and transnational illegal markets. Crime, Law, Soc Change 41:79–94
3. CFR (2017) The growing U.S. opioid crisis: lessons from around the world. https://www.cfr.org/event/growing-us-opioid-crisis-lessons-around-world
4. Canadian Community Epidemiology Network on Drug Use (CCENDU) (2015) Deaths involving fentanyl in Canada, 2009–2014. Ottawa, Ontario
5. Canadian Pharmacists Association (CPhA) (2017) Environmental scan: access to naloxone across Canada. Ottawa, Ontario
6. Carney A (2017) The current national opioid crisis: conceptualizing a plan of action. http://inquirepublication.com/the-current-national-opioid-crisis-conceptualizing-a-plan-of-action/
7. Centers for Disease Control and Prevention (CDC) (2017a) Fentanyl: preventing occupational exposure to emergency responders. https://www.cdc.gov/niosh/topics/fentanyl/risk.html
8. Centers for Disease Control and Prevention (CDC) (2017b) Opioid data analysis. https://www.cdc.gov/drugoverdose/data/analysis.html
9. Chester KL (2017) Fentanyl: the next wave of the opioid crisis. Subcommittee on oversight and investigations committee on energy and commerce United States house of representatives. http://docs.house.gov/meetings/IF/IF02/20170321/105739/HHRG-115-IF02-Wstate-ChesterK-20170321.pdf
10. Cobaugh DJ, Gainor C, Gaston CL, Kwong TC et al (2014) The opioid abuse and misuse epidemic: implications for pharmacists and health systems. Am J Health Syst Pharm 71:1539–1554
11. DODAF (2015) DODAF V 2.0. Volume II: Architectural data and models. Available at: http://dodcio.defense.gov/Portals/0/Documents/DODAF2/DoDAF%20v2.02%20Chg%201%20Vol%20II%20Final%202015-01-19.pdf [accessed 11 Oct 2015]
12. Drewes AM, Jensen RD, Nielsen LM, Droney J et al (2013) Differences between opioids: pharmacological, experimental, clinical and economical perspectives. Br J Clin Pharmacol 75:60–78
13. Duijn PAC, Sloot PMA (2015) From data to disruption. Digit Invest 15:39–45
14. Duijn PAC, Klerks PPHM (2014) Social network analysis applied to criminal networks: recent developments in Dutch law enforcement. In Masys AJ (ed) Networks and network analysis for defence and security. Springer
15. European Monitoring Centre for Drugs and Drug Addiction (EMCDDA) (2016) Preventing OPIOID overdose deaths with take-home Naloxone. http://www.emcdda.europa.eu/system/files/attachments/2092/News%20release_1_EN_2016_Naloxone_Final.pdf
16. Farrell M (2014) Drug legalisation. BMJ 349:g5233
17. Feng Y, He X, Yang Y, Chao D et al (2012) Current research on opioid receptor function. Curr Drug Targets 13:230–246
18. Fischer B, Russell C, Murphy Y, Kurdyak P (2015) Prescription opioids, abuse and public health in Canada: is fentanyl the new centre of the opioid crisis? Pharmacoepidemiol Drug Saf 24:1334–1336
19. Florence CS, Zhou C, Luo F, Xu L (2016) The economic burden of prescription opioid overdose, abuse, and dependence in the United States, 2013. Med Care 54:901–906
20. Frank RG, Pollack HA (2017) Addressing the fentanyl threat to public health. NEJM 376(7):605–607
21. Friesen KJ, Woelk C, Bugden S (2016) Safety of fentanyl initiation according to past opioid exposure among patients newly prescribed fentanyl patches. Can Med Assoc J 188:648–653
22. Gladden RM, Martinez P, Seth P (2016) Fentanyl law enforcement submissions and increases in synthetic opioid-involved overdose deaths—27 states, 2013–2014. Centers Dis Control Prev MMWR 65(33):837–843

23. Goldin I, Mariathasan M (2014) The butterfly defect: how globalization creates systemic risks and what to do about it. Princeton University Press, Princeton, NJ
24. Halliday JEB, Hampson K, Hanley N, Lembo T, Sharp JP, Haydon DT, Cleaveland S (2017) Driving improvements in emerging disease surveillance through locally relevant capacity strengthening. Science 357(6347):146–148
25. International Narcotics Control Board (2016) Narcotic Drugs 2016. https://www.incb.org/documents/Narcotic-Drugs/Technical-Publications/2016/Narcotic_Drugs_Publication_2016.pdf
26. Janssen Inc. (2017) Product monograph: Duragesic. Markham, Ontario
27. Kanjhan R (1995) Opioids and pain. Clin Exp Pharmacol Physiol 22:397–403
28. Kaundert M, Masys AJ (2018) Mass migration, humanitarian assistance and crisis management: embracing social innovation and organizational learning. In: Masys AJ (ed) Security by design. Springer Publishing
29. Kerensky T, Walley AY (2017) Opioid overdose prevention and naloxone rescue kits: what we know and what we don't know. Addict Sci Clin Pract 12:4
30. Levi M, Maguire M (2004) Reducing and preventing organised crime: an evidence-based critique. Crime, Law Soc Change 41: 397–469
31. Lotsch J, Walter C, Parnham MJ, Oertel BG et al (2013) Pharmacokinetics of non-intravenous formulations of fentanyl. Clin Pharmacokinet 53:23–36
32. Marquardt KA, Tharratt RS, Musallam NA (1995) Fentanyl remaining in a transdermal system following three days of continuous use. Ann Pharmacother 29:969–971
33. Masys AJ (ed) (2015) Applications of systems thinking and soft operations research in managing complexity. Springer Publishing
34. Masys AJ (2016) Disrupting terrorist and criminal networks: crime script analysis through DODAF applications. In: Masys AJ (ed) Exploring the security landscape—non-traditional security challenges. Springer, Berlin
35. Meda Pharmaceuticals Inc. (2012) Product monograph: Onsolis. Solna, Sweden
36. Moore M, Westley F (2011) Surmountable chasms: networks and social innovation for resilient systems. Ecology and Society 16(1):5. [online] URL: http://www.ecologyandsociety.org/vol16/iss1/art5/
37. Mounteney J, Giraudon I, Denissov G, Griffiths P (2015) Fentanyls: are we missing the signs? Highly potent and on the rise in Europe. Int J Drug Policy 26:626–631
38. Murray R, Caulier-Grice J, Mulgan G (2010) The open book of social innovation. Nesta
39. Nicholls A, Simon J, Gabriel M (2015) Introduction: dimensions of social innovation. In: Nicholls A, Simon J, Gabriel M (eds) New frontiers in social innovation research. Palgrave Macmillan
40. O'Connor S (2017) Fentanyl: China's deadly export to the United States. U.S.-China economic and security review commission. https://www.uscc.gov/sites/default/files/Research/USCC%20Staff%20Report_Fentanyl-China%E2%80%99s%20Deadly%20Export%20to%20the%20United%20States020117.pdf
41. Paladin Labs Inc. (2017) Product monograph: abstral. St-Laurent, Quebec
42. Patton MQ (2011) Developmental evaluation: applying complexity concepts to enhance innovation and use. Guilford Press, New York
43. Pereira J, Lawlor P, Vigano A, Dorgan M et al (2001) Equianalgesic dose ratios for opioids: a critical review and proposals for long-term dosing. J Pain Symptom Manage 22:672–687
44. Phills JA, Deiglmeier K, Miller DT (2008) Rediscovering social innovation. Stanford Soc Innov Rev Fall 2008: 34–43
45. Prekupec MP, Mansky PA, Baumann MH (2017) Misuse of novel synthetic opioids: a deadly new trend. J Addict Med 11(4):256–265
46. Richmond B (2001) An introduction to systems thinking (iThink software). High Performance Systems, Inc. NH: Hanover
47. Roberts N, Everton SF (2011) Strategies for combatting dark networks. J Soc Struct 12(2). http://www.cmu.edu/joss/content/articles/volume12//RobertsEverton.pdf
48. Roxburgh A, Burns L, Drummer OH, Pilgrim J et al (2013) Trends in fentanyl prescriptions and fentanyl-related mortality in Australia. Drug Alcohol Rev 32:269–275

49. Sandoz Canada Inc. (2013) Product Monograph: Fentanyl Citrate Injection. Boucherville, Quebec
50. Simonsen KW, Edvardsen HM, Thelander G, Ojanpera I et al (2015) Fatal poisoning in drugs addicts in the Nordic countries in 2012. Forensic Sci Int 248:172–180
51. Stanley TH (2014) The fentanyl story. J Pain 15:1215–1226
52. Suzuki J, El-Haddad S (2017) A review: fentanyl and non-pharmaceutical fentanyls. Drug Alcohol Depend 171(2017):107–116
53. Trescot AM, Datta S, Lee M, Hansen H (2008) Opioid pharmacology. Pain Physician 11: S133–S153
54. UNODC (2017) Global smart update: fentanyl and its analogous—50 years on. https://www.unodc.org/documents/scientific/Global_SMART_Update_17_web.pdf

Risk-Based Aviation Security—Designing-Out Terror?

Simon Bennett

Abstract Stung by a wave of terror attacks in the 1960s and 1970s, commercial aviation introduced a range of hard and soft security measures to mitigate the risk of unlawful interference. Terrorists responded by devising new forms of attack, such as the use of shoulder-launched surface-to-air missiles in the vicinity of poorly-surveilled airport perimeters, and reviving old tactics, such as the kamikaze attack. The terror attacks of September 11th, 2001, spurred further anti-terror measures, including the risk-based vetting of passengers. This chapter reviews the pros and cons of risk-based aviation security. Advantages include the opportunity to use scarce resources to best affect and—for most airline customers—a speedier passage. Disadvantages include greater operational complexity and additional capital and revenue cost. There is also the matter of political sensitivity: a risk-based aviation security system that is perceived to be discriminatory risks losing the support of those it is designed to protect—with unpredictable consequences. Finally, if the assumptions that inform the computerised vetting algorithms are wrong, the innocent may be inconvenienced, and the malevolent waved through. Referencing the Israeli experience, the author concludes that realising the benefits of risk-based aviation security requires planning, intelligent design and the will and ability to deliver a professional vetting service *in perpetuity*.

Keywords Terrorism · September 11th, 2001 · Risk-based aviation security
Benefits · Costs

S. Bennett (✉)
Civil Safety and Security Unit, University of Leicester,
University Road, LE1 7RH Leicester, UK
e-mail: sab22@leicester.ac.uk

A. J. Masys (ed.), *Security by Design*, Advanced Sciences and Technologies for Security Applications, https://doi.org/10.1007/978-3-319-78021-4_13

1 Introduction

Compared to the number of people killed in road traffic accidents (RTAs), or the number of lives lost to medical error, relatively few people are killed in terror attacks. According to the National Consortium for the Study of Terrorism and Responses to Terrorism [1] "In 2015, a total of 11,774 terrorist attacks occurred worldwide, resulting in more than 28,300 total deaths and more than 35,300 people injured". According to the World Health Organisation (WHO), in 2013 around 1.25 million people were killed in RTAs, with 20–50 million sustaining non-fatal injuries. In 2013, over 1,800 died on Britain's roads and over 34,000 on America's roads. "Road traffic injuries are the ninth leading cause of death globally" explains the WHO [2]. Despite the significant loss of life on our roads, in our hospitals, in the workplace and even in the home, governments feel morally compelled to focus on the terrorist threat.

Counter-terrorism operations are costly. According to Britain's Secret Intelligence Service (SIS), the number of Britons capable of mounting a Daesh-inspired terror attack numbers around 3,000. Given that it takes up to 30 officers to mount a 24-h surveillance operation on a suspect, the cost of mitigating the risk of Daesh-inspired terror attacks is significant. Other costly mitigations include the Prevent strategy (a civic counter-radicalisation intervention) and efforts to stop the radicalisation of prison inmates. Despite Britain's determined counter-terror effort, on 22 May, 2017, 22 concert-goers were murdered, and 116 injured by a suicide bomber in Manchester. Sixty ambulances and four hundred police officers attended. Troops were put on the streets. Shortly after they were stood down, terrorists murdered 8 and injured 48 in London. One of the attackers was known to MI5. The attack was low-tech, the terrorists using a van and hunting knives as weapons.

This chapter discusses the costs and benefits of risk-based security in commercial aviation. Given aviation's importance to the world economy (it supports 56.6 million jobs and contributes $2.2 trillion to global GDP [3]) and likelihood that a terrorist attack would kill a large number of people (the Airbus A380 carries 853 passengers in its all-economy configuration), significant resources are devoted to securing commercial air services.

For much of the post-war period, aviation security was relaxed and low-key. While this made flying pleasurable, it provided a window of opportunity for terrorists, malcontents and the mentally ill. Relaxed security facilitated aviation's nadir—the 1970 hijacking by Popular Front for the Liberation of Palestine (PFLP) terrorists of three airliners to Dawson's Field, a remote, sand-blasted ex-Royal Air Force airstrip in Jordan (Fig. 1). When Switzerland, Germany and the United Kingdom refused to agree to a prisoner-swap, the PFLP destroyed the aircraft.

Since the security nadir of the 1960s and 1970s, considerable intellectual, scientific, technological, political and financial resources have been devoted to securing aviation. Despite this effort, attacks are still mounted. In October 2015, Metrojet Flight 9268, en-route from Sharm El Sheikh to St. Petersburg, fell from an

Fig. 1 In September, 1970, the PFLP invited the press to witness the destruction of three airliners

altitude of 31,000 feet (9,400 m), killing all 224 on board. Daesh claimed responsibility. The Egypt-led investigation is ongoing (June, 2017). In March 2016, Daesh recruits detonated two nail bombs at Brussels Airport. Seventeen were killed (including the terrorists) and eighty-one injured. Only if those responsible for aviation security innovate, and invest in hardware, software and liveware, will they stay one step ahead of the terrorist, malcontent and psychopath.

2 Definition of Terms

2.1 Terrorism

The European Union Council defines terrorism as: "[I]ntentional acts that are committed with the aim of seriously intimidating a population, or unduly compelling a government or international organization to perform or abstain from performing any act, or seriously destabilizing or destroying the fundamental political, constitutional, economic or social structures of a country or an international organization" [4]. The US Code of Federal Regulations defines terrorism as: "[T]he unlawful use of force and violence against persons or property to intimidate or coerce a government, the civilian population, or any segment thereof, in furtherance of political or social objectives" [5]. This chapter references the latter, pithier definition.

2.2 *Risk-Based Security*

Risk-based security may be defined as the intelligence and/or information-led focusing of security resources—hardware, software and liveware—for maximum effect [6, 7]. According to the Deputy Director of the George Washington Centre for Cyber and Homeland Security [7], the Transport Security Administration's (TSA's) risk-based security regime has helped address four challenges:

(a) The TSA's poor reputation with the travelling public
(b) The TSA's poor reputation with Congress
(c) A fluid threat environment
(d) The Federal Government's post-2008 financial crash austerity measures.

3 Methodology

Deductive and inductive, the methodology is a hybrid. The chapter uses an established body of knowledge—systems theory—to highlight factors that determine the effectiveness of aviation security, while drawing on understandings gained by the author during his eighteen year aviation research career. Developing a sociology of the flight-deck has seen the author spend 1470 h on the jump-seat and *circa* 200 h on the ramp. He has observed: 232 × A319-operated sectors; 66 × A320-operated sectors; 62 × A321-operated sectors; 132 × B737-operated sectors; 181 × B757-operated sectors; 7 × A300-operated sectors. A sector is a point-to-point flight.

4 A Model for Risk-Based Security?

At Israel's only international airport, Ben Gurion, a system of risk-based security supported by enhanced perimeter checks help keep the airport's 16.5 million passengers safe. "By the time a traveller is finally allowed to board a plane departing from Israel, he or she will have cleared no less than 12 layers of security" observes the *Daily Telegraph's* David Blair [8]. The first perimeter check is conducted by armed guards over two miles from Ben Gurion's check-in desks. The check is democratic, every visitor being quizzed.

At check-in, every passenger is questioned. A passenger's answers determine whether s/he is allowed airside or subjected to further questioning away from check-in. Regarding his staff's questioning of passengers, Ben Gurion's Managing Director has said: "We are looking for suspicious signs or abnormal behaviour, in order to increase our chances of detecting a terrorist as far as possible from the main target, the aeroplane" [9]. Regarding the sometimes random and out-of-place questions put to passengers by security staff, the Managing Director of consulting firm Leading Edge Strategies has said: “Once you understand how people respond to ‘normal’ questions that they should know the answers to, and have no reason to lie about, you’ll be able to spot the lie when asked other questions” [10]. Normal questions cover subjects like the amount of luggage carried by a passenger, why the passenger chose flying over other forms of transport, intended length of stay and reason(s) for visiting.

While the Ben Gurion regime has proved successful in mitigating the risk of a terrorist attack—no passenger has been killed or wounded inside the airport, or on board an aircraft on the airport’s ramps and runways, for 44 years—it is not without its critics. Israeli newspaper *Haaretz* has accused the airport’s security staff of obsessive behaviour, alleging “the religion of security drives the authorities out of their senses” [11]. Responding to allegations of religious and ethnic profiling by security staff, the Airports Authority’s legal adviser told a Knesset hearing: “The inspection is not done according to population groups” [12]. Nevertheless, the Airports Authority has conceded that, given the numbers transiting the airport, “there are extraordinary events that we regret” [13].

In one ‘extraordinary event’, British-Cypriot Jack Angelides, general manager of Israeli soccer team Maccabi Tel Aviv, “travelling with a lap-top and a stack of printed reading material … was told to part with one or the other, due to unspecified security concerns” [14]. Angelides described the episode as “very unpleasant, very uncomfortable” [15]. Federman [14] observes: “[Ben Gurion’s] tough security … is stirring debate. On one side stand those concerned about Israel’s good name, tourism potential and moral standing. On the other are those for whom security arguments can seem close to sacrosanct in a country hit with decades of attacks by Palestinian terrorists, a series of hijackings in the 1960s and 70s, and whose travellers abroad are targeted in terrorist attacks” (Fig. 2).

Fig. 2 Before the storm: an El Al Lockheed constellation in 1951

5 Risk-Based Security—A Critique

Although effective in the case of Ben Gurion, risk-based security regimes are problematic. This critique draws on systems theory, specifically Challenger, Clegg and Robinson's [16] conceptualisation of a socio-technical system (Fig. 3).

Systems theory recognises the mutability of systems. It concerns itself with the system-as-found rather than the system-as-designed. It documents systems phenomena like tight coupling, interactive complexity, translation, alignment, common-mode failure, reactive patching, practical drift, safety migration and emergence. It recognises that sub-optimal employee behaviour (for example, soldiering, or use of an authority position to indulge prejudices) can degrade system performance, prevent attainment of organisational goals and damage image and reputation. Systems theory documents *lived reality* and *praxis*. It recognises that factors without a socio-technical system, for example social, economic and political forces, can impact system topography and functioning [17–25].

Perceived issues with risk-based security include:

5.1 Civil Liberties

A full-blown risk-based security system requires a database of low-risk passengers that is maintained to a high standard, in perpetuity. Passengers judged low-risk, being subject to fewer on-airport checks, are able to access airport facilities and aircraft more quickly. Civil libertarians have expressed concerns about the volume and security of the data required to build and maintain a database of low-risk

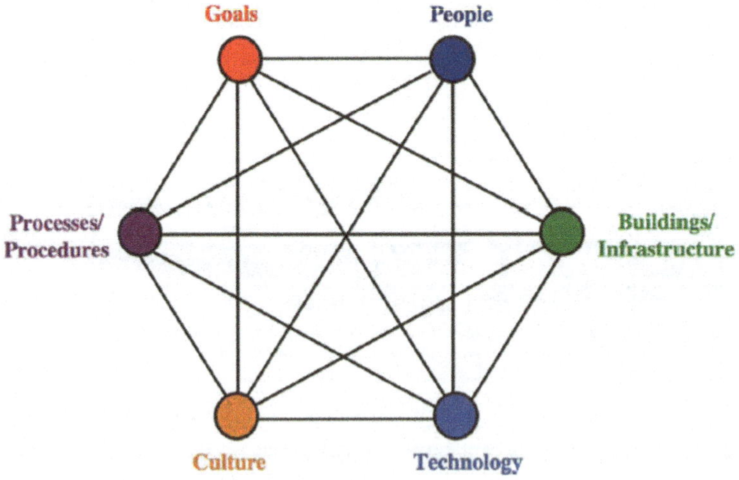

Fig. 3 Socio-technical systems are multi-faceted and dynamic [16]

passengers. The May, 2017 global ransomwear attack and confirmation by the CIA and FBI that Russia used cyber-espionage to try to influence the outcome of the 2016 US Presidential Election, speak to the vulnerability of information technology systems. The ransomwear attack affected at least 200,000 companies worldwide, including giants such as FedEx and Renault. In the UK, it compromised the majority of National Health Service computers. Given such incidents, it is reasonable to assume that *no* computer system is secure.

In November 2001, the Air Transport Association (ATA) called on Federal authorities to introduce a voluntary system of biometric ID cards for air travellers. The American Civil Liberties Union's (ACLU's) Steinhardt [26] warned that those travellers unwilling to participate in the scheme "would be increasingly subjected to intrusive, humiliating and time-consuming searches—in short, treated like second-class citizens". He concluded: "In a short time, Americans will be, for all practical purposes, forced to acquire a card and to submit to whatever procedures are used to issue them" [26]. Steinhardt highlighted an unpalatable truth—that governments can make life for those who refuse to participate in voluntary ID schemes so uncomfortable that non-conformity is no longer a realistic option.

5.2 *Prejudice*

There are concerns that risk-based security provides an opportunity for the expression of prejudice. Specifically, it would be possible for an immigration official to assess risk not on the basis of an objective analysis of a prospective passenger's profile, but on the basis of a passenger's dress code, skin colour, native tongue, accent, country of birth, non-verbal communication or demeanour. A risk-based security system would offer prejudiced police officers, immigration officers, airport and airline staff the opportunity to express those prejudices. Prejudice is widespread. It afflicts every profession, including academia (where it generally manifests as anti-Semitism) and security (where, in the West, it generally manifests as racism or Islamophobia).

Systems theory teaches that culture, which may be defined as 'the way we do things around here', is a key element in any system, influencing how it performs and how it is perceived by funders/shareholders and users/customers. A work ethic contaminated by prejudice and/or incompetence represents what Professor James Reason [23] would call a latent error or resident pathogen. It is a liability, especially in countries whose legal codes proscribe discrimination on the basis of race or religion.

5.3 *Soldiering and Satisficing*

The consequences of soldiering ("deliberately working slowly so as to avoid doing a full day's work" [27]) and satisficing ("a process whereby employees settle for adequate rather than optimal results" [28]) in the operationalisation of a risk-based security system could be catastrophic. While soldiering and satisficing are damaging in any environment, they are especially dangerous traits in aviation, because the consequences of not getting it right every time could be severe (for example, the loss of hundreds of lives and disruption of the global aviation system). Assessing a possibly fractious and abusive passenger, at the head of a long line of passengers, in a hot and noisy terminal building, on the basis of information presented on a small screen, demands of an official significant mental stamina, physical endurance and dedication to the security mission. Sub-optimal performance on the part of the official performing the risk-based assessment would create in the security system a latent error or resident pathogen (see [23] for a definition), that could morph into an active error (a security lapse). Booming passenger numbers and pressure to reduce costs put security systems and those who administer them under pressure. Because it speaks to the problem of having to do more with less, Hollnagel's [29] theory of a trade-off between efficiency and thoroughness (his efficiency-thoroughness trade-off (ETTO) principle) is relevant to aviation security.

In 2004, the British Broadcasting Corporation (BBC) aired a documentary that examined security at Manchester Airport [30], an international facility in the north of England. The BBC uncovered evidence of satisficing amongst security staff: "Evidence provided by the [BBC] journalist's hidden camera and tape-recorder confirmed the following: (1) The falsification of records by some security officers and security team leaders; (2) Inadequate staffing levels at security points; (3) Malfunctioning, non-functioning, or unavailable security equipment (including walk-through metal detectors and hand-held scanners); (4) A lack of security awareness by security staff (team leaders sometimes bypassed security checkpoints); (5) Lax airside security that allowed relatively easy access to parked aircraft (door seals were sometimes not applied, stairs were left at aircraft doors and aircraft were left unlocked overnight)" [31]. The BBC journalist established:

(a) that security staff overstated the number of bags searched, and
(b) that some supervisors were complicit in the falsification of records.

Said one supervisor: "It's just a paper exercise" [30]. These behaviours evidenced both satisficing and practical drift, defined by Snook [20] as "[T]he slow, steady uncoupling of local practice from written procedure". Practical drift happens when: "… locally practical actions within subgroups gradually drift away from originally established procedures" [20]. According to Bennett [32], employees' belief that their primary goal was throughput (rather than thoroughness) influenced the *praxis* of security: "Employees' perception that management prioritised throughput influenced their behaviour. Processing passengers quickly was perceived to be more important than adhering to rules and regulations".

Systems theory confirms the importance of investing in quality components. The human element is crucial in any system. Employees shape an organisation's culture [19], which in turn influences how it performs. An organisation's culture can either frustrate or facilitate the attainment of corporate goals. An organisation whose employees demonstrate integrity and application will do better than an organisation whose employees—for whatever reason(s)—demonstrate indifference and incompetence. Synergy is more likely in an organisation whose employees demonstrate integrity and application than in an organisation whose employees demonstrate indifference and incompetence.

5.4 False Negatives

'Never judge a book by its cover'. 'Appearances can be deceptive'. 'All that glisters is not gold'. These and similar pithy aphorisms speak to the problem of the misreading of cues and substitution of effortful, *in loco* fact-based assessments with effortless, prejudice-informed assessments. Not all Daesh recruits/sympathisers are of the Muslim faith. Not all Daesh recruits/sympathisers are non-white. Consider the case of Samantha Lewthwaite, dubbed the White Widow by Britain's tabloid press. Destined to become a senior figure in Islamist terror group Al-Shabaab, Lewthwaite's early life was conventional. Samantha Lewthwaite was born into a military family, her father serving in Northern Ireland. Following her parents' separation, Lewthwaite converted to Sunni Islam—not in itself an indication of subversive intent. She was seventeen. Lewthwaite studied politics and religion at London University. She lived in Aylesbury, a leafy, prosperous Buckinghamshire town.

Her life changed direction when she married Germaine Lindsay (also known as Abdullah Shaheed Jamal). Lindsay participated in the July 7, 2005 London bombings (that killed 52 people and injured nearly 800). Following the 7/7 attacks, Lewthwaite fled the country, attracting an Interpol Red Notice. Lewthwaite's mostly conventional upbringing camouflaged her murderous intentions. In security terms, Lewthwaite was a false negative.

The case of Samantha Lewthwaite reconfirms the dangers of stereotyping. While risk-based security brings advantages, such as reducing the time it takes passengers to negotiate on-airport checks, it also brings problems, such as allowing those who sympathise with Islamist terror groups, but who have done nothing to draw attention to themselves, relatively easy access to aviation infrastructure. If the assumptions that underpin risk-based security are incorrect, the system fails. Assuming that all intelligent, university-educated, prosperous Home Counties Caucasians with United Kingdom passports are on-side in the War on Terror creates in our aviation security system what Professor James Reason [23] would call a latent error or resident pathogen. Erroneous assumptions create weaknesses—blind-spots or lacunae—that give terrorists room to work. Connectivity (globalisation) ensures that system weaknesses are instantly propagated, creating multiple opportunities for off-the-radar terrorists. Mindful of Reason's Swiss Cheese model of system failure [23], an accumulation of erroneous assumptions, in combination with pathogenic employee behaviour (soldiering and satisficing), could degrade a security system to the point where it is no longer effective.

5.5 Capital and Revenue Costs

To be effective, a risk-based security system requires state-of-the-art information technology, a data set that is up to date and accurate, and staff who are motivated, resilient and trained to correctly interpret passenger responses and behaviour. Systems and staff require support (e.g. hardware maintenance, software development, initial and recurrent training) *in perpetuity*. The axiom 'garbage in, garbage out' is pertinent to risk-based security systems.

A 2015 evaluation of the US Transport Security Administration's risk-based security system PreCheck [7] suggests that a proficiently-managed and well-funded risk-based security system can deliver both cost savings and an improved public image: "[C]urrently [circa 2014] around 45% of air travelers in the United States (more than 276 million in 2014) are receiving expedited screening. This growth in PreCheck has allowed TSA to reduce its screener workforce, and the agency has estimated that these staffing efficiencies will allow it to save at least $100 million in the current fiscal year". Regarding public image, the evaluation notes: "The rapid early growth of PreCheck has undoubtedly been a success story for TSA, one that has improved the agency's image with Congress and with the general public" [7].

5.6 Markers Perceived to Be Discriminatory or Unlawful

A risk-based security system uses markers to create a risk profile for each passenger. Markers can be low or high-fidelity. A low-fidelity marker might reference a passenger's country of birth. A high-fidelity marker might reference a passenger's international travel history. Markers perceived by the public to be discriminatory may cause reputational damage to organisations and their bureaucratic and political masters. On January 27, 2017, US President Donald Trump signed an executive order denying Iraqi, Syrian, Iranian, Sudanese, Libyan, Somalian and Yemeni passport holders entry into the United States [33]. Because it focused on countries with Muslim-majority populations, some considered Trump's executive order (that articulated an election promise to 'Keep our country safe') discriminatory. The executive order impacted airlines as well as travellers. Consequences for the world's airlines included:

(a) Lost business

Measures that reduce load factors are bad for the airline industry.

(b) Roster disruption

As global employers, international airlines do not discriminate on the basis of nationality. An applicant's competence determines whether s/he is hired. German carrier Lufthansa prides itself on the multi-culturalism of its long-haul crews. A representative from the German cabin crew union Unabhängige Flugbegleiter Organisation (UFO) commented: "Lufthansa has always ensured it has very diverse crews, with staff of different nationalities, and that means that we are, for the first time in decades, having to look at where people come from" [34]. According to aviation consultant John Strickland [35], Trump's ban kept airline rostering departments busy: "I cannot think of anything comparable. This brings a mix of administrative confusion, impact and uncertainty for many travelers as well as practical operational headaches and complexities for airlines in planning their flight programs". Emirates, the world's largest long-haul carrier with roughly 23,000 flight attendants and 4,000 pilots, faced a major challenge in adjusting to the new reality.

Reaction

Political reaction to the travel ban was swift. Affected by the executive order, Iran retaliated by banning all US citizens from entering the Islamic Republic. Although not included in the ban, Indonesia, home to the world's largest Muslim population, reacted strongly. "We have deep regrets about the policy" said Indonesia's foreign minister. Germany, France and Canada were critical of the executive order [36].

The Trump executive order saga provokes five observations:

(a) Risk-based security markers that are perceived to be discriminatory may cause a political backlash
(b) Risk-based security markers that are believed to be unconstitutional may be challenged in the courts
(c) Risk-based security markers that affect airline employees disrupt international airline rosters, incurring costs
(d) Risk-based security markers that are too coarse/broad may reduce passenger numbers, affecting load factors and profitability
(e) Ideology-inspired interventions in matters pertaining to national security may produce unintended and undesirable outcomes. Reading Trump's January 27 executive order as unconstitutional, civil libertarians protested at airports (Fig. 4). Following a series of high profile legal challenges and interventions (for example, by former US President Barack Obama), the Trump administration backed down. On March 6, 2017 the administration introduced a revised travel ban that "excluded Iraq from the list of Muslim-majority countries whose citizens were temporarily blocked" [37]. This second ban was blocked on March 15, 2017, "US District Court Judge Derrick Watson … concluding that the order likely violates the Establishment Clause of the Constitution by

Fig. 4 Do public perceptions of risk-based security markers matter? Post-ban protest at a US airport

> disfavouring Muslims" [37]. The President framed Judge Watson's ruling as "an unprecedented judicial overreach" [38]. During the early stages of the executive order saga, Trump had denigrated the judiciary. Sages told the President that, for the sake of public order, it was unwise for the head of state to question the legitimacy of the judiciary, however unhelpful its decisions.

Trump began problematising immigration some two years before his ill-fated executive order. At a December 7, 2015 campaign rally in Mount Pleasant, South Carolina, candidate Trump mooted a ban on Muslim immigration: "Donald J. Trump is calling for a total and complete shutdown of Muslims entering the United States until our country's representatives can figure out what the hell is going on" [39].

In 1968, systems thinker Horst Rittell [40] observed: "Design is an activity, which aims at the production of a plan, which plan—if implemented—is intended to bring about a situation with specific desired characteristics without creating unforeseen and undesired side and after effects". Judged against Rittell's measure of good design, the Trump administration's 2017 security model was a design failure. First, the new measures were overturned in the courts not once, but twice. Secondly, the new measures drew criticism at home and abroad from all strata. Thirdly, any impediment to the free movement of talent and ideas is economically, as well as politically disruptive. To thrive, corporations must be able to hire the best talent without impediment. The ban undermined the Trump Administration's efforts to restore the fortunes of the US economy. It foregrounded the new Administration's lack of coherence on policy.

6 Discussion and Conclusions

The terror attacks of September 11, 2001, in which jihadists used commercial airliners as flying bombs, provoked new thinking in aviation security circles [41]. While airborne kamikaze attacks had been mounted by the Japanese in World War Two (Fig. 5), never before had such attacks been mounted by terrorists against civilians. The 9/11 watershed provoked high and low-fidelity responses. For example:

- the industry-wide installation of armoured, lockable flight-deck doors (low-fidelity)
- the banning of sharp objects from the cabin (low-fidelity)
- the introduction of risk-based passenger vetting systems such as the TSA's PreCheck scheme (high-fidelity).

In their efforts to 'design-out' terrorism, western nations found themselves having to play catch-up with countries like Israel. Finding itself a focus of international terrorism [42], Israel improved its aviation security. Innovations included armed sky marshals and steel-reinforced cockpit doors [43]. At Ben Gurion, innovations like security sweeps remote from the terminal area (physical defence-in-depth) and risk-based passenger vetting have delivered over four decades of incident-free operation.

Fig. 5 "Those who fail to learn from history are doomed to repeat it" observed Winston Churchill. On 14 May, 1945, this kamikaze fighter dived into the aircraft carrier USS enterprise, killing fourteen

Following the kamikaze attacks of September 11, 2001, the United States overhauled its aviation security system. There were bureaucratic innovations (for example, the Transportation Security Administration, established on November 19, 2001) and on-airport security innovations such as PreCheck. If we look at efforts to design-out terrorism post-9/11 through a systems-thinking lens, a number of lessons can be drawn.

First, for the system to deliver the best results, every element must be perfected. In the case of risk-based security, this means capital investment in hardware, software and liveware, and revenue investment in software and hardware maintenance, upgrades, data gathering, recruitment and training. The importance of hiring capable, committed staff cannot be overstated. Employee attitudes and behaviours shape organisational culture, which influences organisational performance [19, 44].

Secondly, a risk-based security system must be socially, as well as politically acceptable. Systems that are perceived to be discriminatory or overly-intrusive risk losing the support of those they are designed to protect, with unpredictable consequences. Witness the disruptive protests at US airports following President Trump's January 27, 2017 executive order concerning immigration markers. Systems theory reminds us that factors without a system can influence its structure, organisation and functioning. To a greater or lesser degree, systems have a reciprocal relationship with the environment in which they operate [45]. In 2017, several factors, including court decisions, foreign leaders' opprobrium, retaliation, mass protests and negative publicity, bore down upon America's re-worked system of passenger vetting. At the time of writing it was unclear whether the Trump administration's reactive patching (see [46] for a definition) of its passenger vetting policies would appease the US courts and public.

Thirdly, a risk-based security system is only as good as the intelligence that feeds its risk-assessment algorithms. Actor-network theory (a forensic systems theory) [47, 48] shows us how the quality of intra and inter-system co-ordination influences system effectiveness. In the argot of ANT, effectiveness is partly a function of actant and systems *translation* and *alignment*. A risk-based security system is most effective:

(a) when its component elements complement and support one another, generating synergies
(b) when it is respected by the public it serves. Respect bestows legitimacy
(c) when it is effectively supported by third parties such as the security and intelligence services, the Legislature and the Executive.

A 2007 systems-thinking-informed analysis of events leading up to 9/11 [49] suggested that failings within, and friction between, America's security agencies may have:

(a) led to an incorrect framing of the threat posed by global terror networks such as Al Qaeda
(b) stymied or distorted the flow of intelligence prior to 9/11.

Fig. 6 The consequences of intelligence gridlock ('bad design') and inadequate security can be catastrophic: Ground Zero

The August, 2007 publication *Office of the Inspector General Report on CIA Accountability With Respect to the 9/11 Attacks* [49] talked of "systemic problems". For example: "Agency officers … did not always work effectively and co-operatively …. [There were] failures to … properly share and analyse critical data. If IC [intelligence community] officers had been able to view and analyse the full range of information available before 11 September, 2001, they could have developed a more informed context in which to assess the threat …". Prior to 9/11, networking between the CIA, the FBI, the Immigration and Naturalisation Service (INS) and the State Department was not as good as it could have been [50, 51]. Applying an actor-network lens, the intelligence community suffered from *weak actant translation and alignment*. As 9/11 demonstrated, in the world of security, the consequences of sub-optimal actant translation and alignment can be catastrophic (Fig. 6).

References

1. National Consortium for the Study of Terrorism and Responses to Terrorism (2016) Country reports on terrorism 2015, national consortium for the study of terrorism and responses to terrorism. University of Maryland, MD
2. World Health Organisation (2016) World Health Statistics 2016. World Health Organisation, Geneva
3. Air Transport Action Group (2017) Social and economic benefits of aviation. Available at: http://www.atag.org/our-activities/social-a-economic-benefits-of-aviation.html. Accessed 10 May 2017
4. European Union Council cited in Meloy JR, Roshdi K, Glaz-Ocik J, Hoffmann J (2015) Investigating the individual terrorist in Europe. J Threat Assess Manag 2(3–4):140–152
5. Federal Bureau of Investigation (2001) Code of Federal Regulations (28 CFR Section 0.85) relating to terrorism
6. McLay LA, Lee AJ, Jacobson SH (2010) Risk-based policies for airport security checkpoint screening. Transp Sci 44(3):333–349
7. Beckner C (2015) Risk-based security and the aviation system: operational objectives and policy challenges, Centre for Cyber and Homeland Security. The George Washington University, Washington DC
8. Blair D (2016) Israel's risk-based approach to airport security 'impossible' for European airports. Available at: http://www.telegraph.co.uk/news/2016/05/20/. Accessed 11 May 2017
9. Zakay cited in Blair D (2016) Israel's risk-based approach to airport security 'impossible' for European airports. Available at: http://www.telegraph.co.uk/news/2016/05/20/. Accessed 11 May 2017
10. Price cited in Federman J (2014) Stern security at Ben-Gurion airport questioned. Available at: http://www.timesofisrael.com/. Accessed 11 May 2017
11. *Haaretz* cited in Federman J (2014) Stern security at Ben-Gurion airport questioned. Available at: http://www.timesofisrael.com/. Accessed 11 May 2017
12. Shaham cited in Federman J (2014) Stern security at Ben-Gurion airport questioned. Available at: http://www.timesofisrael.com/. Accessed 11 May 2017
13. Airports Authority cited in Federman J (2014) Stern security at Ben-Gurion airport questioned. Available at: http://www.timesofisrael.com/. Accessed 11 May 2017
14. Federman J (2014) Stern security at Ben-Gurion airport questioned. Available at: http://www.timesofisrael.com/. Accessed 11 May 2017
15. Angelides cited in Federman J (2014) Stern security at Ben-Gurion airport questioned. Available at: http://www.timesofisrael.com/. Accessed 11 May 2017
16. Challenger R, Clegg CW, Robinson M (2010) Understanding crowd behaviours, vol 1: practical guidance and lessons identified. The Stationery Office, London
17. Perrow C (1983) The organizational context of human factors engineering. Adm Sci Q 28(4): 521–541
18. Perrow C (1984) Normal accidents. Basic Books, New York
19. Vaughan D (1997) The challenger launch decision. Risky technology, culture and deviance at NASA. University of Chicago Press, Chicago
20. Snook S (2000) Friendly fire: the accidental shootdown of U.S. Black Hawks over Northern Iraq. Princeton University Press, Princeton, NJ
21. Johnson CW (2005) What are emergent properties and how do they affect the engineering of complex systems? Department of computing science. University of Glasgow, Glasgow
22. Black J, Koopman P (2009) System safety as an emergent property in composite systems. Carnegie Mellon University, Pittsburgh, PA
23. Reason JT (2013) A life in error. Ashgate Publishing Ltd., Farnham
24. Hollnagel E, Leonhardt J (2013) From safety-I to safety-II: a white paper, European organisation for the safety of air navigation, Brussels

25. Shorrock S, Leonhardt J, Licu T, Peters C (2014) Systems thinking for safety: ten principles. Eurocontrol, Brussels
26. Steinhardt B (2001) ACLU says 'traveller's IDs' for airline passengers will create dangerously false sense of security. Available at: https://www.aclu.org/news/aclu-says-travelers-ids-airline-passengers-will-create-dangerously-false-sense-security. Accessed 11 May 2017
27. Taylor FW (1911) Principles of scientific management. Available at: https://www.marxists.org/reference/subject/economics/taylor/principles/ch01.htm. Accessed 13 May 2017
28. Bennett SA (2015) The benefits of a systems-thinking approach to accident investigation. In: Applications of systems thinking and soft operations research in managing complexity. Springer, pp 203–226
29. Hollnagel E (2009) The ETTO principle: efficiency-thoroughness trade-off. Why things that go right sometimes go wrong. Ashgate Publishing Ltd., Aldershot
30. British Broadcasting Corporation (2004) Whistleblower: Manchester Airport. [television broadcast] BBC Channel 1, 7 September 2004, 21:00
31. Bennett SA (2009) Aviation security: at risk from practical drift? In: Protecting airline passengers in the age of terrorism. Praeger Security International, pp 33–62
32. Bennett SA (2015) Aviation safety and security: the importance of teamwork, leadership, creative thinking and active learning. Libri Publishing Ltd., Faringdon
33. Cornwell A (2017) Emirates changes pilot, crew rosters on US flights after Trump order. Available at: http://www.reuters.com/article/us-usa-trump-immigration-emirates-idUSKBN15D0JM. Accessed 17 May 2017
34. Baublies cited in Cornwell A (2017) Emirates changes pilot, crew rosters on US flights after Trump order. Available at: http://www.reuters.com/article/us-usa-trump-immigration-emirates-idUSKBN15D0JM. Accessed 17 May 2017
35. Strickland cited in Cornwell A (2017) Emirates changes pilot, crew rosters on US flights after Trump order. Available at: http://www.reuters.com/article/us-usa-trump-immigration-emirates-idUSKBN15D0JM. Accessed 17 May 2017
36. VOA News (2017) International reaction to Trump's refugee ban begins. Available at: http://www.voanews.com. Accessed 17 May 2017
37. Almasy S, Simon D (2017) A timeline of President Trump's travel bans. Available at http://www.cnn.com. Accessed 17 May 2017
38. Trump cited in Almasy S, Simon D (2017) A timeline of President Trump's travel bans. Available at http://www.cnn.com. Accessed 17 May 2017
39. Trump cited in Johnson J (2015) Trump calls for 'total and complete shutdown of Muslims entering the United States'. Available at: http://washingtonpost.com. Accessed 18 May 2017
40. Rittell cited in Lee AJ (2016) Resilience by design. Springer, Cham
41. Tamasi G, Demichela M (2011) Risk assessment techniques for civil aviation security. Reliab Eng Syst Saf 96:892–899
42. Zalman A (2017) 1968 PFLP hijacking of El Al Flight. Available at: https://www.thoughtco.com/1968pflp hijacking of el al flight 3209270. Accessed 20 May 2017
43. Ginsburg M (2014) How to thwart a gunman at 29,000 feet, by the only pilot who ever did. Available at: http://www.timesofisrael.com/how to defeat airplane terrorists from the only pilot who ever foiled a skyjacking. Accessed 20 May 2017
44. Armstrong M (1996) A handbook of personnel management practice, 6th edn. Kogan Page, London
45. Payne R (2002) Organisations as psychological environments. In: Psychology at work, 5th edn. Penguin, pp 375–398
46. Weir DTH (1996) Risk and disaster: the role of communications breakdown in plane crashes and business failure. In: Debates in risk management. UCL Press, pp. 114–126
47. Callon M, Latour B (1981) Unscrewing the big Leviathan: how actors macro-structure reality and how sociologist help them to do so. In: Advances in social theory and methodology: towards an integration of micro and macro-sociology. Routledge, pp 277–303

48. Callon M, Law J (1997) After the individual in society: lessons on collectivity from science, technology and society. Can J Sociol 22(2):165–182
49. Office of the Inspector General (2007) Office of the inspector general report on CIA accountability with respect to the 9/11 attacks, office of the inspector general, Langley VA
50. Freedman D (2010) What Really Led to 9/11. Available at: https://www.forbes.com/2010/09/08/september-11-al-qaeda-government-opinions-columnists-daniel-freedman.html. Accessed 23 May 2017
51. National Commission on Terrorist Attacks Upon the United States (2004) The 9/11 commission report, national commission on terrorist attacks upon the United States, Washington DC

Customizing Web-EOC Crisis Management Software to Facilitate Collaboration and Increase Situational Awareness: Advances in Business Resource Center (BRC) Design for Business Continuity Management

Jason Levy and Ross Prizzia

Abstract The Hawai'i Emergency Management Agency embarked upon a project with the University of Hawai'i at West O'ahu to determine the best design and use of a Business Resource Center (BRC) to increase the disaster management collaboration among the public, private and non-governmental sectors. The visitor industry was identified as the first sector that will be engaged. The project includes five major tasks: creation of a BRC Annex and Standard Operating Guide; the development of a BRC WebEOC portal; the design of small businesses preparedness materials; the identification of emergency communications requirements; and emergency management outreach, training and exercises. This paper focuses on the second task listed above: development of a BRC WebEOC portal for effective disaster management collaboration, information sharing and situational awareness. This involves the design of data fields for the WebEOC crisis management system. Finally, focus groups in Hawaii are examined to learn more about current levels of disaster engagement, gaps in disaster information or services, and opportunities for increased success with disaster management through a central BRC. In general, there was a strong sense of support for the development of a collaborative BRC space for the public and private sectors. Investigations pertaining to business disruptions are undertaken with an emphasis on modeling, understanding and characterizing the complex causality that defines these incidents. In so doing this chapter uncovers creative, timely and important strategies improving business continuity management and disaster resilience in Hawaii.

Keywords Business continuity · WebEOC · Emergency preparedness
Vulnerability · Business resource center · Supply chain disruption

---

J. Levy (✉) · R. Prizzia
Disaster Preparedness and Emergency Management Program,
Public Administration Division, University of Hawaii-West Oahu,
91-1001 Farrington Highway, Kapolei, HI 96707, USA
e-mail: jlevy@hawaii.edu

A. J. Masys (ed.), *Security by Design*, Advanced Sciences and Technologies for Security Applications, https://doi.org/10.1007/978-3-319-78021-4_14

1 Disaster Resilience and Business Continuity Planning in Hawaii

As an isolated archipelago, the US island state of Hawaii is susceptible to a range of hazards since much critical infrastructure is located in vulnerable coastal regions. Hawaii is also heavily dependent on imported goods (from fuel and food to consumer goods) and transportation costs are extremely high. Accordingly, supply chain disruptions could have severe ramifications in Hawaii. Many businesses in Hawaii are extremely small. In particular, 86% of Hawai'i businesses have less than twenty employees [2] while 94% of Hawaii's companies have fewer than fifty workers. Given the dominant role of the visitor industry in Hawaii's economy, the closure of airports and harbors to tourists would have significant impacts across multiple sectors. Harbors on the island of O'ahu are also needed to provide critical resources and supplies to the neighboring islands of Maui, Kauai, Hawaii, Molokai and Lanai islands (and smaller islands). A majority of businesses in Hawaii lack business interruption insurance and many possess no disaster coverage at all. This is a major source of vulnerability since many small businesses anticipate insurance as their primary means of disaster recovery assistance. Other businesses may employ workers that use English as a second language (or do not speak English at all). This can lead to a plethora of challenges including difficulty finding recovery assistance and resources; they may also find it difficult to complying with documentation requirements for insurance claims. Particularly vulnerable are businesses in Hawaii with less than 20 employees and those that work from home [2]. According to Hawaii's Department of Business Economic Development [3] 38% of businesses surveyed in Hawaii (primarily companies with fewer than 25 workers) did not have a Business Continuity Plan.

Larger businesses tend to fare better than smaller businesses in the event of a disaster, due to their increased access to resources and economy of scale. Smaller businesses may be devastated by disaster as they typically have fewer financial assets (line of credit, cash reserves, capital assets, short term operating funds) and may lack an extended network of employees and technical services located outside of the area of impact (to provide response assistance). On the other hand, large corporations such as Starbucks can rely on extensive emergency plans, including the use of backup generators, activation of internal emergency operation centers, and the option to switch computing control to locations outside the region [4]. There are a number of reasons why it is essential for smaller businesses to implement disaster recovery plans: they often are located in a single site (compared to more spatially distributed enterprises); they typically depend on a small number of key employees and are more likely to suffer temporary, or even permanent closure, after a challenging disaster [6, 7].

The smallest of Hawaii's businesses lack knowledge about disaster resilience and may be poorly informed about available disaster recovery resources or the benefits of disaster preparedness. For example, emergency management training in

Hawaii is not common for smaller enterprises and many small businesses are entirely dependent on others for assistance, such as land owners or property managers for recovery. In addition many small businesses do not have emergency supplies at their office location and lack off-site data back-ups [3]. A range of institutional and social pressures encourage small businesses to remain at their same location, even though the company may face grave risks [5, 8]. Yoshida et al. [9] found that small businesses are acutely lacking in disaster preparedness and planning knowledge, particularly in reference to insurance. It has been shown that retail businesses that rent space are more vulnerable than owners [1, 7]. There are a range of factors that increase the probability of small business recovery, including experience with hazards, emergency preparedness and the degree of financially stability. The rising intensity and frequency of disasters and the location of critical assets in vulnerable geographic locations can increase business risks.

The State of Hawaii Emergency Management Agency (HIEMA), formerly the "State Civil Defense" agency, supports County and State Agencies in their efforts to respond and recover from natural and man-made disasters. To support emergency preparedness for small businesses in Hawaii the twenty-eighth Hawaii State Legislature enacted, and the Governor of Hawaii signed, House Bill (HB) 1343 that appropriates $328,000 of state funds. This was matched by $1.3 million US dollars in Federal funds from the US Department of Commerce over the three year life of the project (2016–2018). The Business Resource Center (BRC) seeks to address emergency preparedness and disaster recovery needs as defined local businesses and to facilitate operational coordination and information sharing between emergency management and private industry. The Construction component of the project involves the renovation of approximately 2000 ft^2 of space sited within HI-EMA's Building 303 located at 4204 Diamond Head Road, Honolulu, HI within the Diamond Head State Monument. The property is owned fee simple by the State of Hawaii. The constructed Diamond Head space will house the Business Resource Center.

The authors of this paper are helping the Hawaii Emergency Management Agency (HEMA) to undertake an initiative called the Business Resource Center (BRC) to formalize collaboration between the private sector and emergency management. The visitor industry was identified as the first sector that will be engaged. The project includes five major tasks:

1. Creation of a BRC Annex and Standard Operating Guide
2. Development of a BRC WebEOC portal
3. Small businesses preparedness materials
4. Emergency communications requirements
5. Outreach, Training and Exercises.

This paper focuses on the second task listed above: creation of a BRC WebEOC portal to enhance information sharing, collaboration and situational awareness. This involves identifying new fields for the WebEOC crisis management system.

2 Business Recovery and Information Dissemination

Developing a plan is not the final step in the business continuity planning process. It is essential to inform all members of a business and the surrounding community about the plan that has been developed. The contents of the disaster response plan must also be communicated clearly to employees and the community so that they understand the contents of the plan including the location of emergency supplies, procedures for obtaining updates on a disaster situation, what they are expected to do in an emergency (such as how to evacuate or shelter-in-place) and where to obtain notifications on whether or not to report for work. The BRC will use a number of methods to ensure that this information is disseminated to relevant parties. For example, the BRC can help companies to ensure that new employees are informed of disaster plans during their orientation and that periodically thereafter employees are trained on their responsibilities in a disaster. Accordingly, business continuity information should be part of new employee orientation, manager training, and regularly scheduled meetings.

The BRC will work with Economic Development Organizations (EDOs), Chambers of Commerce, business leaders, Small Business Development Centers (SBDC), Community Development Financing Institutions (CDFIs), academics and other organizations to disseminate important business recovery information ranging from business interruption insurance to continuity of operations planning. It is expected that the BRC will hold workshops at convenient times in local communities to allow the maximum number of small business owners and community members to participate. Webinars, conferences and other vehicles for training and outreach should be employed. Each member of the Hawaii Business Recovery Center's Business Recovery Team will likely have their own network of community stakeholders that should be engaged for information dissemination purposes.

The BRC will help businesses in Hawaii, both large and small, to regularly execute business recovery and continuity of operations plans. The BRC will encourage business owners to plan for resiliency in the event of a large disruption through the provision of free workshops and/or webinars that provide tools that small businesses can use to carry out a comprehensive situational assessment of what organizational processes and procedures will be affected by a disaster. This training will consider the fact that small business owners have limited time to engage in such activities. The BRC can help organizations in Hawaii to undergo a formal evaluation of their business continuity processes and procedures: there are extensive approval processes and qualifying protocols to ensure that Hawaii's businesses achieve excellence in Business Continuity Management (BCM), including an onsite inspection of an organization's BCM program, credentials, ethical practices, and training records. In this way the BRC will help Hawaii's businesses to not only plan for the unexpected and uphold the highest standards of business recovery planning but also to become resilience leaders within their industry.

2.1 Disaster Preparedness Training and Financial Incentives

Exercises constitute a valuable approach to identify any flaws in business continuity planning. Even a well thought-out disaster response can be found to be missing important pieces: exercises ensure that disaster responses work as intended and provide an opportunity to reinforce training and expected response roles and identify any needed changes. It is recommended that exercises include all key employees so that everyone can see and hear what actions others will be taking. The BRC will work with management and employees to discuss how they and their business leaders would react to hypothetical situations associated with an incident. A schedule of training/exercises should be included in business continuity plans (and adjusted following every training session). The BRC will address disaster preparedness training in many ways. For example, members of the Business Recovery Team can attend company meetings to discuss business recovery plans, train employees and community members on their roles and responsibilities, and answer questions. In particular, "tabletop" exercises can be scheduled to "talk" through various scenarios to ensure employees are clear on their roles and responsibilities and expected responses under various circumstances. "Walk-throughs" or "Functional Drills" allow employees to physically act out their roles and responsibilities. (Evacuation drills fall into this category.) The BRC will seek to help employees to become familiar with the following key procedures, processes and responsibilities:

- Emergency response procedures during an incident
- Evacuation, shelter-in-place, and accountability procedures
- Individual roles and responsibilities
- Location and use of company emergency equipment
- Shutdown procedures (if there is time) of critical business equipment
- Restart of critical business equipment.

The Hawaii BRC will approach financial institutions and insurance companies about providing incentives for small businesses that successfully complete business continuity efforts. These incentives may take the form of interest rate or insurance discounts for local businesses. These discounts provides a financial incentive for local business to successfully engage in disaster preparedness activities (business continuity planning, employee training, mitigation, data back-up, etc.), which will speed recovery for the entire community. The training should address the affordability of insurance and costs associated with disaster preparedness activities.

2.2 *Business Recovery and Information Dissemination*

Developing a plan is not the final step in the business continuity planning process. It is essential to inform all members of a business and the surrounding community about the plan that has been developed. The contents of the disaster response plan must also be communicated clearly to employees and the community so that they understand the contents of the plan including the location of emergency supplies, procedures for obtaining updates on a disaster situation, what they are expected to do in an emergency (such as how to evacuate or shelter-in-place) and where to obtain notifications on whether or not to report for work. The BRC can help companies to ensure that new employees are informed of disaster plans during their orientation and that periodically thereafter employees are trained on their responsibilities in a disaster. Accordingly, business continuity information should be part of new employee orientation, manager training, and regularly scheduled meetings.

The BRC will work with Economic Development Organizations (EDOs), Chambers of Commerce, business leaders, Small Business Development Centers (SBDC), Community Development Financing Institutions (CDFIs), academics and other organizations to disseminate important business recovery information ranging from business interruption insurance to continuity of operations planning. It is expected that the BRC will hold workshops at convenient times in local communities to allow the maximum number of small business owners and community members to participate. Webinars, conferences and other vehicles for training and outreach should be employed. The BRC will encourage business owners to plan for resiliency in the event of a large disruption through the provision of free workshops and/or webinars that provide tools that small businesses can use to carry out a comprehensive situational assessment of what organizational processes and procedures will be affected by a disaster. This training will consider the fact that small business owners have limited time to engage in such activities.

2.3 *Credentialing and Tiered/Phased System of Re-entry*

Following a wide-scale evacuation, managers, owners and employees of local businesses may be restricted from returning to their property/operations depending on the nature of the incident. The longer the delay in granting individuals access safely back into to their community, the more devastating the effects on the businesses and the affected community including decreased employment opportunities for residents and a significant decline in the tax revenue base, thereby extending the recovery time. Specifically, as a result of re-entry delay, perishable inventory may spoil, employees may find alternative work and business closures can disrupt the operation of critical businesses needed to restore infrastructure, provide essential services and goods in the impacted area (i.e. food and water, gas, daycare, health services, sustainable living conditions, etc.) and jumpstart the local economy. To respond to this issue, the BRC can work with city and county emergency management personnel to develop a tiered system of re-entry following a disaster. The purpose of this tiered system is to allow for the expedited, safe, orderly return of re-entry of:

- tier 1 agencies/groups involved in emergency response and restoring normal operations following a disaster (search and rescue; emergency healthcare staff; utilities and infrastructure repair personnel; damage assessment teams; and pre-designated government staff and other critical personnel);
- tier 2 groups such as relief workers; healthcare agencies and suppliers; insurance agents; business operators such as important food and building material retailers, fuel distributors and stations, debris management, financial institutions (not allowed under tier 1). This may include financial institutions, business anchors for key industries, employers with a large workforce, and/or a major tax revenue source for the community and select businesses with unique circumstances (fragile inventory, hazardous waste, large workforce, global distribution, etc.);
- remaining tier 3 residents and business operators (not allowed under tier 2) that can prove they live, own, rent, or lease in the restricted area; licensed contractors and other repair service providers; and possibly family and friends who re-enter with an eligible resident.

While not guaranteeing re-entry, a credentialing program with ID cards or passes as well as training for local law enforcement to recognize the cards/passes (and allow access when the appropriate 'tier' is activated) can facilitate the timely re-entry of critical businesses and emergency personnel to assist in the community's recovery effort. Working proactively with the business community in advance of a disaster is important for making the re-entry plan relevant to business considerations. There should be clear procedures of securing a re-entry ID cards or passes such as requiring that only business owners, operators, and managers apply for such cards/passes for their employees. Partnering closely with emergency management and local law enforcement personnel to garner support of this plan will help to ensure its adoption by the public sector. Preparing a tiered/phased re-entry system which accommodate business re-entry needs assists community leaders and emergency personnel in administering recovery efforts in a more timely and organized manner; without a re-entry plan, local businesses may suffer further challenges to reopen in a timely manner, thereby hampering the local economic recovery engine.

Staggering business re-entry after a disaster and evacuation when conditions on the ground include damaged infrastructure (roadway, sewer, water and electricity lines) promotes safety and ensures expedited re-entry for the critical businesses. Once safety concerns are addressed, the remaining legitimate business operators should be allowed to quickly return to their community to secure property and re-establish business operations, thereby facilitating recovery and promoting the economic vitality of the local community. The BRC could create an online website for qualifying businesses to apply for re-entry vehicle passes, and learn about Hawaii's evacuation re-entry plans. The re-entry plan should indicate which businesses qualify for each re-entry tier based on the type of specific service(s) they provide in disaster response and recovery). The remainder of the chapter examines DBEDT [3] survey results which explore the following policies and issues.

3 State Legislative Action to Change a Statute or Ordinance, or Appropriate Funding (NDERS Poll Results)

The comprehensive list of strategies pertaining to "State Legislative Action to Change a Statute or Ordinance, or Appropriate Funding" are provided below based on DBEDT [3] poll results:

- A-1 Establish a Recovery Coordinator at the State level
- A-2 Create incentives for businesses with Disaster Plans
- A-3 Create subdivision improvement districts
- A-4 Appropriate public funds to support monitoring programs
- A-5 Advocate to amend the federal U.S. Agricultural Act
- A-6 Establish a disaster revolving fund to bridge loan gaps
- A-7 Establish a State "Build it Back Better" program
- A-8 Use micro-grids to improve local energy self-reliance
- A-9 Review facilities to better protect public utilities.

3.1 State/City or County Agencies and Large Business Breakout Group Responses

The composition of the "State/City or County Agencies and Large Business" breakout group for discussing issues pertaining to "State Legislative Action to Change a Statute or Ordinance, or Appropriate Funding" is provided in Fig. 1. Note that governmental agencies constitute 80% of participants this break group: in particular, Federal Agencies comprise 0% of participants, State Agencies represent 60% of participants while City and County Agencies constitute 20%. The remaining 20% participants discussing "State Legislative Action to Change a Statute or Ordinance, or Appropriate Funding" are non-governmental organizations. Here, private industry represents 20% of the focus group (with all representation

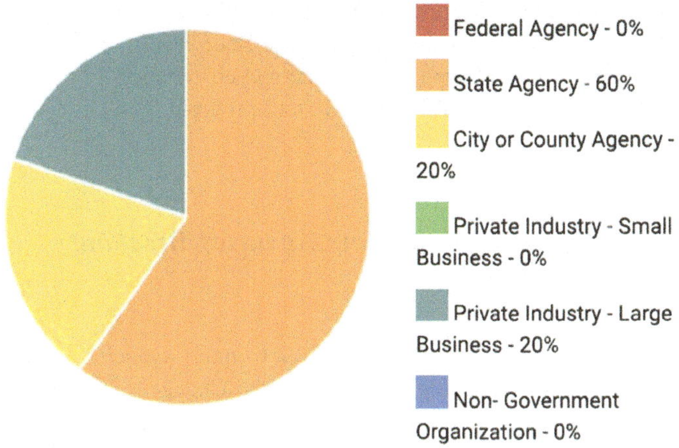

Fig. 1 Composition of the "state/city or county agencies and large businesses" focus group for issues pertaining to "state legislative action to change a statute or ordinance, or appropriate funding"

from large business). More details about the composition of the State/City or County Agencies and Large Businesses" breakout group are provided in Fig. 1.

The detailed responses of the "State/City or County Agencies and Large Businesses" focus group to "State Legislative Action to Change a Statute or Ordinance, or Appropriate Funding" issues are shown in Fig. 2. Here, agency and business preferences are broken down into low, medium and high preference categories.

For "State/City or County Agencies and Large Businesses" the top four strategies (from most to least preferred) with respect to "State Legislative Action to Change a Statute or Ordinance, or Appropriate Funding" are as follows:

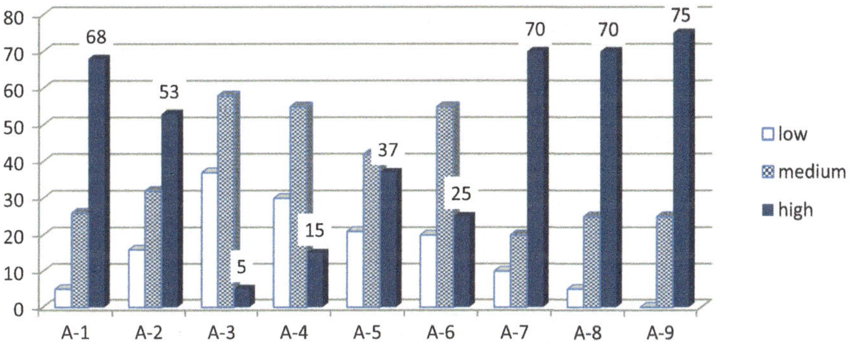

Fig. 2 Responses from state/city or county agencies and large business to "state legislative action to change a statute or ordinance, or appropriate funding" questions

- A-9 Review facilities to better protect public utilities
- A-1 Establish a Recovery Coordinator at the State level
- A-8 Use micro-grids to improve local energy self-reliance
- A-7 Establish a State "Build it Back Better" program.

3.2 Federal/State/City or County Agencies Breakout Group Responses

The composition of the "Federal/State/City or County Agencies" breakout group for issues pertaining to "State Legislative Action to Change a Statute or Ordinance, or Appropriate Funding" is provided in Fig. 3. Note that this focus group is comprised entirely of government agencies. In particular, Federal Agencies comprise 38% of participants while State Agencies constitute 13% of respondents. Finally, fully half of participants discussing "State Legislative Action to Change a Statute or Ordinance, or Appropriate Funding" in this focus group are drawn from City and County Agencies. Note that there is no representation from sectors outside the government: neither the private sector (small or large businesses) nor from non-governmental organizations are included in this focus group.

The detailed responses of the "Federal/State/City or County Agencies" focus group to "State Legislative Action to Change a Statute or Ordinance, or Appropriate Funding" issues are shown in Fig. 4, where agency preferences are broken down into low, medium and high preference categories.

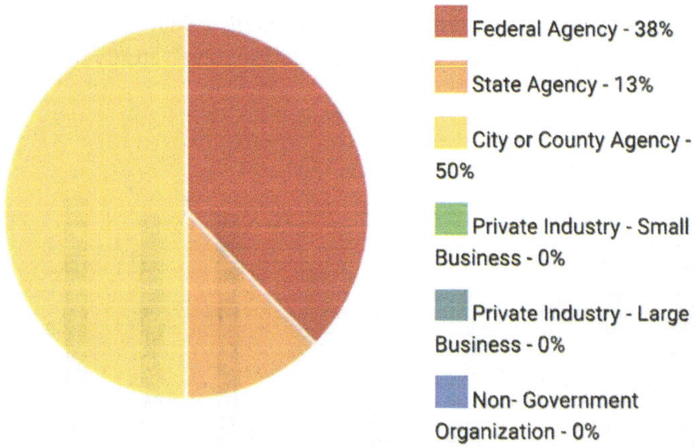

Fig. 3 Composition of federal/state/city or county agencies breakout group for issues pertaining to "state legislative action to change a statute or ordinance, or appropriate funding"

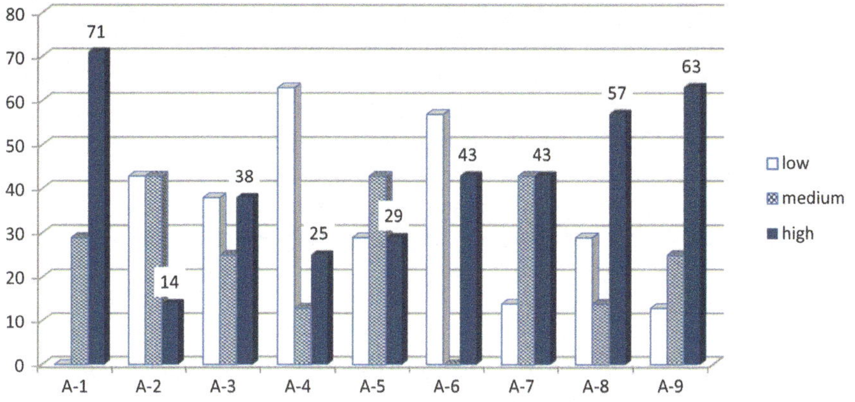

Fig. 4 Responses from federal/state/city or county agencies to "state legislative action to change a statute or ordinance, or appropriate funding" questions

For "Federal/State/City or County Agencies" the top four strategies (from most to least preferred) with respect to "State Legislative Action to Change a Statute or Ordinance, or Appropriate Funding" are as follows:

- A-1 Establish a Recovery Coordinator at the State level
- A-9 Review facilities to better protect public utilities
- A-8 Use micro-grids to improve local energy self-reliance
- A-7 Establish a State "Build it Back Better" program.

4 State or County Government Agency Action to Change Administrative Rules, Policies, or Programs (NDERS Poll Results)

The comprehensive list of strategies pertaining to "State or County Government Agency Action to Change Administrative Rules, Policies, or Programs" are provided below:

- B-1 Establish incentives for business continuity planning
- B-2 Educate businesses about insurance products
- B-3 Increase inspections to ensure preparedness compliance
- B-4 Develop marketing materials for outreach and education
- B-5 Add small business specific info to disaster resources
- B-6 Assign a State-level small business disaster coordinator
- B-7 Assist with outreach to non-English speaking businesses
- B-8 Request that UH offer a continuity planning class
- B-9 Research assistance tools in other states to use in HI
- B-10 Expand and maximize DOL grants for recovery

- B-11 Expand usage from Housing and Urban Dev. recovery fund
- B-12 Establish a State Recovery Coordinator Position
- B-13 Ensure that energy planning supports disaster recovery
- B-14 Develop unified strategies for high risk disaster areas
- B-15 Create a Food Self-Sufficiency Baseline by island
- B-16 Keep records of local food production and consumption
- B-17 Ensure warning and alert systems are working and heard
- B-18 Develop a marketing strategy targeting small businesses.

4.1 Federal/State/City or County Agencies and Small/Large Businesses

The composition of the "Federal/State/City or County Agencies and Small/Large Businesses" breakout group for issues pertaining to "State or County Government Agency Action to Change Administrative Rules, Policies, or Programs" are provided in Fig. 5. Note that governmental agencies constitute 45% of this focus group: in particular, Federal Agencies comprise 10% of participants in this breakout group, State Agencies represent 25% of participants while City and County Agencies constitute 10%. Non-governmental organizations make-up the remaining participants discussing "State Legislative Action to Change a Statute or Ordinance, or Appropriate Funding". In particular, the private sector represents 35% of the focus group (with 5% representation from small business and 30% representation from large business). Finally, 20% of the focus group comes from non-governmental organizations.

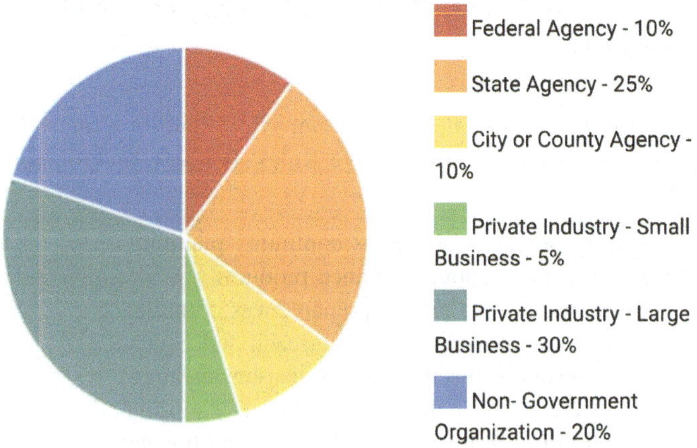

Fig. 5 Composition of "state/city or county agencies and small/large businesses" focus group for issues pertaining to "state legislative action to change a statute or ordinance, or appropriate funding"

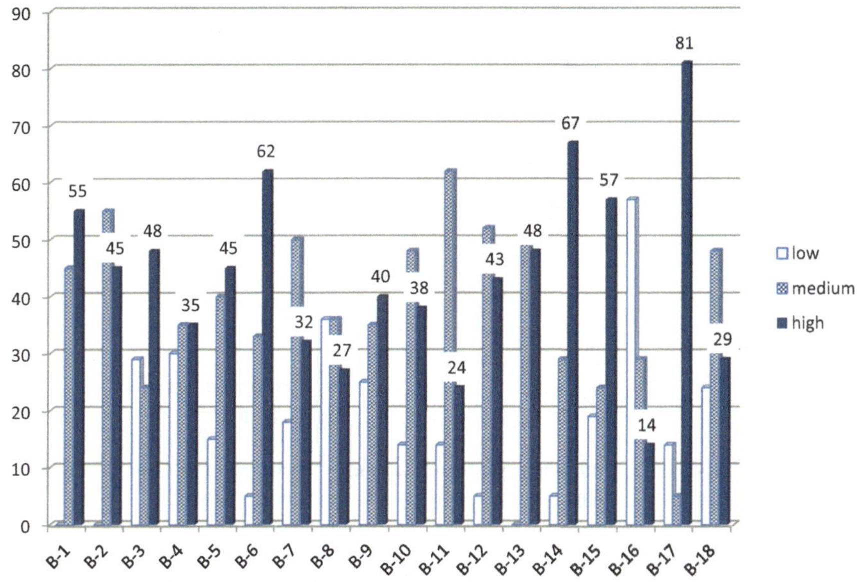

Fig. 6 Responses of the "state/city or county agencies and small/large businesses" focus group for issues pertaining to "state legislative action to change a statute or ordinance, or appropriate funding"

The detailed responses of the "State/City or County Agencies and Small/Large Businesses" focus group to "State Legislative Action to Change a Statute or Ordinance, or Appropriate Funding" issues are shown in Fig. 6, where agency preferences are broken down into low, medium and high preference categories. The most popular strategies and options for the "State/City or County Agencies and Large Businesses" breakout group for issues pertaining to "State Legislative Action to Change a Statute or Ordinance, or Appropriate Funding" are provided below, from most to least preferred:

- B-17 Ensure warning and alert systems are working and heard
- B-14 Develop unified strategies for high risk disaster areas
- B-6 Assign a State-level small business disaster coordinator
- B-15 Create a Food Self-Sufficiency Baseline by island.

4.2 *Federal/State/City or County Agencies*

The composition of the "Federal/State/City or County Agencies" breakout group for issues pertaining to "Federal/State/City or County Agencies and Small/Large Businesses" are provided in Fig. 7. Note that participants in this focus group are

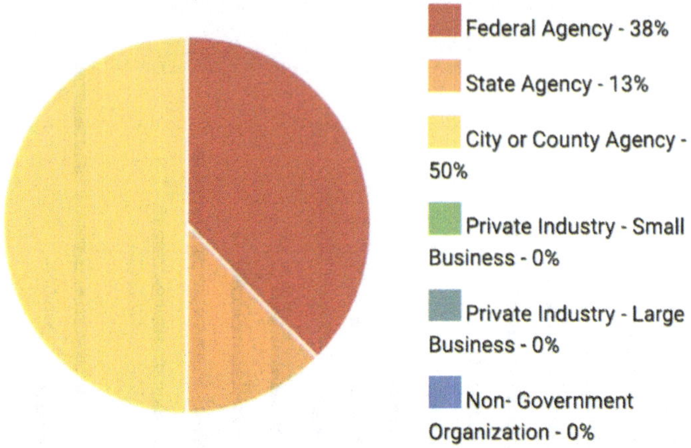

Fig. 7 Composition of "federal/state/city or county agencies" focus group for issues pertaining to "state legislative action to change a statute or ordinance, or appropriate funding"

drawn entirely from governmental entities: in particular, Federal Agencies comprise 38% of participants while State Agencies represent 13% of discussants. Finally, fully half of the breakout group participants come from City or County Agencies. Note that there is no representation from sectors outside the government: neither the private sector (small or large businesses) nor from non-governmental organizations are included in this focus group.

The detailed responses of the focus group comprised of "Federal/State/City or County Agencies" in relation to "State Legislative Action to Change a Statute or Ordinance, or Appropriate Funding" issues are shown in Fig. 8. Here, agency preferences are broken down into low, medium and high preference categories. The most popular strategies and options for participants in the "Federal/State/City or County Agencies" breakout group for issues pertaining to "State Legislative Action to Change a Statute or Ordinance, or Appropriate Funding" are provided below, from most to least preferred:

- B-17 Ensure warning and alert systems are working and heard
- B-12 Establish a State Recovery Coordinator Position
- B-14 Develop unified strategies for high risk disaster areas
- B-18 Develop a marketing strategy targeting small businesses.

5 Public-Private Partnerships (NDERS Poll Results)

A comprehensive list of strategies pertaining to "Public-Private Partnerships" are provided below:

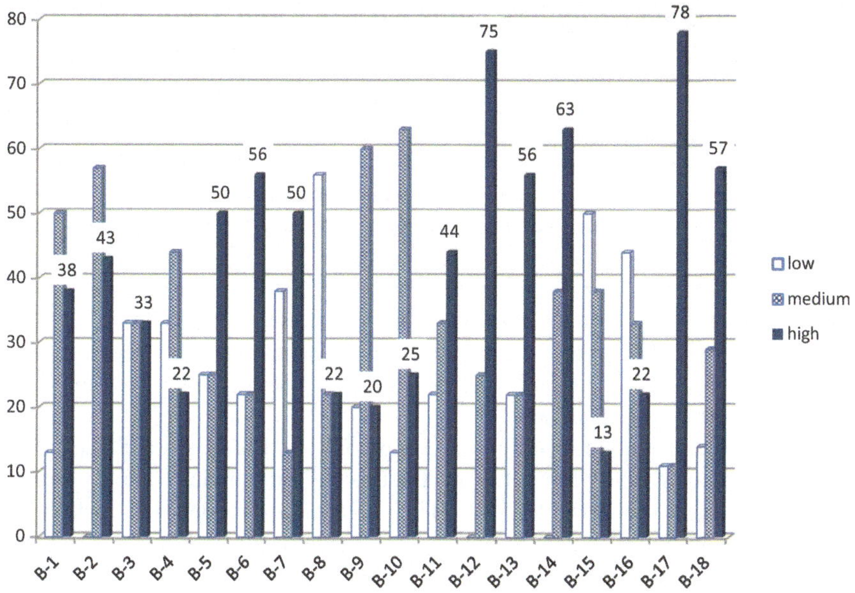

Fig. 8 Responses of the "federal/state/city or county agencies" focus group to "state legislative action to change a statute or ordinance, or appropriate funding" issues

- C-1 Utilize and strengthen unifying events and initiatives
- C-2 Facilitate collaboration to better use federal aid money
- C-3 Partner with SBA Development Centers for more assistance
- C-4 Work with neighborhood-based preparedness groups
- C-5 Study the need for a real-time response virtual platform
- C-6 Find business opportunities in disaster related areas
- C-7 Develop a framework for multi-business partnerships
- C-8 Have pilot programs for community-driven disaster plans
- C-9 Form critical infrastructure task forces on each island
- C-10 Develop technological solutions for resilience
- C-11 Teach businesses about extended interruption mitigation
- C-12 Encourage State and County agreements with businesses
- C-13 Convene response and recovery workshops statewide
- C-14 Develop business and emergency shelter partnerships.

5.1 Federal/State/City or County Agencies and Small/Large Businesses

The composition of the "Federal/State/City or County Agencies and Small/Large Businesses" breakout group for issues pertaining to "Public-Private Partnerships" are provided in Fig. 9. Note that governmental agencies constitute 45% of this focus group: in particular, Federal Agencies comprise 10% of participants while State Agencies represent 25% of participants and City or County Agencies constitute 10%. The remaining participants discussing "Public-Private Sector" issues are non-governmental organizations. Here, the private sector represents 35% of the focus group (with 5% representation from small business and 30% representation from large business). Finally, 20% of the focus group comes from non-governmental organizations.

The detailed responses of the focus group comprised of "Federal/State/City or County Agencies and Small/Large Businesses" discussing "Public-Private Partnerships" are shown in Fig. 10. Here, preferences of decision makers are broken down according to their responses (low, medium and high preferences). The most popular strategies and options for these participants discussing "State Legislative Action to Change a Statute or Ordinance, or Appropriate Funding" are provided below, from most to least preferred:

- C-12 Encourage State and County agreements with businesses
- C-9 Form critical infrastructure task forces on each island
- C-4 Work with neighborhood-based preparedness groups
- C-2 Facilitate collaboration to better use federal aid money.

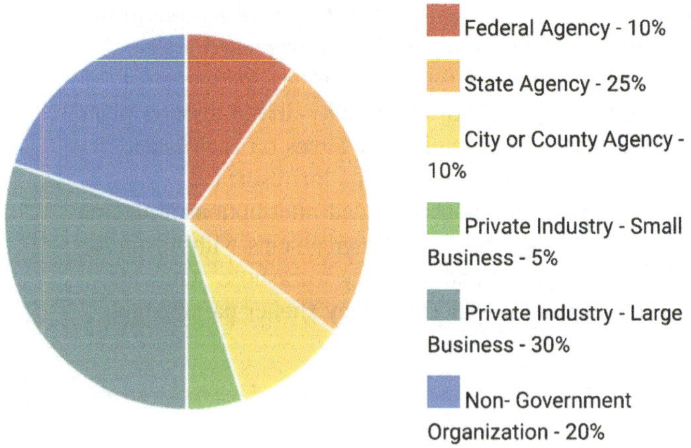

Fig. 9 Composition of the "federal/state/city or county agencies and small/large business" focus group for issues pertaining to "public-private partnerships"

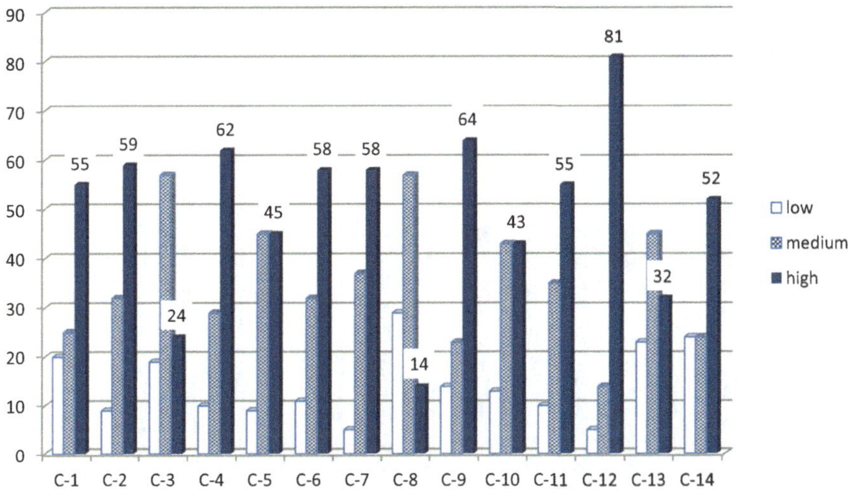

Fig. 10 Responses of "federal/state/city or county agencies and small/large business" for issues pertaining to "public-private partnerships"

5.2 *Federal/State/City or County Agencies*

The composition of the "Federal/State/City or County Agencies" breakout group for issues pertaining to "Public-Private Partnerships" are provided in Fig. 11. Note that participants in this focus group are composed entirely of decision makers from the public sector: in particular, Federal Agencies comprise 38% of participants while State Agencies represent 13% of discussants. Finally, fully half of the

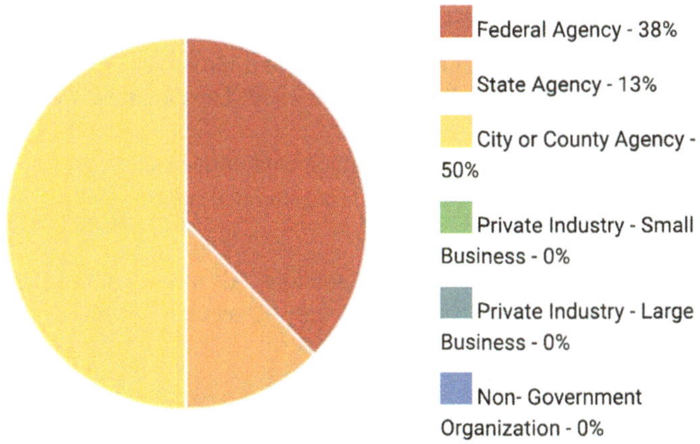

Fig. 11 Composition of "federal/state/city or county agencies" focus group for issues pertaining to "public-private partnerships"

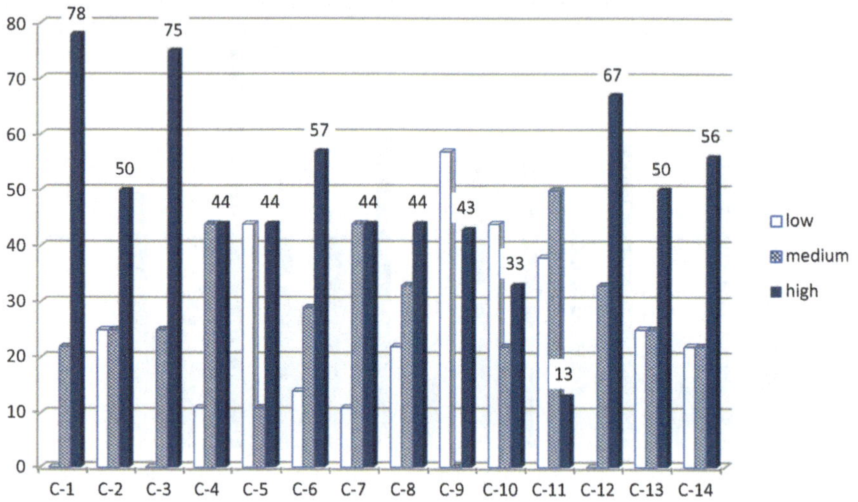

Fig. 12 Preferences of "federal/state/city or county agencies" for issues pertaining to "public-private partnerships"

breakout group participants come from City or County Agencies. Note that there is no representation from sectors outside the government: neither the private sector (small or large businesses) nor non-governmental organizations are included in this focus group.

The detailed responses of the focus group comprised of "Federal/State/City or County Agencies" discussing "Public-Private Partnerships" are shown in Fig. 12. Here, preferences of respondents are categorized according to their preferences (low, medium and high values). Figure I highlights that the most popular strategies and options for participants discussing "Public-Private Partnerships" are provided below, from most to least preferred:

- C-1 Utilize and strengthen unifying events and initiatives
- C-3 Partner with SBA (Small Business Agency) Development Centers for more assistance
- C-12 Encourage State and County agreements with businesses
- C-6 Find business opportunities in disaster related areas.

The full details are provided below.

Note that in the two breakout sessions dealing with "Public-Private Partnerships", encouraging State and County agreements with businesses (i.e. strategy C-12) falls within the top three choices

6 Private Sector Initiatives (NDERS Poll Results)

A comprehensive list of strategies pertaining to "Private Sector Initiatives" are provided below:

- D-1 Business associations and professional networks should designate a disaster point person or committee.
- D-2 Develop incentives and educational materials to encourage and reward client preparedness and accountability.
- D-3 Establish a certification program for businesses.
- D-4 Explore business models and for-profit opportunities to support business preparedness, continuity planning, and recovery.
- D-5 Seek out mutually beneficial businesses relationships.
- D-6 Actively involve and engage other businesses in individual business continuity planning.
- D-7 Consider a mentorship with other businesses to share information and experiences in disaster planning and recovery.
- D-8 Work with other businesses, residents, neighborhood boards and community associations to develop a community natural disaster strategy.

6.1 Federal/State/City or County Agencies and Small/Large Businesses

The composition of the "Federal/State/City or County Agencies and Small/Large Businesses" breakout group for issues pertaining to "Public-Private Partnerships" are provided in Fig. 13. Note that governmental agencies constitute 45% of this focus group: in particular, Federal Agencies comprise 10% of participants while State Agencies represent 25% of participants and City or County Agencies constitute 10%. The remaining participants discussing "Public-Private Sector" issues are non-governmental organizations. Here, the private sector represents 35% of the focus group (with 5% representation from small business and 30% representation from large business). Finally, 20% of the focus group comes from non-governmental organizations.

The detailed responses of the focus group comprised of "Federal/State/City or County Agencies and Small/Large Businesses" discussing "Private Sector Initiatives" are shown in Fig. 14. Here, the preferences of decision makers are broken down according to their responses (low, medium and high preferences). The most popular strategies and options for these participants are provided below, from most to least preferred (Fig. 14):

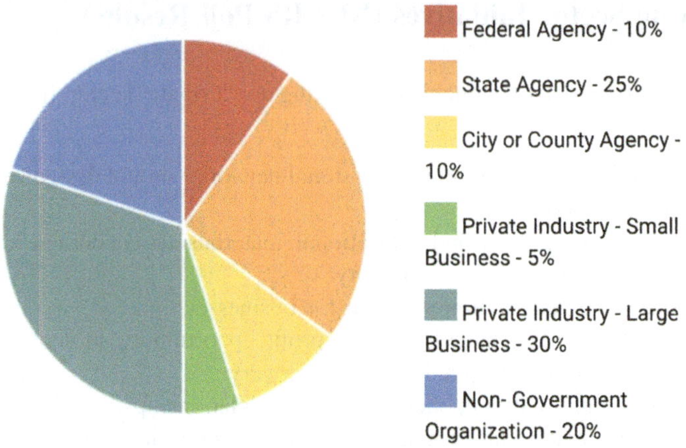

Fig. 13 Composition of the "federal/state/city or county agencies and small/large business" focus group for issues pertaining to "private sector initiatives"

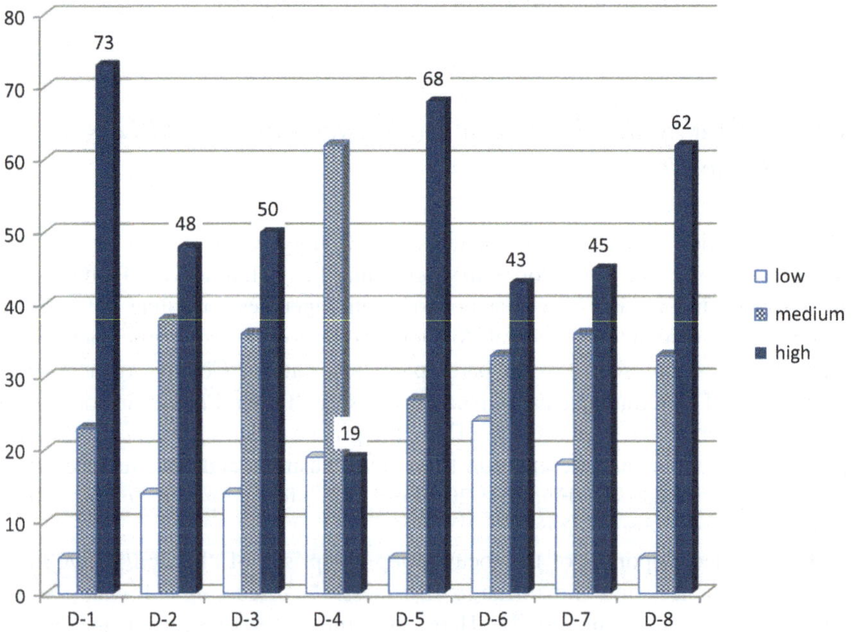

Fig. 14 Responses of "federal/state/city or county agencies and small/large businesses" discussing "private sector initiatives"

- D-1 Designate a disaster point person in every association
- D-5 Seek out mutually beneficial business relationships
- D-8 Develop diverse preparedness efforts in all communities
- D-3 Establish a preparedness certification program.

6.2 Federal/State/City or County Agencies

The composition of the "Federal/State/City or County Agencies" breakout group for issues pertaining to "Public Sector Initiatives" are provided in Fig. 15. Note that participants in this focus group are composed entirely of decision makers from the public sector: in particular, Federal Agencies comprise 38% of participants while State Agencies represent 13% of discussants. Finally, fully half of the breakout group participants come from City or County Agencies. Note that there is no representation from sectors outside the government: neither the private sector (small or large businesses) nor non-governmental organizations are included in this focus group.

Figure 16 provides the detailed responses of the focus group comprised of "Federal/State/City or County Agencies". Here, preferences of respondents are provided (broken down into low, medium and high values). Figure 16 highlights that the most popular strategies and options for participants drawn from "Federal/State/City or County Agencies" discussing "Public-Sector Initiatives" are as follows (from most to least preferred):

- D-1 Designate a disaster point person in every association
- D-7 Consider mentor relationships in disaster planning

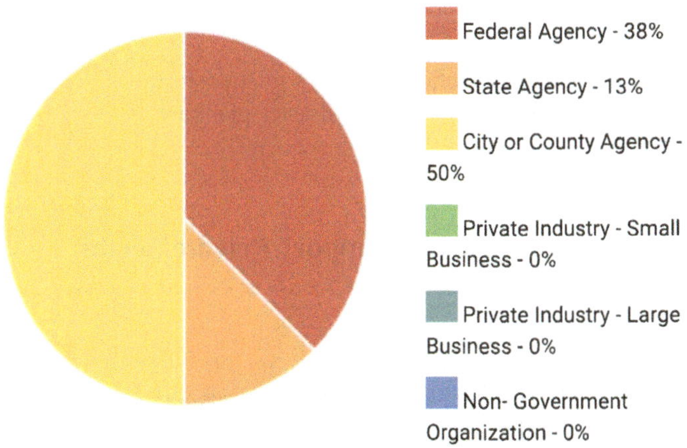

Fig. 15 Composition of "federal/state/city or county agencies" focus group for issues pertaining to "public sector initiatives"

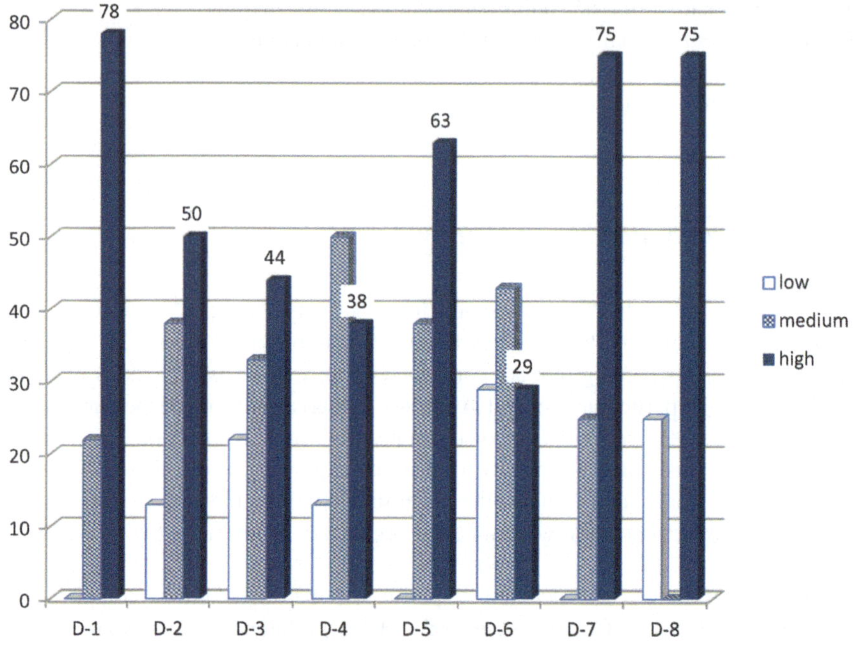

Fig. 16 Responses of "federal/state/city or county agencies" discussing "private sector initiatives"

- D-8 Develop diverse preparedness efforts in all communities
- D-5 Seek out mutually beneficial business relationships.

For both breakout sessions, strategy "D-1. Designate a disaster point person in every association" is the top choice. The second tier selection for both breakout sessions involves the following two strategies (each of which are roughly equally preferred):

- D-5 Seek out mutually beneficial business relationships
- D-7 Consider mentor relationships in disaster planning.

7 WebEOC Data Fields to Support Private Sector Recovery

Key data fields to support small businesses in the event of disaster are now described below with a focus on the following key categories: transportation fields, school fields, critical infrastructure fields, business access fields, donation management fields, business waivers, re-entry protocols and credentialing.

7.1 Transportation Issues

Are there transportation routes closed by this disaster? If so, are there alternate routes. What is that routing? Is there a prioritized list of what trucks must be carrying to be allowed through the alternate routes? If there are detour routes on minor county or city roads, are there weight restrictions or other requirements for trucks allowed through the detours as we may need to use smaller vehicles or load our trucks differently in order to be allowed through. Is there required identification that our trucks must carry to be allowed through? Specific transportation related fields include:

- Road Closures
- Alternative Routes
- Detour Restrictions (weight limits, etc.)
- Identification requirements.

7.2 School Issues

Are schools open or closed? Employees will need to arrange alternate child care or some companies may offer it on site.

7.3 Critical Infrastructure Issues

- Water challenges: Are there areas where water systems have been contaminated? Where are these areas?
- Power Infrastructure challenges: What areas are without power? What is the location of downed powerlines? What is the expected time of re-entry?

7.4 Business Access Issues

Are there disaster areas where access to my business is restricted due to safety concerns? Will I be able to get critical workers into assess the damage and get my business up and running? Is there special ID required? What.

7.5 *Donations*

Who do I call if I have these goods to donate? Is there a list of needed donation items?

7.6 *Waivers*

Are there federal regulations that will be temporarily waived to allow drivers to move goods faster to disaster areas? (such as the number of hours drivers are required to work). Who in government (county, city, state, federal) can temporarily waive various regulations to expedite recovery of my mission critical business function?

7.7 *Re-entry Protocols*

Is there a tiered/phased re-entry system which accommodates business re-entry needs? For example, the purpose of this tiered system is to allow for the expedited, safe, orderly return of re-entry of the following three tiers:

- tier 1 agencies/groups involved in emergency response and restoring normal operations following a disaster (search and rescue; emergency healthcare staff; utilities and infrastructure repair personnel; damage assessment teams; and pre-designated government staff and other critical personnel);
- tier 2 groups such as relief workers; healthcare agencies and suppliers; insurance agents (not allowed in tier 1); food and building material retailers, fuel distributors and stations, debris management, financial institutions, anchors for key industries, major employers with a large workforce, and/or a major tax revenue source for the community and businesses facing high risk circumstances (fragile inventory, hazardous waste, perishable goods, etc.);
- remaining tier 3 residents and business operators (not allowed under tier 2) from the restricted area as well as licensed contractors and other repair service providers.

7.8 *Credentialing/Re-entry Identification*

Are there clear procedures of securing re-entry ID? Is there a list of recognized credentials to re-enter the facility?

8 Conclusions

The Hawai'i Emergency Management Agency embarked upon a project with the University of Hawai'i at West O'ahu to determine the best use of a Business Resource Center (BRC) to increase the collaboration and connectivity of public and private sectors with respect to disaster management. Focus groups of key business and government sectors in Hawaii are examined to learn more about current disaster engagement, gaps in information or services, and opportunities for increased success with disaster management through a central BRC. This feedback helps determine the format, design, and precise purpose of a Business Emergency Operating Center (BEOC) for the State of Hawai'i. In general, there was a strong sense of support for the development of a collaborative space for the public and private sectors around emergency management.

As an isolated archipelago, the US island state of Hawaii is susceptible to a range of hazards since much critical infrastructure is located in vulnerable coastal regions. Hawaii is also heavily dependent on imported goods (from fuel and food to consumer goods) and transportation costs are extremely high. Accordingly, supply chain disruptions could have severe ramifications in Hawaii. Many businesses in Hawaii are extremely small. In particular, 86% of Hawai'i businesses have less than twenty employees [2] while 94% of Hawaii's companies have fewer than fifty workers. Given the dominant role of the visitor industry in Hawaii's economy, the closure of airports and harbors to tourists would have significant impacts across multiple sectors. Harbors on the island of O'ahu are also needed to provide critical resources and supplies to the neighboring islands of Maui, Kauai, Hawaii, Molokai and Lanai islands (and smaller islands). A majority of businesses in Hawaii lack business interruption insurance and many possess no disaster coverage at all. This is a major source of vulnerability since many small businesses anticipate insurance as their primary means of disaster recovery assistance. Other businesses may employ workers that use English as a second language (or do not speak English at all). This can lead to a plethora of challenges including difficulty finding recovery assistance and resources; they may also find it difficult to complying with documentation requirements for insurance claims. In summary, Hawaii is critically dependent upon business resilience. Particularly vulnerable are businesses in Hawaii with less than 20 employees and those that work from home [2]. According to Hawaii's Department of Business Economic Development (DBEDT) Natural Disaster Economic Recovery Plan [3] 38% of businesses surveyed in Hawaii (primarily companies with fewer than 25 workers) did not have a Business Continuity Plan

This chapter seeks to overcome economic disruptions in Hawaii for small, medium and large sized businesses by proposing timely and important data fields for information sharing in the Web-EOC crisis management software. In order to promote business continuity planning and disaster forensics in Hawaii, the twenty-eighth Hawaii State Legislature enacted, and the Governor of Hawaii has signed, House Bill (HB) 1343 which provides funds for a new BRC, a joint partnership between the State of Hawaii Emergency Management Agency (HIEMA), the State of Hawaii Department of Business, Economic Development and Tourism (DBEDT), the State of Hawaii State Procurement Office (SPO), and the University of Hawaii West Oahu (UHWO). The designated BRC hub will provide both outreach and dissemination of business recovery resources in addition to serving as a center for understanding disaster risk in small businesses. Investigations pertaining to business disruptions are undertaken with an emphasis on modeling, understanding and characterizing the complex causality that defines them. In so doing this chapter uncovers creative, timely and important strategies for reducing disaster risk and that impacts the economy of Hawaii. The BRC center will help inform businesses of the importance of disaster preparedness; assist with post-disaster business recovery efforts; and create a robust business recovery network that shares the highest-level of management and governance with business leaders and strives for best disaster management practices and continuous improvement.

References

1. Chang SE, Falit-Baiamonte A (2002) Disaster vulnerability of businesses in the 2001 Nisqually Earthquake. Environ Hazards 4(2–3):59–71
2. DBEDT (2012) Department of business, economic development and tourism. Research & Economic Analysis. DBEDT
3. Department of Business, Economic Development and Tourism (2014) http://files.hawaii.gov/dbedt/op/spb/2014_nders_final.pdf
4. Freitag R (2002). The business of earthquakes—what we learned from the Nisqually earthquake. Paper presented at the CREW conference
5. Graham LT (2007) Permanently failing organizations? Small business recovery after. Econ Dev Q 21(4):299–314
6. Insurance Institute of Business and Home Safety (2012) https://www.disastersafety.org/commercial_maintenance/commercial-vulnerability-assessment_ibhs/. Accessed 20 Aug 2015
7. Tierney KJ (1997) Business impacts of the Northridge earthquake. J Contingencies Crisis Manag 5(2):87–97
8. Vale LJ, Campanella TJ (eds) (2004) The resilient city: how modern cities recover from disasters. Oxford University Press, Oxford
9. Yoshida K, Deyle RE (2005) Determinants of small business hazard mitigation. Nat Hazards Rev 6(1):1–12

Advances in Cybersecurity Design: An Integrated Framework to Quantify the Economic Impacts of Cyber-Terrorist Behavior

JiYoung Park, Jason Levy, Minsu Son, Changkeun Park and Ha Hwang

Abstract An integrated framework to quantify the economic impacts of cyber-attacks on the U.S. airport security systems is put forth. A cyber-attack on the United States would involve complex strategic behavior by terrorists, who could invade an airport electronic system without any U.S. border or entry point. At the same time, defending entities must consider the complex processes that may cause any turmoil. The possibility of simultaneous threats from cyber-attacks makes it more difficult for defending entities to secure their airports, highlighting the need for improved and integrated inter-governmental collaboration. Collaborative networking requires close inter-governmental coordination to overcome such cyber-terrorism to protect critical infrastructure systems. Constructing a new model for strategic cyber-terror security requires a combination of competitive and cooperative game situations to develop specific strategies against cyber-terrorism. In addition, an airport shutdown would have ripple effects throughout domestic and international economies; these effects must be analyzed with a spatially disaggregate economic model. Combining competitive and cooperative game situations with an economic impact model, this study proposes the Game Theoretic National Interstate Economic Model (G-NIEMO) framework. G-NIEMO identifies which airport may be most vulnerable in the event that an airport electronic system is shut down. Using the probabilistic costs of airport closures, the model identifies the economic importance of cyber security by event location and industry type.

J. Park
Department of Urban and Regional Planning, University at Buffalo, Buffalo, USA

J. Levy (✉)
University of Hawaii-West Oahu, 91-1001 Farrington Highway, Kapolei, HI 96707, USA
e-mail: jlevy@hawaii.edu

M. Son
Korea Institute of Civil Engineering and Building Technology, Goyang, Korea

C. Park
Korea Environment Institute (KEI), Seoul, Korea

H. Hwang
University at Buffalo, The State University of New York, Buffalo, USA

A. J. Masys (ed.), *Security by Design*, Advanced Sciences and Technologies for Security Applications, https://doi.org/10.1007/978-3-319-78021-4_15

From G-NIEMO, equilibrium strategies to protect U.S. airport systems can be developed and a general guideline for the evaluation of resource allocation can be provided to U.S. government agencies.

Keywords National aviation security · Terrorism · G-NIEMO framework
Economic impacts

1 Introduction

The world is currently generating large volumes of information and "information containers", far outpacing the growth of the information technology professionals required to manage it. Ensuring data integrity, confidentiality and availability are among the most challenging and urgent issues for businesses, governments and non-profit organizations in the twenty-first century. The number of cybersecurity incidents reported by United States (U.S.) federal agencies has increased over 500% in recent years: a host of assets from classified defense networks to critical infrastructure systems have been targeted [1, 2]. There has been a multi-faceted approach to dealing with cybersecurity problems. Specific counter-measures include employee cybersecurity education, innovative security policies and the establishment of industry, national and international security standards. The results have been mixed as the scale of cyber-attacks has increased. In some cases organizations have failed to keep pace with the evolving nature of cyber-attacks. In other cases, cyber-standards and policies have been adopted, although organizational compliance has often been reactive and insufficient.

The U.S. economy increasingly relies on the security of computer-based information assets. While cybersecurity professionals have made great strides in confronting information security risks, ensuring data integrity, availability and confidentiality remains a serious challenge. Information systems security is vital to protect private sector organizations, governmental agencies and individuals—as well as society at large and the global economy. For example, a cyberattack on the US aviation sector might affect US economic prosperity and the integrity of socio-economic systems: the airline industry is absolutely critical for the regional and global economy (including international trade, importing food, fuel and other essential goods). Such an attack could also create global risks to food, transportation, manufacturing, retail trade and other industries.

As commerce and business increasingly interacts, organizes, and creates communities through cyberspace, cybersecurity represents one of the most serious challenges to the U.S. economy. Accordingly, governments at all levels (from the local to the national) are beginning to grapple with the enormity of this threat. For example, U.S. critical infrastructure networks are impacted by a wide plethora of information security risks from denial of service attacks the theft of valuable intellectual property. Online crime and fraud are also growing concerns.

Networked information systems and interconnected critical infrastructure are essential to the U.S. economy. Accordingly, professionals have devoted considerable efforts to protecting cyber-critical systems from cyber-threats in order to prevent a loss of business continuity and socio-economic upheaval.

Serious concerns about cybersecurity and the vulnerability of the U.S. aviation sector remain with grave threats to business continuity, national security, and civilian critical infrastructure: the need for increased information security is greater than any other modern technology [3]. One key challenge is that increasingly sophisticated cyberattacks now requires exponentially less intruder knowledge then in previous decades. In particular, phishing, denial-of-service attacks, SQL injections, cross-site scripting, viruses, Trojan Horses and worms can lead to catastrophic impacts in the aviation sector. A major attack could not only cause massive social dislocation but also severely degrade economic infrastructure and threaten life-sustaining services. While progress has been made towards preventing sophisticated attacks and identifying attackers, there is an urgent need to protect the aviation industry against cyber incidents. Collective aviation cybersecurity and the protection of computer systems in the broader transportation sector requires better understanding the economic impact of disasters in this sector. Specifically, Fig. 1 highlights the phenomenon of increasingly sophisticated cyberattacks which can be launched with low cost by relatively inexperienced attackers. In particular, phishing, denial-of-service attacks, SQL injections, cross-site scripting, viruses, Trojan Horses and worms can lead to catastrophic impacts in the aviation sector. The increasing complexity of a 'hyperconnected' world is exacerbating information security threats to the U.S. aviation sector.

There are several approaches to cyber-security risk management: checklists, risk analyses, formal methods, and bow-tie diagrams [4–7]. Each approach has unique strengths and weaknesses and the prescriptive and descriptive implications of each method has been examined in detail. For example, checklists constitute the first category of information systems security. Often intended as security evaluation guidelines, the earliest checklists, including IBM's 88 point security assessment questionnaire and the *Computer Security handbook* focused on a wide range of security issues from disaster planning and encryption to off-site backup and physical security. They also provided a straightforward taxonomy of threats. Computer center audits have been carried out with SAFE (*Security Audit and Field Evaluation*) checklists for computer facilities and information systems [8]. The American Federation of Information Processing Societies (ARIFP) provides a *Checklist for Computer Centre Self Audits*. This sophisticated tool involves a kernel style framework of hazards and associated defenses.

Based on quantitative risk assessments, risk-based approaches help identify the most serious threats and hazards, evaluate whether risks exceed accepted thresholds. They can also be used to selecting risk treatment strategies based on the expected probability distributions and consequences of loss. Risk analysis methods can also overcome the problems involved with evaluating and implementing a large number of controls. Risk analysis cybersecurity approaches include the Central Computer and Telecommunications Agency's (CCTA) CRAMM (CCTA Risk

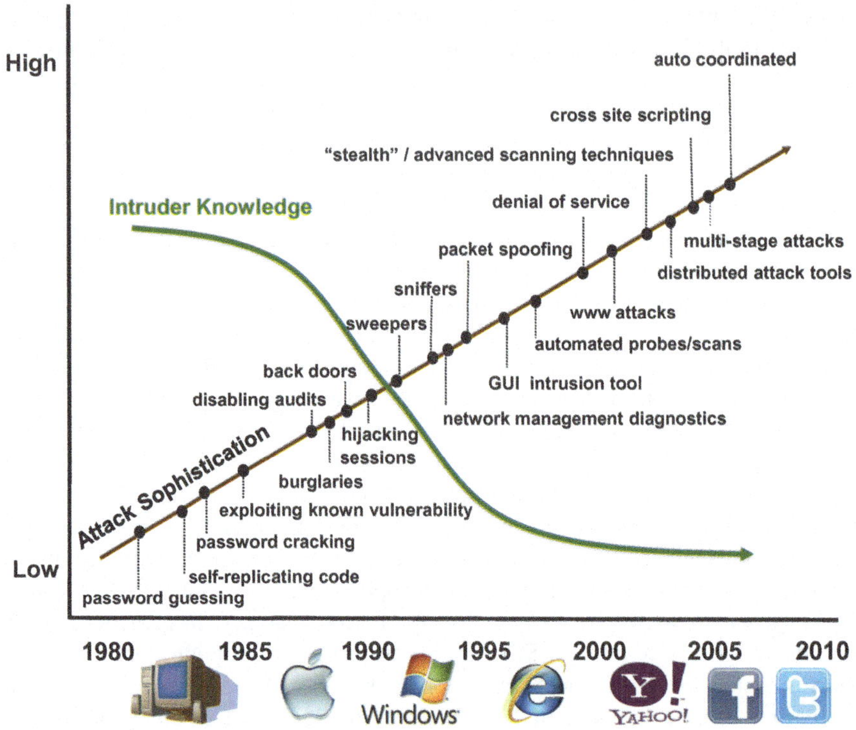

Fig. 1 Cybersecurity attack sophistication versus intruder knowledge (modified from CMU [63])

Analysis and Management Method) tool and the Methodologie d'Analyse des Risques Informatiques et d'Optimisation par Niveau (MARION) [9]. Enterprise risk management standards support strategic and tactical risk management processes with a particular focus on organizational governance (e.g. leadership and accountability), risk treatment, and risk analysis.

A third approach for managing cybersecurity was popularized in the 1970s and 1980s: formal security models and methods. These discrete event-oriented approaches can overcome, at least in part, the weaknesses of information systems security testing. An interesting body of work involves formal models for cyber-security. Traditionally directed at the systems-software level of computer security (i.e. [10, 11] they now often incorporate fuzzy logic [12] and neural networks. Finally, Bow-ties can be used to supplement the guidance provided by checklists, risk analyses and formal models. For example, Fig. 2 shows that bow-ties involve a "top event" (an undesirable incident) where a cyber intrusion has occurred. Figure 2 also highlights multiple layers of safeguards (i.e. 'controls') that attempt to prevent the cyberattack from causing widespread damage. Such bow-ties are a combination

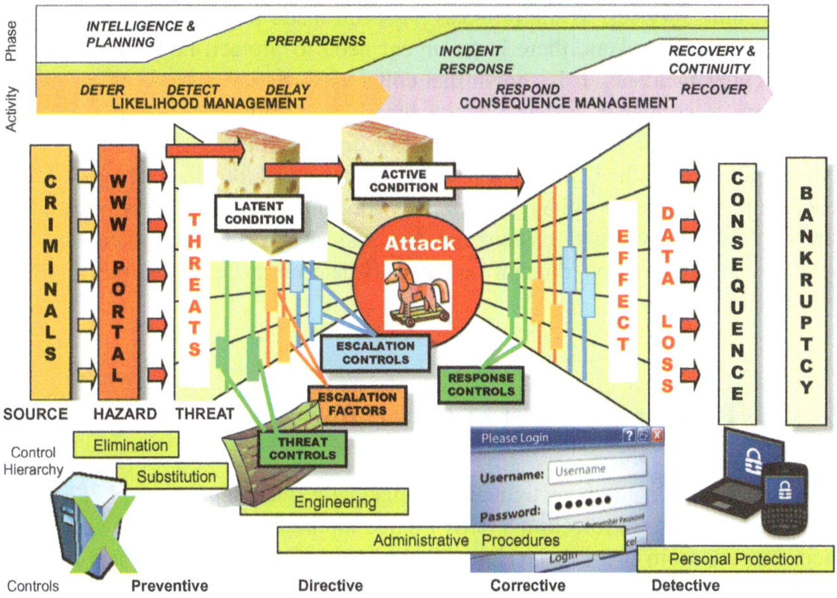

Fig. 2 Bow-tie risk management framework for cybersecurity threats

of fault trees and event trees: note that the fault tree constitutes the left had side of a bow-tie (before the cyber-incident) and the event tree the right hand side (after the event). The bow-tie methodology can also be used to highlight the relative vulnerability of bow-tie branches. This is discussed in more detail in the BowTie Pro™ the "Swiss cheese model" of accident causation [13].

The rest of this study is organized as follows. Section 2 discusses how cyber-threats to the aviation sector pose a grave national security concern, and why it is important to measure the direct and indirect impacts on the economy. Cyberspace must be viewed as an essential domain in the aviation sector, with associated responsibilities, implications and consequences for examining the economic impacts of attacks (as well as enhanced cyber-training, exercising, response and recovery). Section 3 discusses both competitive and cooperative strategic behaviors in a game theoretic situation (Sect. 3.1), a competitive game example (Sect. 3.2) and a spatially disaggregated economic impact model for the United States (Sect. 3.3). Section 4 introduces an integrated modeling framework for strategic cyber-security against terrorism that combines both competitive and cooperative game situations with the spatially disaggregated economic model applied to measure probabilistic economic costs. The final section (Sect. 5) concludes with a brief summary and a discussion of further research. It is concluded that a major attack on the aviation sector could not only cause massive social dislocation but also severely degrade economic infrastructure and threaten life-sustaining services. While progress has been made towards preventing attacks and identifying attackers, there is an urgent need to protect the aviation industry against cyber incidents. It is shown that collective aviation cybersecurity (and the protection of computer systems in the broader transportation sector) requires a better understanding the economic impact of disasters in this vital sector. Current cyber-challenges in the U.S. aviation sector include a lack of adequate cybersecurity leadership, technologic vulnerabilities and workforce shortages. It is necessary to not only create dynamic, customized and robust information systems security solutions in the aviation domain but also to determine the economic impact of cybersecurity attacks.

2 The Economic Impact of Cyberattacks

U.S. border security has been tightened dramatically since September 11, 2001. U.S. Customs and Border Protection's (CBP) stated top priority is "to keep terrorists and their weapons from entering the United States" [14]. Enhanced tightening of the U.S. border for national security has been implemented to prevent physical invasions. However, it appears that the Department of Homeland Security (DHS) may ineffectively plan for and fail to successfully manage possible physical invasions across U.S. borders [15], and this is much clearer in the case of cyber-attacks.

Many economic impact studies on infrastructure damages have focused only on evaluating physical terrorist attacks, which may potentially cause severe economic consequences stemming from infrastructure damage and business interruption in the areas affected, in terms of the direct and indirect impacts on the economy [16–19]. Major U.S. airports have been provided with useful information on possible catastrophes, and public and private industry sector decision-makers have been advised to evaluate their expenditures to mitigate and respond to emergent events. However, such economic impact analyses have been limited to measuring physical economic damage under some restricted assumptions and scenarios without considering cyber-events or the probability of a resilient process after an event.

A cyber-attack on U.S. airports through the invasion of an airport electronic system bypassing any U.S. border or entry point would require complex strategic behavior by terrorists. This is because defending entity should consider the turmoil stemming via complex processes. Any simultaneous cyber-attack further makes the defending entities difficult in securing their airports. Indeed, the improved security collaboration between local governments, between local and federal governments, and between countries subject to an attack type is essential to overcome cyber-terrorist threats. For example, cyber-terrorists successfully invaded could cause various problems in the U.S. airport and aviation systems including operational software and the database because they involve valuable information. According to Oxford Economics [20], the air transport industry contributes to the global economy up to $1540 billion annually. Indeed, a single event could generate not only a region's economy but also other countries that are connected to the region's airport domestically and internationally. An advanced model design for strategic cyber-security against terrorism requires to including both competitive and cooperative game situations to be applied for the simulations of specific strategies against cyber-terrorism. Spatial ripple impacts need to be analyzed via a spatially disaggregated economic model.

Developing a new design framework to quantify the possible economic impacts of breached strategic airport security requires combining the probability of cyber-attacks with the consequent economic impacts. The new probabilistic economic impact model suggested in this study can provide regional and industrial estimates of differentiated economic impacts. A full integration of both competitive and cooperative games with a spatially disaggregated economic model proposes a new framework of a Game Theoretic extension of the National Interstate Economic Model (NIEMO) [16, 17, 21–24]. This G-NIEMO, therefore, can be used to (1) identify which U.S. airport and aviation systems would be most vulnerable when an airport electronic system is shut down; and (2) determine economically important airports and aviation systems for cyber security. The results collected via simulations of this research will help us understand how cyber-attacks affect the real U.S. economies. G-NIEMO can identify equilibrium strategies to protect the U.S. airport system against cyber-terrorism and provide a general guideline for the resource allocation strategies for U.S. governments and related agencies. Finally, this study is expected to provide a basis for communication among policy makers, the general public, and various local entities associated with aviation security and policy in the U.S.

3 Game Theory and the Economic Impact Model

In summer 2011 of New Delhi, Indira Gandhi International (IGI) Airport's Terminal 3 experienced a virus attack on the Common Use Passengers Processing System with a simple technical failure. The system was down for half a day and manual check-in was used during that time; about 50 flights were delayed for 15–20 min [25]. More recently, the U.S. experienced a cyber-attack on airport information systems via a malware insertion that might have caused serious malfunction in diverse systems from air traffic control to air conditioning pipes on airplanes [26]. Such events indicate that cyber-attacks are more than simple threats to airport control systems because they may cause serious physical disruptions at the targeted airport and result in airport shutdown as a result [27].

There is no doubt that an analytic approach of the cost-effectiveness to heightening the level of border security is useful to understand the risk of terrorist attacks. This analysis requires not only applying a benefit–cost analysis for each border but also understanding the defender's strategy in determining the vulnerability of airport and air systems in terms of the terrorists' perspective. When considering cyber-attacks that disable or delete critical infrastructure system data [28], the strategic game situation is applicable to the case of cyber-terror attacks. A physical attack planned by terrorists to disrupt critical infrastructure systems of the U.S. usually starts with complex strategic behaviors by terrorists, who may intend to cross a U.S. border. At the same time, a defensive entity should consider the complex processes that may lead to catastrophic results should such an attack happen. Likewise, if cyber-terrorists are able to successfully invade a U.S. airport system to cause a problem with the operational software that manages airplane schedules, one attack may generate not only the target airport, but also the other airports connected domestically and internationally. An airport shutdown will make ripple effects throughout the domestic and international economic links.

In the modern technologically advanced era, the following fundamental questions need to be raised in a cyber-terror policy analysis procedure: how can we effectively analyze defensive scientific and technological improvements to address terrorists' strategies? If we consider both economic damage and risk probability, how can we advance an analysis tool that considers them effectively and simultaneously? Part of the answer involves a clearer understanding of cyber-attack procedure and the economic impact of cyber-terror events. It requires the analysis of strategic behaviors among groups involved in cyber-attacks. In addition, after an event has occurred, the costs estimated need to consider additional impacts that extend from the direct impacts that occur in airport system disruptions. The estimated costs of counter-terrorism usually consider the policy benefits that hinge on the economic losses that can be prevented; these costs need to be extended to measure other regional losses.

It is certain that U.S. anti-terrorism policy has relied tremendously on improving its scientific and technological assets. At the same time, modern scientific and technological innovation has advanced the strategic options that terrorists may consider in planning an invasion of the U.S. Certainly, there are extensive records of recent cyber-attacks on modernized scientific transportation sectors, and the number of such attacks has increased, indicating that increasing computerization in transportation systems means more potential vulnerability to attack from cyber-hackers. For example, cyber-hackers may be able to disable substantial automobile and subway control systems because modern transportation network systems are largely computerized and can be attacked without spatial and temporal restrictions [29, 30].

It is clear that more air, water, and ground networks will be affected by cyber-terrorists in the future. Further, it is increasingly likely that cyber-attacks can simultaneously target domestic and international air system-related infrastructures because visible and invisible threats have increased dramatically with ease of access to the internet. Simultaneous attacks highlight the need for improved, integrated collaboration between local governments, between local and federal governments, and between the countries involved in the attacks. Collaborative networking, connected horizontally and vertically in terms of governmental cooperative integration, requires close inter-governmental coordination to overcome the threat of cyber-terrorism.

Traditionally, game theory has been widely applied to analyze competitive, non-cooperative interactive decision processes, especially since 1970. Based on the strategies evolved from "Dawkins' *The Selfish Gene*," the competitive, evolutionary dynamic game process is more effective in finding the best agreement under restricted conditions [31]. However, agreements formed through the competitive process are unstable and difficult to maintain because they are more effective to problem identification or diagnostic process. A proposed public solution requires public support that satisfies interest groups and organizations; the public policy process requires this type of public solution to a problem that is publically identified [32]. This type of process suggests publically enforceable agreements. A game theory tool applicable to collaborative interaction provides the various mechanisms that can encourage cooperation to overcome disastrous risk [33, 34]. As Benkler [31] clearly stated, cooperative, collective interdependent actions among groups are more effective and play a more important role in the threat of a disaster than the best action plan determined via a competitive and non-cooperative strategic behavior process because of public agreement.

How can cooperative, collective action be applied to the defense of an unarmed civilian society threatened by cyber-terrorist attacks? Civilian society works to improve cyber security and reduce economic damage with the rapid growth of the internet, which removes geographical and temporal restrictions and helps create more established and developed societies. A strategic approach to modeling cooperative and collective actions among governments and the complex competition actions between hackers and governments requires integrated, structural super-cooperation in order to include horizontally and vertically connected but adverse actions. Figure 3 demonstrates the integrated structure of complex behavioral actions in cyber-terrorism.

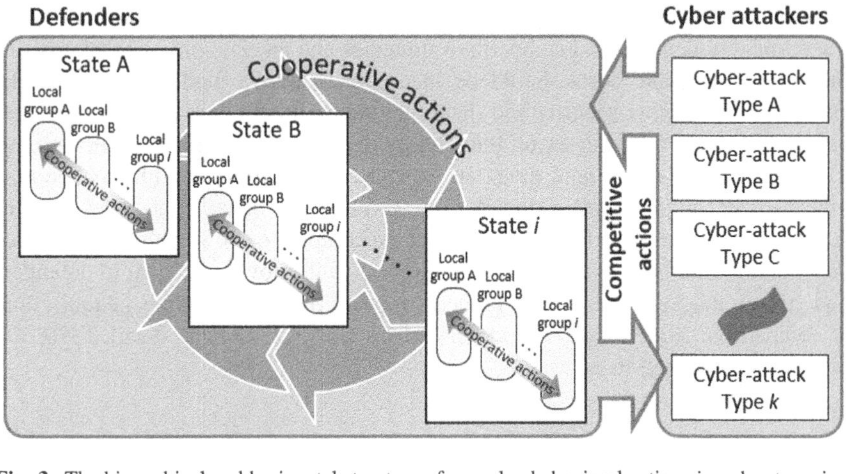

Fig. 3 The hierarchical and horizontal structure of complex behavioral actions in cyber-terrorism

3.1 The Competitive and Cooperative Game Process

Competitive game theory has been widely applied to the study of strategic interactions between attackers and a defending government [35]. Terrorists are intelligent and adaptive; they may determine where border security measures are weak and utilize diverse, illegal network channels to transport money, weapons, and personnel. For example, when considering the border security of a country in terms of a physical terrorist invasion to explain the competitive strategy of national security's response to terrorist attacks, the game theoretic situation begins with the behavioral strategy that terrorists may intend to cross a U.S. border. As a defender, the U.S. may respond to possible terrorist attacks when making defensive decisions concerning, for example, the amount of resources that need to be allocated to prevent any possible attack [36]. Having resolved the competitive game probabilities, by connecting to an economic impact model as a spatially disaggregate macro-economic model, U.S. federal policy makers can benefit from suggestions on which borders should be considered first for increased protection. Specifically, it is expected that the effectiveness of U.S. Customs and Border Protection can be significantly enhanced by the findings suggested with the competitive approach because it eventually delivers a simulated probabilistic economic costs package that can be used by DHS and by other agencies involved in border security. The study conducted by Zhuang and Bier [36] analyzed such strategic competitive interactions between terrorist groups and the defender using game-theoretic models.

While the game-theoretic view of a physical invasion by terrorists is a complex and competitive situation [35], the game view does not consider how local border securities collectively cooperate to protect U.S. borders or how national governments cooperate internationally to protect their critical infrastructure systems from possible terrorist attacks. Furthermore, although several innovative approaches to deterring possible attackers have been suggested (see, for example, [37–39], it is rare to find the studies that suggested the strategic allocation of defensive resources to improve border security by reducing the expected costs of a potential attack. Therefore, cooperative and competitive strategies and an economic impact analysis measuring economic costs should be incorporated into a model to evaluate the current level of cyber-security and determine the optimal allocation level of international and federal resources for the security of the aviation infrastructure systems. The procedure of combining the strategic situation of terrorists with an economic impact model is presented in Fig. 4. The lower panels demonstrate how the economic impact model can be combined with the complex game situation in the upper panels. The total probabilistic economic impacts will provide a signal to defenders and cyber-attackers, and the process repeats until it reaches equilibrium. A competitive game example is described in Sect. 3.2, and the detailed NIEMO process is explained in Sect. 3.3.

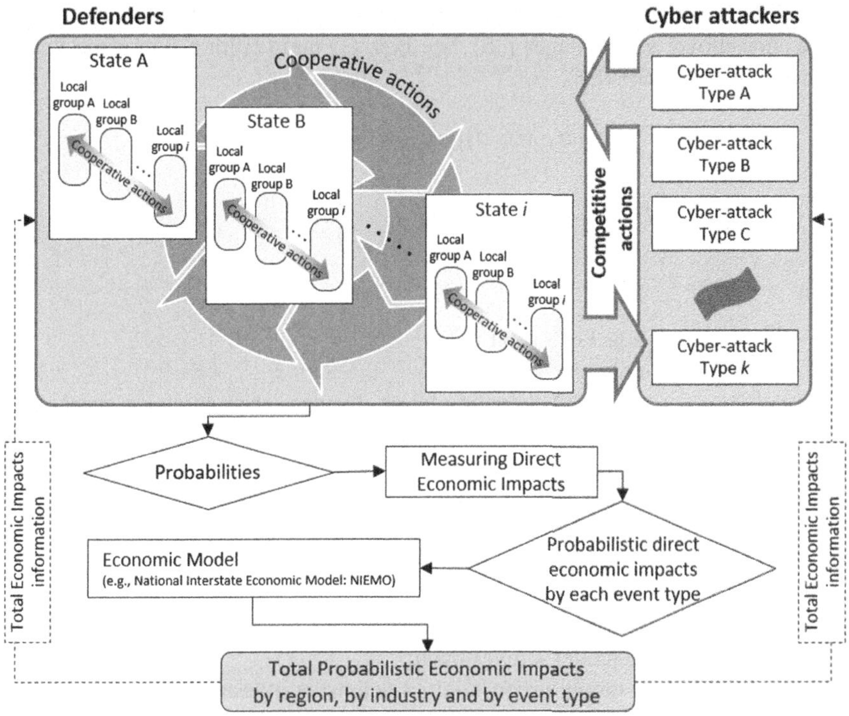

Fig. 4 Procedure to measure economic impacts via an integrated game theoretic economic impact model

3.2 A Competitive Game Example

Based on the existing literature dealing with a competitive game theory for attacks only, various assumptions were made as following. First, all entities, for example, local and regional airport agencies, the private sector to protect a regional airport, local to federal and international governments will collectively cooperate together and protect the global aviation system from cyber-terror threats. Second, even though strategic allocation of defensive resources can improve airport security by reducing the expected costs of a potential attack, defenders collectively behave and do not apply any additional strategic behaviors among defenses. Third, both an attacker and a defender's total utilities are the difference of target valuation and disutility, which is the cost of attack or defensive behavior. This is to help not only evaluate existing security measures, but also simulate possible allocations of local, regional, national and international levels, and find optimized resource allocation that minimizes the threat of attacks.

With assumptions described above, the game goal was to find a probability function of a security attack corresponding to government investment level on aviation infrastructure improvement for a target airport. To maximize a defender's total utility, a sequential game situation was applied in which a defender moves first due to first-mover's advantages [36]. See Eqs. (1) and (2) for a Sub-game Perfect Nash Equilibrium (SPNE).

$$U_A = u_A[w(a,d)] - g_A(a) = \frac{aw}{a+d+c} - \alpha a \tag{1}$$

$$U_D = u_D[v(\hat{a},d)] - g_D(d) = \frac{(d+c)v}{\hat{a}+d+c} - \beta d \tag{2}$$

where

U_A	attacker's total utility;
U_D	defender's total utility;
$u_A[w(a,d)]$	utility of expecting target damage when an attack succeeds;
$u_D[v(a,d)]$	utility of expected target damage when a defense fails;
$w(a,d)$	attacker's valuation of a target airport;
$v(a,d)$	defender's valuation of a target airport;
$g_A(a)$	attacker's disutility for attack effort;
$g_D(d)$	defender's disutility for defensive investment;
a	attacker's effort;
â	attacker's best response;
d	defender's investment;
c	a target airport's initial defensive investment level;
α	marginal utility of attack effort;
β	marginal utility of defensive investment; and
$\frac{(d+c)v}{\hat{a}+d+c}$ and $\frac{(d+c)v}{\hat{a}+d+c}$	Contest Success Functions (CSFs) for an attacker and a defender, respectively

The result of the Sub-Perfect Nash Equilibrium with binary strategies (attack/no-attack and defense/no-defense) shows "no-attack if defense" strategy is the equilibrium. Hence, a government should make enough investment on the aviation security infrastructure improvement, so to deter terrorist's attacks. With continuous strategies (a level of attack effort and defense resource allocation), the attacker's best response function is shown below in Eq. (3):

$$\hat{a} = \left\{a : U_A^{(a)}(a,d) = 0\right\}. \tag{3}$$

$$U_A^{(a)}(a,d) = \frac{\partial a}{\partial} \cdot \left(\frac{aw}{a+d+c} - \alpha a\right) = \frac{(d+c)w}{(a+d+c)^2} - \alpha = 0 \tag{3.1}$$

$$\therefore \hat{a} = -(d+c) + \sqrt{\frac{(d+c)w}{\alpha}} \tag{3.2}$$

Through a mathematical solution process demonstrated by Eqs. (3.1) and (3.2), the probability of a successful terrorist attack can be computed in Eq. (4)

$$P_a = \sqrt{\frac{(d+c)\alpha}{w}} \tag{4}$$

By entering the attacker's best response function to the attacker's Contest Success Function (CSF), the cumulative probability function of a security attack corresponding to the level of government investment can be obtained. CSF denotes the probability of a successful attack. By entering $\hat{a}$ to the defender's CSF, $1 - P_a$ is obtained. Optimal government investment level (for example d under $p = 0.5$) and probabilistic direct economic impacts can be computed through this initial model. The simulation would generate total impact (direct and indirect impact) responding different game situations as following: (1) both an attacker and a defender do not know the total impact outcome; (2) only defender knows; (3) both of them know.

A general implication of this approach includes: when a government increases level of investment d, attacker's effort a also would increase, but Δa would decrease for the first and second cases. For the third case, a terrorist would rapidly increase the level of a, but decrease once governments decide to keep increasing the defense investment d. It is expected to present that the pattern in the first two cases would be similar despite information available would be different. This can be explained possibly because governments match the level of d along with terrorist's probability of attack (P_a) that is decided by the level of attacker's effort a. However, comprehensive sensitivity and validation tests should be conducted based on the advanced model of G-NIEMO.

3.3 *The National Interstate Economic Model (NIEMO)*

Many economists and regional scientists have applied an input–output (I–O) or computable general equilibrium (CGE) analysis to measure the socioeconomic impacts of disruptions resulting from diverse disasters. The traditional, one-region type of I–O model that Isard used in 1951 cannot capture interlinked effects among regions. To measure the interregional impacts of terrorist attacks on one country, the economic links among sub-national areas should be clarified. Instead of surveying all interregional economic connections, for which the costs are huge, Chenery [40] and Moses [41] developed a relatively simplified multiregional I–O model (MRIO) framework in response to the "ideal interregional model" suggested by Isard [42, 43]. This can serve as an alternative to measuring the interregional economic connections based on interregional trade flows to avoid some of the problems associated with excessive spatial aggregation. When an I–O model is spatially aggregated, it especially loses spatial information—a terrorist attack may negatively impact the targeted area but may give positive economic boosts to another. Because most politicians have a keen interest in their local constituency, a state-to-state economic model may substantially contribute to providing such information.

The National Interstate Economic Model (NIEMO), an economic MRIO model for the 50 states and the District of Columbia (D.C.) of the United States, is the first operational MRIO model of the United States since 1990 [17]. Most studies that apply NIEMO have focused on regional and national economic impact measures stemming from diverse man-made and natural disasters to suggest plausible public policy alternatives. As a primary tool of application to the U.S. security problems, NIEMO's applications have adopted econometric approaches and quantified the economic losses.

The main reason that this type of model has not been operational for the United States stems from the non-existence or rarity of useful interstate trade data. Even though intraregional inter-industrial data and interregional trade data must be comparable and compatible, current commodity-based trade-flow data between states are only sporadically available, and it is difficult to connect the sector system to other U.S. industrial sector systems such as the North American Industrial Sector Code (NAISC). Building an operational MRIO that includes all the states is difficult because it requires highly detailed interstate trade flow data by industry type. Because of these data problems, only a few MRIO-type models have been constructed for the United States; these include the MRIO for 51 regions and 79 sectors for 1963 by Polenske [44] and the MRIO for 51 regions and 120 sectors for 1977 by Jack Faucett Associates [45]. The latter was updated in 1988 [46].

NIEMO applied Commodity Flow Survey (CFS) data to estimate interstate trade flows [24] and IMPLAN data (from the Minnesota IMPLAN Group; MIG, Inc.) for inter-industrial transaction flows by state. The Bureau of Transportation Statistics (BTS) has released CFS data every five years since 1993 following the discontinuation of the U.S. Commodity Transportation Survey Data on inter-regional trade flows; however, these CFS data do not report complete interstate trade flows. As Fig. 5 shows, based on the currently available CFS data, Park et al. [24] estimated complete trade flow data and updated them to a target period that matches IMPLAN data sets by applying an Adjusted Flow Model (AFM) and a Doubly Constrained Fratar Model (DFM). This approach necessarily requires creating conversion tables that reconcile the sectors between CFS and IMPLAN; detailed procedural bridges are explained in Park et al. [24]. While the current NIEMO relies on 1997 CFS and 2001 IMPLAN data, a systematic approach that is still operational is updatable. The current version of NIEMO tried to reconcile the different sector definitions and classifications of the commodities among multiple data sources using a new commodity sector system of 29 commodity and 18 service sectors. Park and Gordon [47] demonstrated that the interstate trade flows and trade flows between the U.S. states and the rest of world are reliable and even more important for providing spatial information than for providing detailed sectoral information. The modeling and development process of NIEMO is illustrated in Fig. 5.

NIEMO has been applied to various empirical studies that include hypothetical terrorist attacks [21, 16, 48]. NIEMO has also been applied to diverse natural disaster studies [22] and to the closure of all U.S. borders during the international avian influenza epidemic [49]. See Table 1 for various studies NIEMO has estimated.

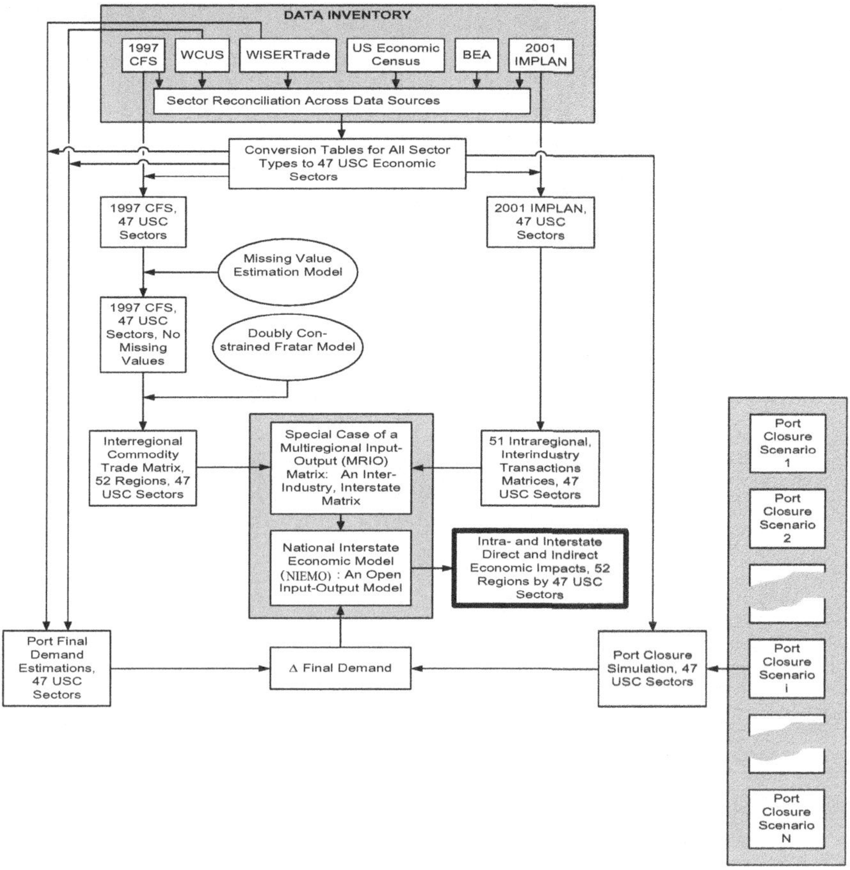

Fig. 5 NIEMO modeling and development process *Source* Park et al. [17], p. 213, "Simulating the State-by-State Effects of Terrorist Attacks on Three Major US Ports: Applying NIEMO (National Interstate Economic Model)"

NIEMO has evolved: it has been extended to the sub-state level (Southern California Interregional Input–Output: SCI–IO) and to temporal models (Flexible NIEMO: FlexNIEMO), and has been combined with transportation network system (Transportation network NIEMO: TransNIEMO) and environmental (EnviNIEMO) models to address local and dynamic issues in a country. The extended NIEMO may investigate the magnitudes and policy implications of negative economic impacts resulting from disruptive events, including losses of infrastructure services (notably transportation) due to natural or man-made disasters at various regional levels. The importance of maintaining social, economic and community development systems in an era of new uncertainties from possible disasters involving climate change or terrorist attacks can be accurately and easily addressed by NIEMO and its extensions. Certainly, terror prevention programs that are

Table 1 Various economic impact studies using NIEMO

Nature of disruption	Targets	Type of economic impact	Total economic impacts ($M)	Base year/duration/ model	Citations	Note
Explosives	LA/LB, Houston, and NY/NW	Ports shut down	23,258	2001/1month/ demand-driven NIEMO	Park et al. [17]	Direct/Indirect state-by-state impacts
Dirty bomb	LA/LB	Ports shut down	26,905	2001/1 month/ supply-driven NIEMO	Park [16]	Direct/Indirect state-by-state impacts
9/11	U.S. Airports	Loss of air passengers	214,347–420,455	Sep. 2001–Aug. 2003/2 years/IMPLAN IO model	Gordon et al. [19]	Direct/Indirect/ Induced U.S. impacts
Mad cow disease	U.S. Bovine animals	Cessation of foreign exports	13,681	Dec. 2003–Apr. 2005/15 months/demand-driven NIEMO	Park et al. [17]	Direct/Indirect state-by-state impacts
Explosives	13 U.S. theme parks	Consumer losses	20,747–24,921	2004/18 months/ demand-driven NIEMO	Richardson et al. [48]	Direct/Indirect state-by-state impacts
Hurricanes Katrina and Rita	PADD III	Disruption of Oil refinery industries	4,849	2005/13 months/ supply-driven FlexNIEMO	Park et al. [22]	Direct/Indirect state-by-state and month-to-month impacts
International avian influenza epidemic	Loss of Air passengers, U. S. seaports closing, loss of cross-border shopping, loss of legal and illegal labors	U.S. border closures	1,734,075–5,408,796	2001/1 year/demand- and supply-driven USIO and NIEMO	Gordon et al. [49]	Direct/Indirect state-by-state or U.S. impacts
Hurricanes Katrina and Rita	Louisiana customs district	Seaports shut down	44,374	Aug. 2005–Mar. 2006/8 months/demand- and supply-driven NIEMO	Park et al. [62]	Direct/Indirect state-by-state impacts

implemented at all levels of government should closely connect national prevention programs with many localized constituencies for political support. In addition, aggregate cost analysis may not be able to deliver localized effects because local impacts may cancel each other out.

4 Game Theoretic NIEMO: G-NIEMO

The distinctive feature of this study is that it incorporates the strategic situation of cyber-terrorism by combining game theory with NIEMO. Constructing a new model for strategic cyber- security against terrorism requires a combination of both competitive and cooperative game situations with a spatially disaggregated economic model; the latter requires the expected input costs in order to estimate the total costs. As demonstrated in Fig. 4, the general conceptual approach to calculating the economic costs of cyber-terror attacks is to multiply the success probability of a cyber-terrorist invasion on important transportation infrastructure, such as airports, and the corresponding costs if those airports are subsequently closed. A metric table that involves the probabilistic costs for all U.S. airports considered will be used to evaluate the airport vulnerability from cyber security. The dual-methodology model of combining a game-theory and NIEMO will generate a new integrated model that we call the Game-Theoretic National Interstate Economic Model (G-NIEMO).

G-NIEMO provides probabilistic economic costs for each airport or for multiple airports targeted simultaneously. G-NIEMO first analyzes the probability of each or multiple airports in the U.S. being attacked. The direct cost of the closure of one or multiple airports is then estimated with disruption scenarios. The product of the probability and the direct cost is used as input information to NIEMO. Via NIEMO, the probabilistic total economic cost (by sector, by region, and by scenario) of the airports closed will be estimated. Finally, the total economic losses will be provided for defenders and cyber-attackers as new information and new economic impacts will be measured. Detailed procedures for this approach are as follows.

Step 1: Modeling Competitive (Defender and Attacker Strategic) and Cooperative (Defender and Defender Strategic) Interactions

For one or multiple airports, the U.S. government chooses the level of defensive investment in terms of finance, equipment, and personnel. An attacker observes (or anticipates) the defender's choices and then chooses the best response (or the probability of launching an attack against each or multiple airports). While competitive strategic interactions at this level may apply different forms of CSFs in order to estimate the probabilities [50, 51], another form of the function requires accommodating local governmental cooperative interactions as well as international cooperative interactions to defend collectively against cyber-terrorist attacks. For example, simultaneous cyber-attacks are plausibly made against John F. Kennedy (JFK) International Airport in New York City in the United States and Toronto

Pearson International Airport in Canada, and it is extremely important that any piece of information to defend the economic activities in these two megacities be shared, as most airports and routes are served by these main hubs. Furthermore, JFK is tightly connected with other airports in the United States, and as such, an attack would further disrupt U.S. economic activities. Hence, local cooperative strategic interactions are critical for preventing cyber-violation. This step requires analysis of competitive and cooperative strategic behaviors.

Step 2: Estimating Direct Costs

For possible individual and cooperative defender strategies concerning cyber-attackers, the hypothetical and historical measures of direct costs for a successful invasion that may occur should be estimated. Numerous historical data can be found from sources such as the Bureau of Economic Analysis (BEA), the Office of Travel and Tourism Industries (OTTI), the Travel Industry Association (TIA), the Bureau of Transportation Statistics (BTS), and the Bureau of Labor Statistics (BLS), and these should be combined with hypothetical invasion scenarios. The collected data will be used to calculate the direct cost of airport closure and provide numerical experimental results to measure the importance of each airport.

Step 3: Estimating Probabilistic Direct, Indirect, and Total Costs

Based on the probability of a successful attack and the direct cost for one or multiple airport closures, a panel graph can be drawn as shown in Fig. 6, which demonstrates where each airport fits in each panel.

High probability (High-P) and high direct economic cost (High-DEC) airports should be considered as a priority for U.S. government investment. Airports in the panel of low probability (Low-P) and low direct economic cost (Low-DEC) require the least consideration. Hence, we need to compare the cases in the panes of Low-P/High-DEC and High-P/Low-DEC. Small direct cost may cause substantial indirect cost if economically connected tightly with other airports. The probability of an attack if succeeded will be used to verify the possible direct cost associated with the airport's closure. Aggregating the probabilistic direct costs of each airport by state provides the expected direct damages that each state may experience. The expected probabilistic direct costs by state will be used to estimate the indirect and total

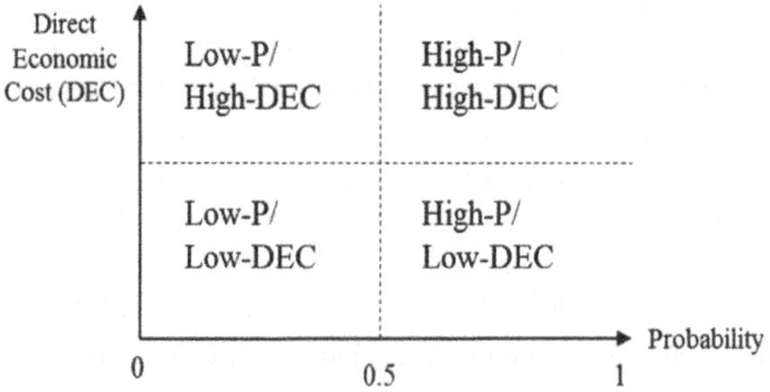

Fig. 6 Probability and direct economic cost (P/DEC) panel for cyber-terror invasion

potential costs for the U.S., considering the inter-industrial and interregional economic relations that each airport supports via NIEMO.

Step 4: Evaluating Equilibrium Probabilistic Impacts and Vulnerable Ranking Metrics

The estimated total impacts will be used for additional information for cyber-terrorists and defenders. Both players will receive this information in deciding their equilibrium strategies, which result in an updated estimation of the total economic impacts as suggested in the right and left up-pointing arrows in Fig. 6. This feedback process will repeat until equilibrium is reached. Finally, the equilibrium total impacts provide a vulnerability ranking score metric that demonstrates the vulnerability level of U.S. airports from cyber-attacks. The vulnerability metric, then, will be used for an analysis by industry type, by scenario, and by geographical boundary. The provided information will be used to determine which states and airports should be first considered for the distribution of defensive resources available in the United States.

Step 5: Evaluating G-NIEMO

Equilibrium strategies for U.S. airport protection will be computed using G-NIEMO. To evaluate G-NIEMO reliability, the computed equilibrium strategies from G-NIEMO are compared to other strategic results via traditional game-theoretic models and/or to the real data. The evaluation results will support the reliability of G-NIEMO and help the U.S. governments allocate the resources.

5 Conclusions

This study provides an effective research design to fight cyber-attacks. Even though aviation security is especially important due to its huge contribution to domestic and international economies once disrupted due to rapid human and freight movements, cyber-terrorists have easily invaded the U.S. governmental cyber security. This is clearer in the transportation network system [18]. Because U.S. airports are varied by the size, it is difficult to protect the U.S. airports using the centralized approach of aviation security by the Transportation Security Administration [52].

An effective way to protect vulnerable airports from cyber-terrorism requires improving the horizontal and vertical collaborations among local and federal entities and integrating various local strategies on cyber-attacks. This effort can improve our systems from the threat of simultaneous attacks on multiple U.S. airports. The collaborative strategies horizontally and vertically connected, should be integrated in order to analyze complex behavioral action strategies.

By missing economic impacts stemming from the successful invasion, the game approach could not be used to understand the regional and industrial vulnerabilities.

The G-NIEMO framework can quantify economic impacts on strategic infrastructure security and differentiate economic impacts by location and by industry targeted, measuring equilibrium strategies needed for U.S. airport protection. This study provides the theoretical strategies toward the integrated modeling of G-NIEMO.

Still a question not responded is to consider the economic resilience in the game process using the economic damage information. Numerous scholars agree that the defensive capacity can decrease maximum potential impacts at a given point in time after experiencing a terrorist attack and increase the ability to recover as quickly as the targets can [21, 23, 53–56]. After being repaired and resuming the full operations of its original airline schedule, the airport resilience can be measured. While the experience with 9/11 attacks and most physical disasters indicates that the economies of the U.S., New York and its surrounding states had substantial resilience [57, 58], the aviation and the operational services cannot recapture the lost services. The ex post lost values also need to be measured. The recent Park et al. 22 FlexNIEMO method will provide a potential to measure this lost values.

Fixed parameters estimated have been widely applied to measure a portion of the lost production in the U.S. economy from physical disruption [59–61]. An alternative way to relax functionalized parameters also has been suggested for each of several industrial sectors [22]. Even though most operational service sectors of infrastructure have usually no resilience, resilience to a cyber-attack may be needed to better estimate business interruptions because an economy will be indirectly impacted by infrastructure service disruptions. In this sense, a general, strategic framework addressing national and international transportation network security including aviation systems will be used to prevent cyber-attacks when being combined with multidirectional future strategies and a spatially disaggregate economic model that can trace local economic activities. Further analytic modeling design considering this resilience is left to future research.

Acknowledgements The authors wish to thank our institutional sponsor for this project: the U.S. Department of Transportation through the University Transportation Research Center, Region II (under the UTRC contract number 49997-38-24, DRTC-G-UTC02). However, any opinions, findings, conclusions, or recommendations in this article are those of the authors and do not necessarily reflect the views of the institute. Also, we wish to acknowledge the helpful inputs received from Professors Peter Gordon (University of Southern California) and the authors are also grateful to Professors Shambhu Upadhyaya, Raghav H. Rao, Jun Zhuang, and Changhyun Kwon at the University at Buffalo for their valuable comments.

References

1. Center for Strategic and International Studies (2011) Significant cyber incidents since 2006. http://csis.org/files/publication/110606_Significant_Cyber_Incidents_Since_2006.pdf
2. General Accountability Office (2011) Cybersecurity: continued attention needed to protect our nation's critical infrastructure and federal information systems, US Government, 16 Mar 2011

3. Goldman J (2010) CIO Insight, federal investment in cyber security to reach $13.3 billion by 2015. Government-CIO Insight. 29 Dec 2010. http://www.cioinsight.com/c/a/Government/Federal-Investment-in-Cyber-Security-to-Reach-133-Billion-by-2015-305281/
4. Thomas M, Dhillon G (2011) Interpreting deep structures of information systems security. Comput J Adv Access
5. Dhillon G, Backhouse J (2000) Technical opinion: information systems security management in the new millennium. Commun ACM 43(7)
6. Dhillon G, Torkzadeh G (2006) Value-focused assessment of information systems security in organizations. Inform Syst:293–314
7. Siponen MK (2005) An analysis of the traditional IS security approaches: implications for research and practice. Eur J Inf Syst 14, 3 (September 2005):303–315
8. Krauss L (1972) SAFE: security audit and field evaluation for computer facilities and information systems. Amacon, New York
9. Broder JF, Tucker E (2012) Risk analysis and the security survey. Elsevier, Waltham
10. Landwehr CE (1981) Formal models for computer security. ACM Computer Survey 13, 3w (Sept), pp 247–278
11. McLean J (1990) The specification and modeling of computer security. Computer 23, 1 (Jan), pp 9–16
12. Yang Y, Zhou Y (2011) Fuzzy logic based method for network information security risk assessment. In: International conference on internet technology and applications. IEEE, Wuhan
13. Reason JT (1990) Human error. Cambridge University Press
14. CBP (2008) About border security. Available at http://www.cbp.gov/xp/cgov/border_security/bs/. Accessed Sept 2008
15. Tirman J (2006) Immigration and insecurity: post-9/11 fear in the United States, MIT Center for International Studies. Available at http://borderbattles.ssrc.org/Tirman/. Accessed Sept 2008
16. Park J (2008) The economic impacts of dirty bomb attacks on the Los Angeles and Long Beach Ports: applying the supply-driven NIEMO (National Interstate Economic Model). J Homel Secur Emerg Manage 5(1)
17. Park J, Gordon P, Moore JE II, Richardson HW, Wang L (2007) Simulating the state-by-state effects of terrorist attacks on three major US ports: applying NIEMO (National Interstate Economic Model). In: Richardson HW, Gordon P, Moore JE II (eds) The economic costs and consequences of terrorism. Edward Elgar, Cheltenham, pp 208–234
18. Richardson HW, Gordon P, Moore JE II (eds) (2007) The economic costs and consequences of terrorism. Edward Elgar, Cheltenham
19. Gordon P, Moore JE, Park JY, Richardson HW (2007) The economic impacts of a terrorist attack on the US commercial aviation system. Risk Anal 27(3):505–512
20. Oxford Economics (2009) Aviation: the real world wide web. Onward Publishing, Inc
21. Park J, Gordon P, Moore JE II, Richardson HW (2008) The state-by-state economic impacts of the 2002 shutdown of the Los Angeles-Long Beach Ports. Growth Change 39(4):548–572
22. Park J, Gordon P, Moore II JE, Richardson HW (2017) A new approach to quantifying the impact of Hurricane-disrupted oil refinery operations utilizing secondary data. Group decision and negotiation. Online published. https://doi.org/10.1007/s10726-017-9537-7
23. Park J, Cho J, Rose A (2011) Modeling a major source of economic resilience to disasters: recapturing lost production. Nat Hazards 58(1):163–182
24. Park J, Gordon P, Moore JE II, Richardson HW (2009) A two-step approach to estimating state-to-state commodity trade flows. Ann Reg Sci 43(4):1033–1072
25. Kakkar M (2011) CBI believes cyber attack led to IGI airport's technical problems in June, Sept 25 2011. Available at http://www.zdnet.com/blog/india/cbi-believes-cyber-attack-led-to-igi-airports-technical-problems-in-june/710. Accessed December 2012

26. Dolgow M (2012) Cyberwars reach a new frontier: the Airport. Guerras cibernéticas alcanza frontera nueva: El aeropuerto), Bloomberg Businessweek, 15. Available at http://www.businessweek.com/articles/2012-08-15/cyber-wars-reach-a-new-frontier-the-airport. Accessed Dec 2012
27. AFP (2010) Airliners fly in face of cyber attack scares: What would happen if all those screens tracking all those flights suddenly went blank? 3 Nov 2010. Available at http://news.discovery.com/tech/cyber-attack-air-traffic.html. Accessed Dec 2012
28. Tafoya WL (2011) Cyber terror, FBI law enforcement bulletin, November. Available at http://www.fbi.gov/stats-services/publications/law-enforcement-bulletin/november-2011/cyber-terror. Accessed Dec 2012
29. Koscher K, Czeskis A, Roesner F, Patel S, Kohno T, Checkoway S, Savage S (2010, May) Experimental security analysis of a modern automobile. In 2010 IEEE symposium on security and privacy (SP), pp 447–462. IEEE
30. Ignelzi L (2012) Hackers manipulated railway computers, TSA memo says. http://www.nextgov.com/cybersecurity/2012/01/hackers-manipulated-railway-computers-tsa-memo-says/50498/. Accessed on 10 June 2012
31. Benkler Y (2011) The unselfish gene. Harvard Bus Rev 89(7/8):77–85
32. McCain RA (2009) Game theory and public policy. Edward Elgar, Cheltenham
33. Benkler Y (2011a) The penguin and the Leviathan: the triumph of cooperation over self-interest, 1st. Crown Business, New York, NY
34. Nowak M, Highfield R (2011) SuperCooperators: altruism, evolution, and why we need each other to succeed. Free Press
35. Sandler T, Arce M (2003) Pure public goods versus commons: benefit-cost duality. Land Econ 79(3):355–368
36. Zhuang J, Bier VM (2007) Balancing terrorism and natural disasters-defensive strategy with endogenous attacker effort. Oper Res 55(5):976–991
37. Frey BS, Luechinger S (2003) How to fight terrorism: alternatives to deterrence. Defence Peace Econ 14(4):237–249
38. Perrow C (2006) Shrink the targets continued. Spectrum 43(9):46–49
39. Keohane NO, Zeckhauser RJ (2003) The ecology of terror defense. J Risk Uncertainty 26:201–229
40. Chenery HB (1953) Regional analysis. In: Chenery HB, Clark PG, Pinna VC (eds) The structure and growth of the Italian economy. US Mutual Security Agency, Rome, pp 98–139
41. Moses LN (1955) The stability of interregional trading patterns and input-output analysis. Am Econ Rev 45(5):803–826
42. Isard W (1951) Interregional and regional input-output analysis: a model of a space-economy. Rev Econ Stat 33:318–328
43. Isard W (1960) Methods of regional analysis: an introduction to regional science. The MIT Press
44. Polenske (1980) The US multiregional input-output accounts and model. DC Health, Lexington, MA
45. Jack Faucett Associates, INC (1983) The multiregional input-output accounts, 1977: introduction and summary, vol. I (Final report), prepared for the US Department of Health and Human Services, Washington
46. Miller RE, Shao G (1990) Spatial and sectoral aggregation in the commodity-industry multiregional input-output model. Environ Plan A 22(12):1637–1656
47. Park J, Gordon P (2005) An evaluation of input-output aggregation error using a new MRIO model. In: Paper presented at North American meetings of the regional science association international 52nd annual conference, Riviera Hotel & Casino, Las Vegas, NV, 10–12 Nov
48. Richardson HW, Gordon P, Moore JE II, Park J, Kim S, Pan Q (2007) Tourism and terrorism: the national and interregional economic impacts of attacks on major US theme parks. In: Richardson HW, Gordon P, Moore JE II (eds) The economic costs and consequences of terrorism. Edward Elgar, Cheltenham, pp 235–253

49. Gordon P, Moore J, Park J, Richardson H (2009) The economic costs of border closure: a state-by-state analysis. In: Richardson HW, Gordon P, Moore JE II (eds) Global business and the terrorist threat. Edward Elgar, Cheltenham, pp 341–374
50. Skaperdas S (1996) Contest success functions. Econ Theor 7(2):283–290
51. Hausken K (2004) Mutual raiding of production and the emergence of exchange. Econ Inq 42(4):572–586
52. Poole RW Jr (2007) Airport security: time for a new model. In: Richardson HW, Gordon P, Moore JE II (eds) The economic costs and consequences of terrorism. Edward Elgar, Cheltenham, pp 67–97
53. McDaniels T, Chang S, Cole D, Mikawoz J, Longstaff H (2008) Fostering resilience to extreme events within infrastructure systems: characterizing decision contexts for mitigation and adaptation. Glob Environ Change 18(2):310–318
54. Rose A (2004) Defining and measuring economic resilience to disasters. Disaster Prev Manage: Internat J 13(4):307–314
55. Rose A (2007) Economic resilience to disasters: multidisciplinary origins and contextual dimensions. Environ Hazards: Hum Soc Dimensions (forthcoming)
56. Adger WN (2000) Social and ecological resilience: are they related? Prog Hum Geogr 24(3): 347–364
57. Park J, Gordon P, Jun E, Moore II JE, Richardson HW (2009b) Identifying the regional economic impacts of 9/11. Peace Econ Peace Sci Public Policy 15(2), Article 6
58. Rose AZ, Oladosu G, Lee B, Asay GB (2009) The economic impacts of the September 11 terrorist attacks: a computable general equilibrium analysis. Peace Econ Peace Sci Pub Policy 15(2), Article 4
59. Federal Emergency Management Agency (FEMA) (1997) HAZUS®99 earthquake loss estimation methodology: user's manual
60. Rose A, Lim D (2002) Business interruption losses from natural hazards: conceptual and methodological issues in the case of the Northridge earthquake. Glob Environ Change B: Environ Hazards 4(1):1–14
61. Rose A, Oladosu G, Liao SY (2007) Business interruption impacts of a terrorist attack on the electric power system of Los Angeles: customer resilience to a total blackout. Risk Anal 27(3):513–531
62. Park JY, Gordon P, Moore JE II, Richardson HW, Kim S, Kim Y (2008b) Estimating the state-by-state economic impacts of Hurricane Katrina. In: Richardson HW, Gordon P, Moore JE II (eds) Natural disaster analysis after Hurricane Katrina. Edward Elgar, Cheltenham, pp 147–186
63. Carnegie Mellon University (2006) CERT/CC vulnerability discovery project, CERT Coordination Center, Software Engineering Institute, Pittsburgh Pennsylvania

Operationalizing 'Deliverology' for Humanitarian Aid and Disaster Risk Reduction: Accelerating Improvement in Delivery

Anthony J. Masys

Abstract During the period 2005–2014, the Asia-Pacific region had over 1600 reported disaster events resulting in approximately 500,000 fatalities and affecting over 1.4 billion people (UN ESCAP in Asia Pacific disaster report-2015: disasters without borders, 2015 [29, 30]). Economic costs stemming from these disasters were upwards of $500 billion. During 2015 alone, the Asia/Pacific region experienced close to half the world's 344 disasters resulting in over 16,000 fatalities. Impact of these disasters in 2015 amassed to more than US$ 45.1 billion in economic damage (UN ESCAP in Disasters in the Asia and Pacific: 2015 year in review, 2015 [29, 30]). As noted in Weick and Sutcliffe [32: 1], 'Unexpected events often audit our resilience' and thereby challenge response and recovery activities. Black swan events are not the only mechanisms of destruction in Asia/Pacific region, such slow onset disasters as droughts, heat waves, forest fires and haze impact the safety and security of the region. Disasters stress the safety and security of the region, impinging on the already fragile disaster response capabilities and affecting the most vulnerable. The challenges associated with humanitarian assistance, disaster response and disaster risk reduction regionally and globally highlight the requirement for a more responsive and impactful 'delivery' model of aid. This chapter examines recent disasters and humanitarian aid efforts in the Asia/Pacific Region through the lens of the Sendai Framework and 'Deliverology'. Deliverology is a 'systematic process for driving progress and delivering results in government and the public sector' [1]. The key principles of Deliverology, rooted in design thinking and reflective practices helps organizations (humanitarian aid) define and execute their highest-priority objectives so that they have the greatest possible impact. In so doing, the chapter walks the reader through 'operationalizing deliverology and design' to accelerate improvements in delivery of humanitarian aid in support of the 'New Ways of Working (NWOW)' [22]. Deliverology combined with design is more than a process model but is a **mindset** for truly impacting communities through disaster risk reduction.

A. J. Masys (✉)
University of South Florida, Tampa, FL, USA
e-mail: tmasys@health.usf.edu

A. J. Masys (ed.), *Security by Design*, Advanced Sciences and Technologies for Security Applications, https://doi.org/10.1007/978-3-319-78021-4_16

Keywords Deliverology · Humanitarian · Disaster risk reduction
Design · Sendai framework · New ways of working (NWOW)

1 Introduction

'Poverty, crisis and risk are intimately linked and mutually reinforcing. Most of the poorest people in the world are living in contexts where disasters and conflict-related crises are more likely. An estimated 87% of people living in extreme poverty are in countries that are considered fragile, environmentally vulnerable or both' [5: 13]. The gap between humanitarian assistance aspirations and achievable results continues to grow. This is forcing us to question how we perceive risks and challenges our current mindset and approach to managing crisis, disasters and emergencies. Recently the global threat landscape has seen the emergence of high impact, low probability events. Events like Hurricane Katrina, the Great Japan Earthquake and tsunami, Hurricane Sandy, Super Typhoon Haiyan, Hurricane Harvey, Hurricane Irma, global terrorist activities, aviation and critical infrastructure disasters have become the new normal. The Asia/Pacific region in particular has seen a disproportionate number of such disasters with significant impact [19]. The Chatham House report 'Preparing for High Impact, Low Probability Events', found that governments and businesses remain unprepared for such events [12]. As global interdependencies continue to increase in number, these high impact low probability events prompt us to consider the need for radical change in how we perceive risks and how we manage and reduce disaster risk.

During the period 2005–2014, the Asia-Pacific region had over 1600 reported disaster events resulting in approximately 500,000 fatalities and affecting over 1.4 billion people [29, 30]. Economic costs stemming from these disasters were upwards of $500 billion. From extreme heat to extreme floods, from crippling earthquakes to mega storms, the Asia-Pacific region in 2015 was the world's most disaster prone region.

In response to these events, each year the international community invests globally on disaster risk management (DRM) initiatives to support capacity building and risk reduction. Masys [15] in his complexity analysis of the Horn of Africa crisis of 2011 describes some of the challenges associated with such investments in humanitarian relief such as being 'consistently late' and the 'ineffectiveness of early warning mechanisms'. Regarding disaster risk reduction (DRR) strategies, it was noted by Levine et al. [14: 4] that many plans associated with humanitarian and disaster risk reduction lacked the following:

- an overall strategy to which the various planned interventions were to contribute;
- rationale for the interventions;
- justification or rationale for the proposed scale of intervention;
- impact targets;

- clear triggers for deciding when to implement;
- anticipated calendar months for implementation;
- what needed to be monitored to know when to implement;
- specific actors given specific responsibilities for which they could be held accountable;
- a situational analysis that included predictions about what would be going on outside the area, e.g. movement of livestock in or out—without which most pastoral livelihood interventions would make little sense;
- discussion of specific locations within the area—given mobile livelihoods, which strategy would be needed in which location? Which areas would be likely to need most/least help? Where could conflict issues be a problem?;
- a link to an assessment of the degree of help needed; and
- (most important of all) a link to preparedness: to actions to be taken before the contingency arose in order to be ready to implement the contingency plan on time [14: 4].

Understanding the links between vulnerability, sustainable development and humanitarian assistance is a key enabler for disaster risk reduction. The Sendai Framework for Disaster Risk Reduction 2015–2030 (Sendai Framework), recognizes that the State has the primary role to reduce disaster risk but that responsibility should be shared with other stakeholders including local government, the private sector and other stakeholders [31]. The Sendai Framework represents a commitment to '...address disaster risk reduction and the building of resilience to disasters with a renewed sense of urgency within the context of sustainable development and poverty eradication, and to integrate, as appropriate, both disaster risk reduction and the building of resilience into policies, plans, programmes and budgets at all levels and to consider both within relevant frameworks' [31: 9]. The four priority areas of the Sendai Framework are:

Priority 1: Understanding disaster risk.
Priority 2: Strengthening disaster risk governance to manage disaster risk.
Priority 3: Investing in disaster risk reduction for resilience.
Priority 4: Enhancing disaster preparedness for effective response and to "Build Back Better" in recovery, rehabilitation and reconstruction.

The priorities of the Sendai Framework provide a vision for enabling resilience and delivering capability to support disaster risk reduction. Essentially much of the issues regarding delivering 'capacity building' and humanitarian assistance to enable resilience and reduce disaster risks stems from a crisis of perception regarding unintended consequences, the lack of focused attention on performance measurement, accountability and ineffective cooperation/collaboration/coordination [15]. Through a holistic and comprehensive approach to disaster risk management, a more purposeful and impactful approach is introduced that supports humanitarian aid and disaster risk reduction facilitated by 'deliverology' and design supporting 'New Ways of Working' (NWOW) as described in OCHA [22].

2 Deliverology

Deliverology, rooted in reflective practices, is an approach for managing and monitoring the implementation of activities that, according to its creator, have a significant impact on outcomes [1]. Deliverology combined with systems thinking, resilience thinking and design [17] is more than a process model but is a **mindset** for truly impacting communities through disaster risk reduction. As such, it embodies core elements of classic results-based management which consists of goal setting, establishing metrics, building plans, tracking performance, building 'robust performance dialogues' and linkages to individual incentives.

Key to Deliverology (Fig. 1) is the establishment of a central unit with a focus on using performance information to encourage change and improvement [26: 5].

The humanitarian relief and disaster risk reduction domains are data rich. By encouraging the mindset of data exploitation, Deliverology can drive an important aspect of managing organizational performance that is often easy to forget- **the discipline of follow-up** [26: 5]. In so doing, the delivery culture (Fig. 1) must recognize the requirement for building organic leadership, shared culture, trust relationships and leveraging talent distributed throughout the delivery system. It is through these key themes that performance information and impact can be monitored and enacted. The performance narrative is one that resides with the different stakeholders (from local to regional to national and global). Stakeholder perspective thereby plays an essential role in deliverology.

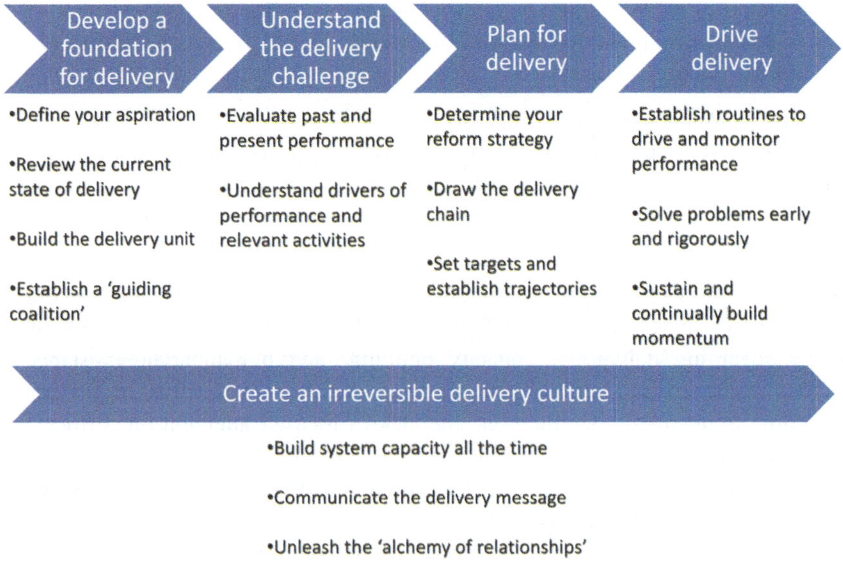

Fig. 1 Deliverology framework [1: xiv]

3 Research Context

The Asia/Pacific region has the highest number of total occurrences, fatalities and effects of natural disaster events (flood and cyclone) and is no stranger to mega-disasters such as the likes of Super Typhoon Haiyan and Indian Ocean Tsunami of 2004. Data from United Nations Economic and Social Commission for Asia and the Pacific (UNESCAP) 2015 shows that in the past 4 decades there has been a growing number of small and medium-scale disasters which have resulted in a total loss of over US$1.15 trillion. However this scale of disaster has received less attention from humanitarian agencies. At the same time the increasing number of natural disasters has increased the demand for humanitarian products and facilities such as shelters, communication devices, health and safety equipment, water and sanitation.

The impact of events such as Super Typhoon Haiyan and lessons learned can be used as a backdrop to help frame the deliverology and design discussion. It captures the devastating effect such a disaster can have and also highlights lessons learned. 'Super Typhoon Haiyan was a city killer' [3: 9]. Typhoon Haiyan is recorded to have had powerful winds up to 200 mph (320 km/h) with gusts up to 225 mph (360 km/h). The islands of Leyte and Samar were hardest hit with 90 percent of the infrastructure destroyed in Leyte's largest urban center, Tacloban City. Table 1 [3: 11] details the impacts of Typhoon Haiyan.

Although the disaster response to Typhoon Haiyan was considered a success, lessons learned that emerged from Typhoon Haiyan response included [3]:

Table 1 Super Typhoon Haiyan impacts [3: 11]

Super Typhoon Haiyan impacts NDRRMC situation report 104, 29 January 2014 0600 PHT	
Number of people dead	6201
Number of people injured	28,626
Number of people missing	1785
Number of families affected	3,424,593
Number of persons affected	16,078,181
Number of families served by evacuation centers	890,895
Number of people served by evacuation centers	4,095,280
Number of totally destroyed houses	550,928
Number of partially damaged houses	589,404
Total number of damaged (totally/partially) houses	1,140,332
Total cost of damages (agriculture)	$445,766,612 (USD)
Total cost of damages (infrastructure)	$430,306,341 (USD)
Total cost of damages	$876,072,953 (USD)

1. Lack of situational awareness and delayed implementation of standard operating procedures and pre-planned responses did not support the optimal use of resources.
2. Information sharing never matured to a more advanced stage due to resource limitations

   a. Lack of commonly accepted information-sharing platform among all major actors continues to confront relief efforts in emergency response.

3. Need to develop more operationally synchronized efforts that bridged the gaps between the government, humanitarians, and militaries. Planning assumptions and products for a multinational relief effort need to be reviewed for cases where the first line of defence—affected state responders—are themselves victims of a disaster.

As described in Jones et al. [10: vi] '...development and humanitarian sectors continue to face criticism over their relative rigidity and short-termism with regards to project funding and delivery'. The question arises, how can we better service communities in disaster scenarios with regards to humanitarian assistance? Through a thematic analysis of case studies, the ensuing discussion will explore how the 'deliverology' and design approach can support disaster management and humanitarian assistance thereby supporting NWOW.

4 Discussion

'Poverty, crisis and risk are intrinsically linked. An estimated 87% of people living in extreme poverty are in countries that are considered environmentally vulnerable, fragile or both. This means that effectively preparing for and responding to crises is critical, not only to save lives and alleviate suffering, but also to achieve a sustainable end to global poverty by 2030' [5: 13]. The way we view the world shapes what we see and hence shapes our decisions and actions. Much of the issues pertaining to humanitarian assistance gaps stems from a 'crisis of perception' or a 'crisis of perspective' [15]. Essentially, this centers on the lack of awareness regarding interdependencies and unintended consequences of actions, inactions and decisions. Lessons learned from Haiyan and other humanitarian and crisis management events [15] have highlighted some key elements regarding the application of perspective to support the delivery of humanitarian assistance and disaster risk reduction. Perspective, through a systems lens reveals striking insights and solutions by helping to frame the problem [21: 15–16]:

- 'First, people see the part to which they are connected but are largely unaware of the bigger system(s) that surrounds it. They miss their impact on others and others on them.
- Second, people tend to lack a time dimension. They see the present but not the past. They are intent on figuring out where a particular system should be in the future. They have some interest in knowing where it is now. But they have little

inclination to understand where it has been. They do not know the history of the present. '…everything that was left unprepared becomes a complex problem, and every weakness comes rushing to the forefront' [32]

- Third, participants miss—and in many cases mischaracterize—many of the key relationships that shape events.
- Fourth, people suffer from process blindness. They do not grasp the process dynamics, especially the deeper ones that are ongoing even within smaller systems. They suggest improvements which do not fit or even acknowledge the way the system actually works. They see individuals or events but not the processes of which they are a part'.

Levine et al. [14: 7] argue that '…a system perspective can often reveal how behaviour that is competent from the standpoint of each individual actor does not contribute to achieving the overall goals which collectively all the actors in the 'system' say they are working towards, in different ways. System problems often result when different actors do not really share the objectives, or when they do not agree on which elements contribute to a single system'.

5 Complex Humanitarian Problem Space

The hallmark of design pertaining to effective deliverology is the application of perspective as part of the problem framing process to support disaster management and humanitarian assistance. This systems view of the humanitarian context supports understanding regarding interdependencies, interconnectivity and unintended consequences from actions and decisions [15]. Changing how we view the complex problem space associated with humanitarian assistance is a key requirement in operationalizing deliverology to support aid. Thompson-Dyck et al. [28: 279] argue with regards to enabling resilience '…there is little chance that planners, experts and engineers can create new and more resilient communities until we transform the way we plan for, prevent, and respond to disasters…the devastation following Hurricane Katrina and Rita in 2005 was amplified by the failure of agents and organizations to work effectively together and with the communities affected before, during and after the hurricanes'. Fundamental to this is the requirement to change the way we see and think.

6 Intervention Strategy and Design Considerations: Understanding Fragility and Vulnerability

'Ongoing and new crises left an estimated 164.2 million people in 47 countries in need of international humanitarian assistance in 2016' [5: 16]. DRR capacity building program and sustainability are often competing agendas. This has been shown in Masys [15] with regards to drought and famine in East Africa. It also resonates with Asia/Pacific challenges [19]. Weick and Sutcliffe [32] argue that 'Unexpected events often audit our resilience, everything that was left unprepared becomes a complex problem, and every weakness comes rushing to the forefront'. With this in mind, poverty, gender and vulnerable populations emerge as key concerns and challenges in the DRR domain [15]. With regards to vulnerable communities, Zoraster [33] argues that '…many high-risk geographical areas have a disproportionately high percentage of marginalized populations; this same population is at a disadvantage for preparation, evacuation, response, and recovery'. This has significant impact on the delivery of humanitarian aid and hence how deliverology is operationalized. Within the Philippines for example, over 25% of the population are living below the poverty line. This characteristic marks a considerable vulnerable population in the face of such disaster events as Super Typhoon Haiyan. As such, it marks a threat to human security in the region. The systems approach in understanding this complexity to support humanitarian aid delivery is an integral phase of the deliverology process: 'Understanding the delivery challenge'.

Understanding this delivery challenge requires seeing the details. Gender is a characteristic that delineates a vulnerable population within the disaster context. In terms of capacity development in support of disaster risk management, IFRC [9: 57–58] argue that '…consideration of how women and men may be affected differently by certain hazards or disaster risks did not appear to have been taken into account. The way in which the design and delivery of a capacity development programme or humanitarian aid intervention is carried out can potentially have differential impacts on men and women'. This becomes a key consideration in the deliverology framework for establishing the vision for delivery and how to design humanitarian assistance with consideration of its impact on the communities as a whole. This is about leveraging 'empathy': a design thinking element. As described in Masys [18: 300]

> *Design is an activity, which aims at the production of a plan, which plan – if implemented – is intended to bring about a situation with specific desired characteristics without creating unforeseen and undesired side and after effects. Hors Rittel, 1968 (cited in Lee, 2016: 19)* [11]

As described in Masys [18], Design Thinking is a solution oriented methodology used by designers to solve complex problems. It draws upon logic, imagination, intuition, and systemic reasoning, to explore possibilities of what could be, and to create desired outcomes. A design mindset is not problem-focused, it's solution focused, and action oriented. It involves both analysis and imagination. Design represents a process that embraces innovation, creativity, opportunity analysis and problem framing and solving.

Through the phases of Inspiration, Ideation and Implementation, Design Thinking is operationalized through an iterative (not linear) 5 step process (Fig. 2). This process is described in detail (http://dschool.stanford.edu/redesigningtheater/the-design-thinking-process/).

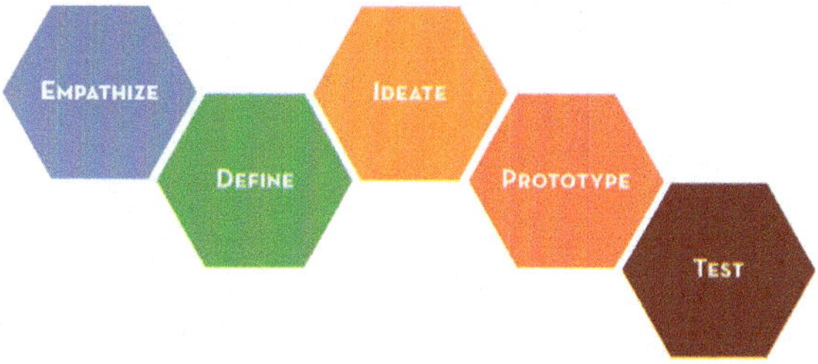

Fig. 2 Design thinking process

Design thinking in supporting humanitarian assistance delivery is certainly applicable. As described in Pourdehnad et al. [24] '…designers have specific abilities to produce novel unexpected solutions, tolerate uncertainty, work with incomplete information, apply imagination and forethought to practical problems and use drawings and other modeling media as means to problem solving. He further argues that designers must be able to resolve ill-defined problems, adopt solution-focusing strategies, employ abductive/productive/appositional thinking and use non-verbal, graphic and spatial modeling media'. As such, the design thinking approach becomes an integral element of the operationalization of deliverology.

In support of design thinking is the requirement for a holistic view of the problem space. Systems thinking [17] provides a means by which one may better understand the nonlinear dynamics and volatility of the humanitarian problem space. Human systems (as complex systems) have inherent feedback processes between interconnected elements that contribute to its dynamic, nonlinear and unpredictable nature. This has a profound effect on how we see the world and our understanding of unintended consequences and butterfly effects. As noted in Ramalingam and Jones [25: 13] '…despite this level of complexity, a bias towards and reliance on simplistic models pervades the aid system. For example, a study on drought-related work in the Sahel identifies that many analyses tend to divide causes into immediate and structural factors, with the structural issues largely ignored in agency responses'. In terms of development, sustainability and humanitarian relief, complex problems require a more holistic and inclusive approach to problem framing and solution navigation. Those being affected need to be included in the dialogue regarding problem definition and solution design.

7 From Analysis to Strategy to Delivery

The challenges associated with humanitarian assistance are manifold. From evidence derived from the literature through cross-case trend analysis, thematic analysis and grounded theory emerged humanitarian aid principles that influence design and delivery of humanitarian assistance intervention strategy. These key intervention strategy principles focus on considerations of **timescales**, **needs assessment** and **dynamic evaluation** (Fig. 3). These are critical considerations in the deliverology process pertaining to humanitarian aid in disaster scenarios.

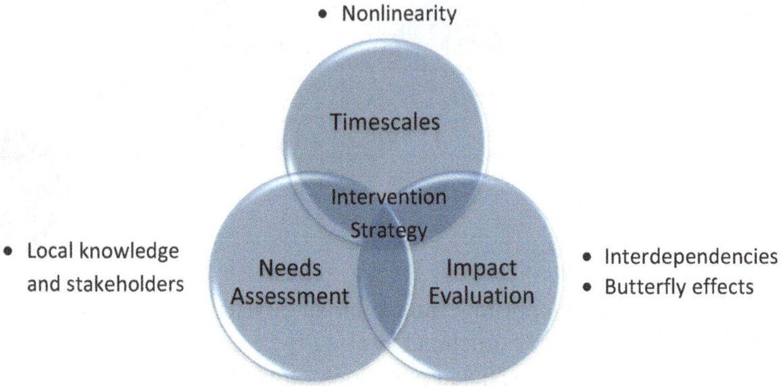

Fig. 3 Core deliverology humanitarian aid principles

8 Timescales

The outcome of intervention strategies regarding humanitarian aid have had mixed results in terms of timelines and impact [15]. It has been documented that the timescales for DRM capacity development programmes are too short. Matters pertaining to sustainability and capacity development for individuals, communities and organizations are generally a long-term endeavour. Within Disaster Risk Management, the shift from a response focus to a more holistic DRR perspective (mitigation, prevention and preparation) presents an added challenge regarding timescales. In addition, it has been cited [9: 56] that '…decisions regarding the length of a programme appear to be driven more by donor funding cycles than by what is deemed necessary during project design'.

Sustainability is a major concern in disaster risk reduction as noted in the Sendia Framework. By its very nature sustainability has temporal aspects. It has been noted in the literature [9] that sustainability is often insufficiently addressed. IFRC [9: 59] argue that '…a common, if often implicit, assumption is that the capacity raised by a programme will continue to exist after withdrawal of support, but meaningful sustainability planning is rarely undertaken and exit strategies are either poorly developed or, often, entirely absent'. Humanitarian assistance, development and sustainability all work on different timescales. This must be considered when supporting disaster risk reduction.

9 Needs Assessment: Problem Framing

Capacity and capability development intervention requires the conduct of a capacity needs assessment to determine the existing resident skills, resources and knowledge within the organization or community. It is an essential step to ensure that the programme is appropriately designed to build on DRM capacities and is well documented as best practice in the literature [9: 57]. Despite this, it was apparent that DRM capacity development programmes are often designed and implemented without a rigorous capacity needs assessment being conducted. As described in IFRC [9: 57–58] '… there were few examples of capacity needs assessments being undertaken at the very initial stages of programme design. Assessments were more typically made after the programme design had been partially conceived. In programmes in Ethiopia and Pakistan, staff did not conduct prior needs assessments because they felt under pressure to get on and start implementing activities'.

10 Impact Evaluation

Across the spectrum of deliverology from assessment to monitoring and evaluation, the literature highlights inadequacy along these key variables. As detailed in IFRC [9], '…the quality and robustness of programme monitoring and evaluation are generally weak'. Only approximately 50% of programmes had completed or planned for an evaluation to be conducted by an external party. In terms of routine monitoring and evaluation, programmes tend to monitor activities and outputs, with less attention paid to outcomes and impact. IFRC [9: 58] argues that '…monitoring and evaluation programme activities were typically driven by the demands of the donor—if they were not a specific requirement from the donor, then these activities tended to be forgotten. Unfortunately this demonstrates how monitoring and evaluation is currently perceived to be a tool for funding accountability rather than an instrument for improving impact and effectiveness'.

As has already been suggested, explanations of phenomena based on linear cause and effect are often not viable in systems that consist of numerous interdependent relationships. As such, one of the key outputs from design thinking to support deliverology is the articulation of the risk and vulnerability narrative that will shape humanitarian assistance. This heightens the importance of local knowledge and a good understanding of the contexts in which an agent is acting within a complex system. One needs a detailed understanding of the factors and dimensions of the complex social system that affect an area in order to refine one's perspective to see which features are important and which are irrelevant to the context. One must then look to discern the wider patterns that drive these factors. It is quite possible that '… rules governing [the local context are likely to be] only be partially and inadequately understood by the outside actor' [6]. The case for seeking out, transferring and harnessing knowledge of local actors increases importance from the perspective of emergent properties' [25: 23].

Design and Deliverology process attempts to bring clarity and transparency to incremental decisions taken to support DRR in humanitarian assistance to address the nonlinear behavior that can sometimes lead to unintended consequences. Reflective practices [20] figures prominently in the deliverology concept of operations. The humanitarian assistance domain is a volatile and complex changing environment. Reflective practices, as an integrated 'best practice in design and deliverology' recognizes how assumptions and mental models shape perception and decision making. Reflection must be part of the day-to-day management (reflection in action).

11 Deliverology and Goals

The challenge for humanitarian aid delivery is to find ways to define and execute the highest-priority objectives so that they have the greatest possible impact [2: 1]. The complex landscape associated with planning and decision making for humanitarian assistance requires attention to the dynamic complexity of the evolving scenario thereby making long-term forecasting problematic. As described in Leleur [13: 9] 'focusing on snapshots of the parts, looking for cause-and-effect links that are close together in time and space, means missing the slow accumulation of change'. Therefore as cited in Leleur [13: 27] 'when addressing complex strategic choices we need to initially examine different 'ways of seeing', which will then set the background for the process and methodology' for rolling out the solution enabled by deliverology. System thinking and reflective practices support this paradigm shift with regards to new ways of seeing.

Figure 4 depicts the central tenets of deliverology process being informed through reflective practices associated problem framing and solution navigation approaches of systems thinking, design thinking and resilience thinking. These approaches are discussed in Masys et al. [20] and Masys [16–18].

These approaches support the development of deliverology's most effective tools: '**targets**—a prioritized set of measureable, ambitious and time-bound goals—and trajectories, a projected progression toward these goals that creates a tight link between planned interventions and expected outcomes' [2: 35]. These targets become the areas of focus for the delivery team.

12 Operationalizing Deliverology Model

Deliverology represents a New Way of Working. As noted in the Global Assistance Report [5: 35] 'Recent commitments to a New Way of Working, bringing together humanitarian and development efforts, reiterate the need for predictable and sufficient development assistance to "shrink humanitarian needs" and meet the Sustainable Development Goals in crisis-affected contexts'. This New Way of Working helps '…frame the work of development and humanitarian actors, along with national and local counter-parts, in support of **collective outcomes** that reduce risk and vulnerability and serve as installments toward the achievement of the SDGs' [22].

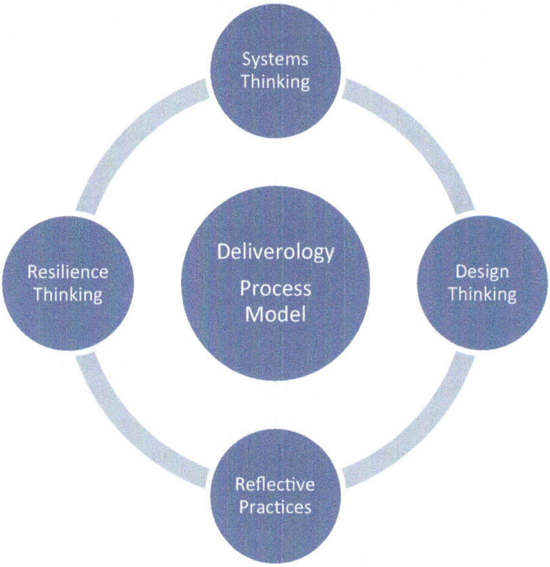

Fig. 4 Supporting approaches to deliverology

Deliverology, as a New Way of Working, provides a delivery mindset rooted on impact, outcomes and accountability. As noted in OCHA [22: 5]:

> *We must bring the humanitarian and development spheres closer together from the very beginning of a crisis to support affected communities, address structural and economic impacts and help prevent a new spiral of fragility and instability. This approach relates to the New Way of Working agreed at the World Humanitarian Summit. To achieve this, we need more accountability, on the level of each individual agency carrying out its mandate, but also its contribution to the work of the United Nations system and of the system as a whole. A strong culture of accountability also requires effective and independent evaluation mechanisms.*

Develop a Foundation for Delivery: The foundation of delivery asks the question 'why'. This foundational piece is the glue that connects the teams to the means to the ends. Any 'how' is doable if you have a strong enough why. Developing the deliverology team requires that 'building networks of organizations committed to a process of continual inquiry, informed action and adaptive learning [4: 335]. These qualities are critical in dynamic disaster scenarios. 'The New Way of Working sets a path for contributing to shared outcomes of reducing humanitarian need, risk and vulnerability through a range of well-aligned short-, medium- and longer-term contributions by humanitarian and development actors' [22: 7].

Define Your Aspiration To have an impact and realize outcomes, the humanitarian assistance aspiration must be clear, sharp and understandable to everyone. Dealing with a diverse group of stakeholders with different perspectives and agendas will require a unifying aspiration that fosters trust relationships, collaboration and cooperation. Common aspirations form the basis for all efforts at delivery because they signify a shared understanding of what success would look like [1: 3].

Identifying 'collective outcomes' focusing on reducing risk and vulnerability is a key requirement in the deployment of deliverology. As described in OCHA [22: 12] '...in some settings, identifying the link between collective outcomes and national plans to pursue the SDGs, in areas such as nutrition, education, health, and others, can offer a clear means to align short- and medium-term programming to bring the most vulnerable onto the path of development progress'.

Understanding the delivery challenge focuses on a reflection of past and present performance. The systems lens provides a holistic view of the problem space such that interdependencies and interconnectivity that lies in the system are understood. A network mapping [8, 17, 20] of the system is essential in order to understand the nonlinear dynamics resident within the humanitarian problem space. Drivers of performance will emerge from this analysis thereby shaping the delivery design considerations and metrics of success. The core deliverology Humanitarian assistance principles figure prominently. As described earlier, establishing the needs assessment is a key step in successful humanitarian aid deployment. This analysis phase examines the interdependencies that reside within the humanitarian crisis scenario to identify opportunities and threats to intervention strategies. Here the systems perspective [14] is essential in order to reveal unintended consequences.

Plan for delivery focuses on the how: it operationalizes the vision and needs assessment. Here is where timelines and resources become a reality in the planning cycle. Planning for delivery will require design thinking prototyping through modeling and simulation to support the design of the delivery and to explore how changes to the delivery strategy will affect outcomes. This will also be used to perform sensitivity analysis to facilitate a foresight/scenario planning exploration [15] to prevent unintended consequences [27]. Planning for collective outcomes requires a review of stakeholder contributions across humanitarian and development domains as well as a review of accountability frameworks in order to address the needs of the most vulnerable. As described in IFRC [9: 52] 'Effective DRM takes place at different levels: national, regional, local and community. For this reason capacity development not only has to occur at each level, but the ability for DRM actors to themselves collaborate and work across levels is itself an important capacity to develop'. The plan for delivery must be inclusive and empower all stakeholders. This is central for plan development.

Drive delivery depends on the 3 other phases. Establishing the drive delivery organization as an HRO [32] is a recommended practice in dealing with uncertainty and dynamic interdependencies. These insights derived from the principles of design thinking bring a relevance and operational 'sensitivity' when deploying humanitarian assistance. Such an operational positioning of deliverology resonates with the five principles of High Reliability Organizations [32]:

1. Preoccupation with failure
2. Reluctance to simplify
3. Sensitivity to operations
4. Commitment to resilience
5. Deference to expertise.

In driving delivery, we can also draw upon the four principles from Resilience Engineering [8]: Response; Monitor; Learn; Anticipate. As described in Hollnagel and Fujita [7: 14]:

1. To respond quickly and effectively to expected and unexpected conditions (disturbances as well as opportunities) either by implementing a prepared set of responses, or by adjusting everyday functioning, and also be able to sustain the response until control of the situation has been regained.
2. To monitor that which is, or can become, a change or disturbance in the near term, covering both what happens in the environment and what happens in the system itself, i.e., its own performance.
3. To learn from experience, in particular knowing how to learn the right lessons from the right experience—successes as well as failures.
4. To anticipate developments, threats, and opportunities further into the future, such as potential changes, novel needs, increased demands, and tighter constraints.

These Resilience Engineering principles resonate and are integrated within the HRO principles. In deliverology, within the context of humanitarian assistance, transformation becomes the goal. Both HRO and Resilience Engineering principles support and operationalize the goal.

As described in Masys [15], it is not enough to have a bold aspiration. Quality execution of tasks supporting humanitarian assistance is required. As described in Barber et al. [1: 3–4], an aspiration answers three questions:

- What do you care about?
- What are we going to do about it?
- How will we measure success?

Figure 5 highlights the relationship between aspiration and quality delivery and shows the region that constitutes successful delivery of humanitarian aid.

Finally, creating an irreversible delivery culture focuses on key factors of: organic leadership amongst the disaster stakeholders and partners, establish trust relationships, leverage distributed talent throughout the delivery system. This resonates with the requirement for stakeholder and community engagement in supporting and designing humanitarian assistance efforts.

Fig. 5 Map of delivery-aspiration and delivery excellence [1: 4]

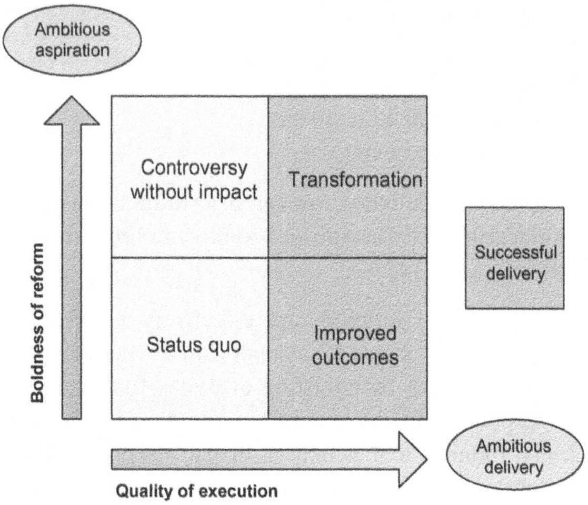

13 Deliverology Design Challenges and Opportunities

13.1 Focusing on Outcomes

The challenge of maintaining an outcomes focus in the face of short term disaster requirements is problematic. The design and deliverology process must balance both long term and short term disaster requirements as part of the intervention strategy. The delivery team must focus on the question: how will what we do make a difference for those we are serving, as soon as possible? And how will we know?

To evolve to this more proactive role, it is important that a deliverology inspired framework for comprehensive disaster risk management be prepared and implemented. This includes bringing together stakeholders from across public agencies, the private sector, and civil society to support problem framing. '***Learning***' to inform actions and interventions that makes a difference becomes a key element of the deliverology paradigm. This involves double loop learning. As described in Patton [23: 11], '…in double loop learning, those involved go beyond the single loop of identifying the problem and finding a solution to a second loop that involves questioning the assumptions, policies, practices, values and system dynamics that led to the problem in the first place and intervening in ways that involve the modification of underlying system relationships ad functioning'. This involves 'experimenting, adapting and developing what they do in response to feedback, changing conditions, new insights, and emergent challenges all around them'. For the deliverology team they must '…assume a world of multiple causes, diversity of outcomes, inconsistency of interventions, interactive effects at every level' [23: 41].

As a NWOW, a focus on collective outcomes is new requiring greater collaboration across stakeholders.

As described in OCHA [22: 14] 'achieving sustained and whole-of-system change will require overcoming political, institutional and structural obstacles at all levels, as well as realizing fundamental changes in attitudes, behaviours and approaches. It will mean supporting new approaches in relevant contexts, measuring concrete results over multiple years, and adapting financing to diversify and layer types of investment to support the achievement of collective outcomes'. This mindset is a key requirement for the delivery team.

14 Measuring the Unmeasurable

Managing disaster risk reduction and humanitarian relief is a wicked problem that will require the deliverology team to tap into new sources of data to make real-time information available. This can include social media and crowd sourcing information. The various stakeholders have different conceptions of DRR, focusing on different dimensions of the disaster cycle: some stakeholders focus on resilience in mitigation and prevention over preparedness, or preparedness over response and recovery. A targeting strategy needs to be flexible enough to adapt to different interventions and different phases as different patterns of vulnerability emerge. Therefore, effective monitoring that enables ongoing assessment of vulnerability is critical. Assessment and targeting approaches must be relevant to the disaster complexity. It is critical that community members be involved in definitions of vulnerability as part of the needs assessment.

15 Conclusion

Disasters stress the safety and security of the region, impinging on the already fragile disaster response capabilities and affecting the most vulnerable. The status quo regarding humanitarian assistance is insufficient and a delivery effort will be needed to achieve a transformational change in the systems performance to provide timely and impactful humanitarian assistance.

The challenges associated with humanitarian assistance, disaster response and disaster risk reduction regionally and globally highlight the requirement for a more responsive and impactful 'delivery' model of aid. The human security and disaster domains locally, regionally and globally will require innovative approaches to disaster risk reduction, humanitarian aid delivery and recovery. Deliverology combined with design is more than a process model but is a **mindset** for truly impacting communities through disaster risk reduction. This mindset will require building organic leadership, trust relationships and leveraging talent distributed throughout the delivery system all with a well-defined mission focus.

The deliverology team must exhibit an enhanced state of situation awareness regarding the uniqueness and instability involved in humanitarian aid. Maintaining to the principles of Deliverolgy can position the team for anticipatory resilience in the face of dynamic complexity and represents a NWOW across the humanitarian and development domains.

References

1. Barber M, Moffat A, Kihn P (2011a) Deliverology 101: a field guide for educational leaders. Thousand Oaks California, Corwin
2. Barber M, Kihn P, Moffit A (2011b) Deliverology: from idea to implementation. https://www.mckinsey.com/~/media/mckinsey/.../TG_MoG_6_Deliverology.ashx
3. CEDMHA (2014) Lessons learned from civil-military disaster management and humanitarian response to Typhoon Haiyan (Yolanda). Sponsored by the Center for excellence in Disaster Management & Humanitarian Assistance
4. Comfort L (2005) Risk, security and disaster management. Annu Rev Polit Sci 8:335–356
5. Global Humanitarian Assistance Report (2017) https://reliefweb.int/sites/reliefweb.int/files/resources/GHA-Report-2017-Full-report.pdf
6. Holland J (2000) Emergence: from chaos to order. OUP, Oxford
7. Hollnagel E, Fujita Y (2013) Fukushima disaster-systemic failures as the lack of resilience. Nucl Eng Technol 45(1)
8. Hollnagel E, Woods DD, Leveson N (eds) (2006) Resilience engineering: concepts and precepts. Hampshire, UK: Ashgate Publishing
9. IFRC (2015) World Disasters Report Focus on local actors, the key to humanitarian effectiveness. https://reliefweb.int/sites/reliefweb.int/files/resources/1293600-World-Disasters-Report-2015_en.pdf
10. Jones L, Ludi E, Beautement P, Broenner C, Bachofen C (2013) New approaches to promoting flexible and forward-looking decision making: insights from complexity science, climate change adaptation and 'serious gaming'. ODI February 2013
11. Lee AJ (2016) Resilience by design. Springer Publishing
12. Lee B, Preston F, Green G (2012) Preparing for high-impact, low-probability events: lessons from Eyjafjallajokull. A Chatham House Report, London
13. Leleur S (2012) Complex strategic choices: applying systemic planning for strategic decision making. Springer, London
14. Levine S, Crosskey A, Abdinoor M (2011) System failure? Revisiting the problems of timely response to crises in the Horn of Africa. Network paper Number 71. Humanitarian Policy Group. Overseas Development Institute
15. Masys AJ (2013) Human security—a view through the lens of complexity. In: Gilbert T, Kirkilionis M, Nicolis G (eds) Proceedings of the European conference on complex systems 2012. Springer Proceedings in Complexity, pp 325–335
16. Masys AJ (ed) (2015) Applications of systems thinking and soft operations research in managing complexity. Springer Publishing
17. Masys AJ (ed) (2016a) Disaster forensics: understanding root cause and complex causality. Springer Publishing, Berlin
18. Masys AJ (2016b) Exploring the Security Landscape- non-traditional security challenges. Springer Publishing
19. Masys AJ, Lin L (eds) (2017) Asia/Pacific security challenges: managing black swans and persistent threats. Springer Publishing, Berlin
20. Masys AJ, Ray-Bennett N, Shiroshita H, Jackson P (2014) High impact/low frequency extreme events: enabling reflection and resilience in a hyper-connected world. In: 4th international conference on building resilience, 8–11 September 2014, Salford Quays, United Kingdom. Procedia Economics and Finance, vol 18, pp 772–779

21. Morgan P (2005) The idea and practice of systems thinking and their relevance for capacity development. European Centre for Development Policy Management. http://portals.wi.wur.nl/files/docs/SPICAD/13.%20Idea%20and%20practice%20of%20systems%20thinking%20(ECDPM).pdf
22. OCHA (2017) New way of working. https://www.unocha.org/sites/unocha/files/NWOW%20Booklet%20low%20res.002_0.pdf
23. Patton MQ (2011) Developmental evaluation: applying complexity concepts to enhance innovation and use. Guilford Press, New York
24. Pourdehnad J, Wexler ER, Wilson DV (2011) Systems & design thinking: a conceptual framework for their integration. http://repository.upenn.edu/cgi/viewcontent.cgi?article=1009&context=od_working_papers
25. Ramalingam B, Jones H (2008) Exploring the science of complexity: ideas and implications for development and humanitarian efforts. Working Paper 285 ODI
26. Richards G, Gallo C, Kronik M (2016) Does deliverology deliver? Can Gov Executive 22(10):5–6
27. Taylor I, Masys AJ (2017) Human Security crisis: complexity and unintended consequences—a system dynamic model of the Refugee migration in Europe. In: Masys AJ (ed) Security by Design. Springer Publishing, Berlin
28. Thompson-Dyck K, Mayer B, Anderson KF, Galaskiewicz J (2016) Bringing people back in: crisis planning and response embedded in social contexts. In Yamagata Y, Maruyama H (eds) Urban Resilience. Springer Publishing, Berlin
29. UN ESCAP (2015a) Asia Pacific disaster report-2015: disasters without borders. http://www.unescap.org/news/asia-pacific-disaster-report-2015-%E2%80%93-disasters-without-borders
30. UN ESCAP (2015b) Disasters in the Asia and Pacific: 2015 year in review. http://www.unescap.org/sites/default/files/2015_Year%20in%20Review_final_PDF_1.pdf
31. UNISDR (2015) Sendai framework for disaster risk reduction 2015–2030. http://www.unisdr.org/files/43291_sendaiframeworkfordrren.pdf
32. Weick KE, Sutcliffe KM (2007) Managing the unexpected: resilient performance in an age of uncertainty, 2nd edn. Wiley, San Francisco, CA
33. Zoraster (2010) Vulnerable populations: Hurricane Katrina as a case study prehospital and disaster medicine. http://pdm.medicine.wisc.edu. Vol. 25, No. 1. http://citeseerx.ist.psu.edu/viewdoc/download?doi=10.1.1.176.2958&rep=rep1&type=pdf

Nuclear Waste Management: Security and Safety Implications

Suzanne Waldman and Anthony J. Masys

Abstract Helbing (Nature 497, [20: 51]) poignantly argues that 'Globalization and technological revolutions are changing our planet'. Along with the benefits and opportunities associated with worldwide collaboration networks comes 'pathways along which dangerous and damaging events can spread rapidly and globally'. Given the significant nuclear footprint throughout the Asia/Pacific, issues pertaining to nuclear waste management resonate with Beck's risk discourse on how distinctively modern manufactured risks are both temporally and spatially displaced. Many Asia/Pacific countries, including Japan and South Korea are facing increasingly large nuclear fuel stocks, with safe space for surface storage growing tight. A key event complicating nuclear power waste storage in the region was the Fukushima nuclear accident, which increased the nuclear waste profile while complicating storage siting by elevating public concern and diminishing trust in nuclear industries and regulators (Rich in Struggling with Japan's nuclear waste, six years after disaster, [45]). At the same time, equally salient manufactured risks from global climate change due to fossil fuel consumption are contending with nuclear risks, adding urgency to the optimization of cradle-to-grave energy life cycle decisions. All in all, the manufactured risk landscape calls for "reflective practices" that leverage the theoretical frameworks of risk perception and risk communication to shed light on issues challenging the Asia/Pacific region. In this chapter, we draw upon theoretical orientations of cultural cognition of risk (Kahan in J Sci Commun 14(3):1–10, [27]), responsible innovation (Guston and Sarewitz in Technol Soc 24 (1–2):93–109, [17]), and design thinking as a lens for reflection and guidance to enable more constructive and conciliatory nuclear waste decision-making within the Asia/Pacific region.

Keywords Nuclear waste management · Risk perception
Risk communication · Security

S. Waldman
Carleton University, Ottawa, Canada

A. J. Masys (✉)
University of South Florida, Tampa, FL, USA
e-mail: tmasys@health.usf.edu

A. J. Masys (ed.), *Security by Design*, Advanced Sciences and Technologies for Security Applications, https://doi.org/10.1007/978-3-319-78021-4_17

1 Introduction

Energy security is a key national security concern. Having reliable, safe and available energy is critical to support national economic and societal prosperity and innovation. Many countries through the Asia/Pacific region are natural resources deficient thereby challenging their energy security posture. In 2014 Japan imported more than 90% of its primary energy supply, giving its nuclear energy plants a key role in enabling energy security. Yet nuclear plants raise their own security concerns, both on the national scale pertaining to management of plants and waste, and on the international scale in monitoring diffuse nuclear material stocks and for weaponry proliferation. With 447 nuclear reactors in 31 countries maintaining and acquiring nuclear power for electricity generation to enable energy security and restrict runaway climate change due to fossil fuel generation, ensuring security around nuclear power and nuclear waste facilities is bound to present significant, and growing, challenges [60]. In the Asia/Pacific region, the question of how to dispose of nuclear waste is growing particularly pressing. The major nuclear-powered countries in the Asia/Pacific region are China, Japan, South Korea, and ROC (Taiwan). All these countries manage nuclear waste according to internationally approved strategies consisting of pools for short term storage, dry casks for medium term storage, and the pursuit of a deep disposal site, whether within their geographical bounds or abroad, for long term storage. In the past decade, however, South Korea had used up to 71% of its temporary nuclear waste storage capacity [8], and "the country's utility fore[saw] a crisis in the next 10 years as the storage at all of the country's four nuclear plants fills up" [31]. Other Asia/Pacific sites present similar limitations such as population density, seismic exposure, and legacies of top-down institutional decision-making exacerbating public distrust. Reprocessing spent fuel remains another solution that has the advantage of minimizing the volume of waste, however some security risks are raised through this approach, given that it creates highly enriched byproducts that are more amenable for nuclear weapon creation than either nuclear generation fuel or waste.

The recent Fukushima Daiichi nuclear accident has brought to the forefront a global reflection on the security and safety risks associated with nuclear energy. The DIET Report Executive summary [9: 9] argues that although the earthquake and tsunami of March 11, 2011 are considered triggers of the cataclysmic event, 'the subsequent accident at the Fukushima Daiichi Nuclear Power Plant cannot be regarded as a natural disaster. It was a profoundly manmade disaster—that could and should have been foreseen and prevented'. This disaster event, to which Beck's [1] notion of 'manufactured risks' applies, changed the nuclear energy landscape globally and particularly in the Asia/Pacific, resonating through public conversations not only about the ongoing viability of nuclear power generation in the region, but also about how to broker nuclear waste storage solutions amidst a global public sphere that is increasingly sensitive to perceived nuclear risks and prepared to be active and vocal in resisting them.

2 Theoretical Lens: Beck Risk Society; Risk Perception/Communication

Nuclear waste management produces what Beck [1] calls "manufactured risks" that are distinctive to modernity and impact economic security, energy security, environmental security, health security and food security over a diffuse geographical area and time frame. Such risks enmesh risk with uncertainty and by being 'more problematic, less easily manageable and more anxiety provoking' than traditional risks, for instance those concerning supplying resources and evading natural disasters. This is because—as Beck characterizes—risk is not synonymous with catastrophe itself, but with the anticipation of catastrophe amidst future occurrences and developments in a world that does not yet exist [2]. Within this paradigm, risk is no longer localized but differentially distributed and borderless, requiring society to deal with persistent, though sometimes vague and unlocalized, insecurities and uncertainties. The manufactured risks that characterize our contemporary "risk society" [1, 2] accordingly engender a preoccupation with debating over, preventing, and managing such risks. Masys [33] argues that manufactured risks move us from risk management to complexity management, revealing the transborder and distributed nature of the risks in the risk society, as well as their tendency to result in unforeseen effects.

Beck [2] argues that 'the dynamic of risk society rests less on the assumption that now and in future we must live in a world of unprecedented dangers; rather we live in a world that has to make decisions concerning its future under the conditions of manufactured, self-inflicted insecurity.' In a twenty-first century context combining climate change alongside increasingly densified and advanced technological development, natural disaster-triggered technological disaster's such as Fukushima Daiichi [35] will continue to heighten the ambiguity associated with manufactured risk.

Given the ubiquity of manufactured risk in modern risk society, the challenge can be posed as how to create reassuring appearances in situations characterized by unpredictable complexity: "how to *feign* control over the uncontrollable—in politics, law, science, technology, economy and everyday life" [3]. The importance of a sensibility of control for diminishing risk dread is indicated by risk perception research that finds involuntary risks to be far more dreaded and resented than voluntary ones over which communities and individuals have decisionary control. Control is extended through the engagement of communities more fully in decision-making about what levels of risk are tolerable and which risk trade-offs they are willing to assume. Until the 1980s, nuclear waste was entirely managed by its producers and, in turn, by governments that sited repositories in what they believed to be optimal geophysical locations without respect to the views of adjacent communities [4: 69–81]. In many cases, this approach aroused local resistance, leading to the indeterminate tabling of many initiated long-term waste disposal projects long-term antipathies to nuclear institutions [30: 85–86]. During the 1990s, the OECD Nuclear Energy Agency observed that progress in nuclear waste disposal was being impeded by poor public confidence in nuclear waste management institutions and processes and recommended greater public involvement to enhance trust [5: 123]. Simultaneously, the norm of public participation in environmental decisions was enshrined in Principle 10 of the UN's Rio Declaration on Environment and Development that held "[e]nvironmental issues are best handled with participation of all concerned citizens" [54]. While this norm of public participation is applied to vastly different degrees and in manifestly different forms ways around the world and in different contexts, there is no doubt that as a general ethos it has been spreading to the Asia/Pacific region during the past decade.

3 Context: Asia/Pacific Region

Currently, progress towards nuclear waste management is delayed in the Asia/Pacific region as compared with Europe and North America. Due to limits on the security of temporary storage solutions, international nuclear scientists and regulatory agencies favour the option of geological burial, especially for spent nuclear fuel [5: 125, 50: 676], with deep geological repositories currently in various phases of testing and construction in countries including Japan and Korea. Yet siting of long term facilities in the Asia/Pacific region is especially sensitive, as many of the countries are densely populated, with few large parcels of land away from urban areas and seismic vulnerabilities placing further limits on suitable sites [12: 82]. Even countries that wish to shutdown their nuclear energy ambitions, the decommissioning of operational reactors produces significant highly radioactive material, usually the used fuel and coolant along with buildings and equipment accounting for thousands of tons of low-level waste.

Over the past few decades, Asia/Pacific countries have attempted to broker deals amongst each other, and with other countries, to store nuclear waste, with patchy success. Exacerbating tensions around nuclear waste has been Russia's long-standing habit of dumping liquid waste in the Sea of Japan [12: 82]. Another solution floated has been the so-called 'Pangea Proposal,' which would see a country with plenty of open space—such as Russia, Australia, or China—host a global repository to absorb the waste of smaller countries such as Japan and South Korea, for a cost [59]. There are, however, significant political barriers to the achievement of a global nuclear waste repository.

Nuclear fuel reprocessing strategies are, meanwhile, also politically sensitive in the politically dense and volatile Asia/Pacific region. Japan for periods of time exported its waste for reprocessing to the UK and France, but this raised the ire of countries such as Indonesia, Malaysia, and Singapore who line its shipping route.

4 China

China is currently pursuing an aggressive campaign to expand its nuclear energy capacity as part of an effort to meet growing energy demands and reduce air pollution from coal-fired plants. In China, near-surface disposal for low and intermediate level waste has been implemented for years on a regional basis (Huiling [22]. China is pursuing a plan to reprocess its nuclear fuel, which would allow it to maximize the energy derived from uranium and minimize the waste [6]. It is also undertaking experiments to validate a geological disposal site for vitrified, high-level radioactive waste in a granite region in the country's Northeast [55].

Typically, China is known for undertaking short, perfunctory public consultations before going ahead with projects. Reportedly due to inroads of social media, effective public resistance to nuclear facilities is growing [25]. In 2013, demonstrations against a planned nuclear waste reprocessing facility protesting its being "rushed through with inadequate public consultation" led to the facility's suspension, suggesting it is becoming increasingly difficult for China to site facilities on a purely top-down basis [53]. Even a nuclear industry executive complained that the 10-day public consultation was "way too short," and that "materials it provided about the project [were] also woefully inadequate" [61]. China appears to be taking the lesson, as the government recently put a draft regulation on the nuclear power industry out for public consultation and is planning public hearings in sites selected for new nuclear power plants [25].

5 Japan

The quest for its own long-term nuclear waste disposal site is proceeding gradually in Japan, though siting remains a contentious issue. Currently in Japan, spent fuel from power reactors is reprocessed to extract reusable uranium (U) and plutonium (Pu) for further power generation with liquids solidified into a stable glass form [24: 92–93]. Under the Japanese disposal concept, vitrified wastes would be disposed of in a stable deep geological environment. Japan is also pursuing reprocessing plans, which, alongside China's and South Korea's reprocessing plans, are raising concerns about the proliferation of plutonium in the Asia/Pacific region [6, 52].

At present, two underground research laboratories (URLs) are testing a range of geologies as well as a disposal system in anticipation of a site being identified [24: 92–93]. As far back as 2002, Japan initially offered communities payment to be surveyed for their suitability for a nuclear waste disposal [48]. However, the mayors that initially agreed to these terms overwhelmed by negative responses from their communities and backed away from the process. Recently, Japan has created a map of geologically suitable sites in the hopes of initiating a more pared-down quest for a host community, which officials acknowledge be long and difficult [38]. However, some stakeholders remain concerned that Japan will ultimately assume a top-down approach to siting a nuclear waste repository, involving the selection of a community and the dedication of significant funds to pressure it into acceptance [48].

6 South Korea

Of the Asia/Pacific countries, South Korea has progressed most along the route to geological nuclear waste disposal. A low-level waste repository near the Wolseong power plant has been in operation since 2015 after a 20-year siting process led to its approval by 90% of the voters in a provincial referendum [58, p. 6, 40]. However, public opinion about nuclear generation has been worsening in South Korea in recent years and there is currently no politically viable plan to deal with the country's high-level nuclear waste, despite the striking of a Public Engagement Commission to advance dialogue and create public consensus around spent fuel management [8]. Meanwhile, the United States and South Korea are working to develop a longer-term solution: a novel "pyroprocessing" generation technology that would create energy without separating plutonium, virtually eliminating the footprint of high-level nuclear waste needing to be buried [8].

7 ROC (Taiwan)

In ROC (Taiwan), the utility Taipower is responsible for disposing of nuclear waste. Taipower is currently in need of more additional storage facilities to deal with its growing amount of spent fuel. Its pools for storing spent fuel in the short term are close to capacity, while medium storage currently occurs on a humid island where casks are prone to disintegration [57]. That the island is home to a large Tao aboriginal community has raised additional concerns siting decisions have not been made in a wholly respectful, consent-based way [46]. Taipower has meanwhile made plans to initiate development of a deep geological disposal repository based on a Swedish design [23]. An undersea repository design is thought to offer another alternative given the shortage of available land mass, though anti-nuclear advocates consider the seabeds of Taiwan too seismically unstable for long-term disposal [7]. Further, the islands identified for pursuing these approaches have registered objections to such developments [15].

Given the constraints it faces in long-term waste disposal, Taiwan has over the past several decades also expressed openness to offshore reprocessing opportunities in countries such as North Korea and Russia. However, opposition to these notions from South Korea, China, and the United States has blocked significant movement in this direction; as Taiwanese reactors and fuel are of U.S. origin, bilateral agreements especially require Taiwan to obtain U.S. consent for reprocessing arrangements [21]. Recently, ROC (Taiwan) has even opened discussions with

China to accept Taiwan's spent fuel inventory, after previously quashing the idea in the 1990s due to political concessions demanded in return by Beijing [15].

8 Discussion

8.1 Risk Perception and Communication

Slovic [51] argues that "it is clear that better information about radiation and its consequences is needed" to expand the acceptability of nuclear facilities with the public. Along these lines, stakeholder groups in the Asia/Pacific region are keen for information on nuclear waste disposal to be shared with the public in more transparent and democratic ways than is observed to be traditional in the Asia/Pacific culture [7]. In the case of the Fukushima Daiichi meltdown, a government-appointed panel found that communication and non-transparent decision-making by the regulator, TEPCO, and the Japanese government itself were major factors exacerbating the incident and undermining the nuclear response and evacuation [14]. Even in China, it has been proposed that due to the effective spread of information and organization of popular resistance on social media, the "nuclear industry, government and scientists must expand their risk communication initiatives to engage more stakeholders" [25].

Yet enhanced risk communication and the circulation of information about nuclear power and waste are not likely in themselves to overcome social divisions around nuclear waste storage. Manufactured risks are sensitive to "technical, social, and psychological qualities of hazard" that are difficult to soothe purely with expert risk analysis and communication [50: 675]. Publics are commonly "aware of commercial imperatives, sceptical about politics, and distrustful of the competence and impartiality of regulatory frameworks" [39: 1142]. Differing degrees of institutional mistrust are, meanwhile, found to be entangled amidst specific socio-cultural worldviews, or attitude packages that distinguish the identities of different sectors of the public [41: 1437]. The broadly left-wing worldview typed as "Egalitarian-Communitarian" features longstanding distrustful of governments and other large institutions with material aims has maintained a persistent and distinctive fear and hostility towards nuclear power in the West for three decades [11, 28].

In the Asia/Pacific region anti-nuclear views have been spreading more widely in recent years through the active advocacy of environmentalist groups connected in international networks. Some of these groups have been active since the 1980s, when nuclear power generation was introduced to the region. In past decades however, these groups have tended to be ignored by technocratic governments bent on quickly scaling up industrial societies, and their perspectives did not achieve currency amongst large-scale publics [56]. The Fukushima Daiichi incident appears to have changed this calculus, leading to abrupt declines in public acceptance of nuclear power plants in Asia/Pacific countries such as ROC (Taiwan) and

South Korea, which have similar seismic profiles [16]. Distrust of institutions has, meanwhile, been exacerbated by persistent revelations of corruption and non-transparency within these counties' technocratic regimes [10]. At the same time, countries in the region are feeling increasingly threatened by North Korea's pugnacious posture and coming to regard their nuclear power plants as vulnerable sites for potential attack [16].

Meanwhile, as the region has been growing more prosperous, a version of the "Egalitarian-Communitarian" worldview has been emerging, mirroring its presence in the West [10]. This regional shift to post-material values is exemplified in the left-wing South Korean President's recent observation that, as "[t]he country's economic status has changed, our awareness on the importance of the environment has changed," and "the notion that the safety and lives of people are more important than anything else has become a firm social consensus" [36]. Responding to this emerging sensibility, governments across the region, in South Korea and ROC (Taiwan) have tabled policies to shutter their nuclear fleets [56]. These decisions remain controversial, however, with experts concerned about impacts on energy security, economic factors such as energy prices and technology export potential, and greenhouse gas emissions [26].

Whichever decisions are taken regarding Asia/Pacific countries long-term nuclear future, decisions about waste storage will need to be made, and strategies enabling the development of secure nuclear waste facilities need to be developed. Theories of risk communication broadly support the need to formulate trust relationships, education and strategies for public engagement upon controversial subjects. With respect to nuclear waste disposal, there is now a widely held understanding that "[f]ormal and substantive involvement in the decision-making process enhances acceptance" of projects [13: 48]. Staged, deliberative processes that integrate technical and lay knowledge are now widely recommended [47: 763].

In particular, consent-based siting approaches offer communities the choice of whether to accept projects, influence over their shapes, veto rights, and authority to shut down facilities [29]. Such consent-based processes are currently becoming the *non plus ultra* of nuclear waste siting in the Western world. Evidence on behalf of the utility of the consent-based approach is provided by the Östhammer repository in Sweden which used a democratic, consent-based approach and is widely viewed as having achieved acceptability to the host community [37]. A consent-based ethic likewise implicitly underlay South Korea's siting of the low and intermediate geological repository at Wolseong, which was gone forward with after a favorable survey of the province. Arguably, countries such as ROC (Taiwan) would be able to obtain more acceptability for siting decisions if they began with the premise that the consent of communities is required to host nuclear facilities in land to which these communities are proximate, rather than assuming that repositories sited on geographically optimal lands will be accepted by the communities around them. However, in small, seismically complex Asian Pacific countries, where there may be a dearth of acceptable sites, the requirement of consent-based siting is a taller order to fulfill.

Further, even community consent-based siting processes are likely to be challenged by sharp disagreements within communities over the acceptability of nuclear waste facilities, whereby different sectors of the community are likely to privilege distinct and apparently irreconcilable goods as paramount. While the competition of pluralist worldviews in modern publics is inevitable and healthy, competing worldviews amongst stakeholders poses a significant challenge to effective dialogue and decision-making on controversial files such as nuclear waste virtually everywhere in the world. This is because positions on controversial issues such as nuclear waste are inevitably refracted by conflicting worldviews held by different sub-sections of communities, or "issue publics" [19]. Given how political affiliations diffract communities around these challenging topics, policy progress on such issues becomes difficult to achieve, as individuals on all sides come to regard the positions they hold as "badges of loyalty" to their political or socio-cultural networks, minimizing openness to new information and perspectives, and ultimately to conciliation [28: 6–7].

Mediating diverse community-based and stakeholder perspectives on topics such as nuclear waste accordingly requires creating effective forms of dialogue across polarities. The Responsible Innovation paradigm holds that the agility of communicative processes about technology must be enhanced in order to forge greater accountability as well as acceptability of technologies to society [18]. The most helpful communicative practices for decision-making around complex problems such as nuclear waste management are observed to be community-based, stakeholder-engaging strategies for exploring and integrating the multidimensional aspects of the problem space, including security and safety, socio-technical, and human factors aspects [44]. Community-based scenario planning, for instance, can help participants with diverse perspectives to collaboratively explore the complexity of a problem space, including the scope for unintended consequences beyond the immediate decision timelines of those making the decision, [32]. A similar strategy is reframing, which "helps people to become mindful of the frame they have been using to make sense of and intervene in the world, as well as what is left out of this frame" [43: 10]. Reframing is enabled "by articulating plausible, often emerging stories" that allow participants with diverse perspectives "to reperceive their world and to bring forth new options for action" [43].

Design thinking is another appropriate methodology that can help reframe the nuclear waste issue in a way supporting the simultaneous pursuit of the multiple goods that different stakeholders may privilege [34]. As described in Masys [33]:

> Design is an activity, which aims at the production of a plan, which plan – if implemented – is intended to bring about a situation with specific desired characteristics without creating unforeseen and undesired side and after effects. Hors Rittel, 1968 (cited in Lee, 2016: 19)

Design thinking is a solution-oriented methodology used by designers that draws iteratively upon logic, imagination, intuition, and systemic reasoning to explore possibilities of what could be and to create desired outcomes. Complex problems are addressed through a 5-step, iterative, non-linear phases invoking innovation,

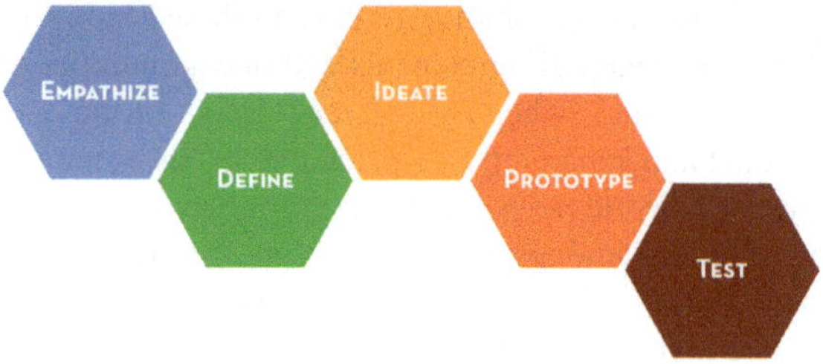

Fig. 1 Design thinking process

creativity, opportunity analysis and problem framing; see Fig. 1 and http://dschool.stanford.edu/redesigningtheater/the-design-thinking-process.

Ideally, design thinking around socially controversial issues such as nuclear waste storage is pursued in a community forum including a diverse range of stakeholders and citizens. Undertaken in this way, it holds promise for enhancing public control, trust, and unity concerning manufactured technological risks. Researchers of technology innovation have suggested that the public should be asked to generate additional design criteria to guide how technologies are implemented in given social contexts [17]. Through such processes, the integration of community-based design input into the nuclear facilities could prospectively help to enhance communities' feelings of control and security in relation to these projects.

Community-based solution design processes have been especially recommended for obstructed policy-making situations, which can be alleviated as impacted stakeholders and citizens work together to co-design specific solutions for local problems [49: 175]. As questions concerning nuclear waste will continue to haunt South East Asian countries whether nuclear power is shuttered or ramped up, engaging citizens in generating design criteria can potentially relieve the social divisions that typically gather around such problems. Cultural risk theorists hold that design-based consultation processes can shift citizens beyond the focus on belief and identity that triggers rigid worldviews by engaging them in the more practical question of how to use collective knowledge to address everyday needs [27: 7–8].

Asia/Pacific countries may especially benefit from the integration of scenario planning, reframing, or design thinking on the community level in nuclear facilities planning, given how their publics remain divided between concerns with economic prosperity on a national scale and community concerns with local harms [10]. These approaches can help to move a risk discourse such as that relating to nuclear waste management from its often polarized focuses towards an integrated paradigm of 'sustainability security,' which is simultaneously concerned with

> protecting, restoring, designing, and implementing a set of integrated natural, industrial, civilian, and security processes that equitably and responsibly meet the biophysical needs of human communities, while maintaining long-term security, respecting financial constraints, meeting ecological limits, and improving institutional arrangements [42].

The need for intensive, stakeholder and community-based reframing and designing processes is especially acute in the Asia/Pacific region, where enacting other emergent norms of public consultation around nuclear waste siting may be challenging. In particular, the norm of consent-based siting, which allows participating communities to retain a binary go/no go option as to whether they go forward with hosting a site, depends on the availability of a surplus of geophysically appropriate sites for nuclear waste disposal. If some communities reject or back out of arrangements to host the site, there will remain others to do so. As the densely populated and seismically vulnerability countries of the Asia/Pacific region may not have the same surplus of appropriate sites as other countries, design-focussed processes can give communities another way to alleviate concerns and exert control in relation to waste disposal facilities.

9 Conclusion

A high level of meaningful community, public, and stakeholder consultation are understood around the world to be required for effective nuclear waste facility siting. This requirement is increasingly pressing in the Asia/Pacific region, where views about nuclear and similar topics are intensifying and pluralizing. Even in China, where plans for nuclear expansion are unabated, experts consider that "[i]f public communication is not done properly, it [will] have a major negative impact on China's future nuclear power development" [61]. Clearly, it is incumbent on nuclear proponents and regulators in the region to develop locally-appropriate processes of intensive and meaningful public engagement around nuclear waste disposal. Public engagement on nuclear waste management could benefit from including elements of community-based scenario planning and design thinking, which may be able to produce more community acceptable integrations of scientific, economic, environmental, and social elements of the nuclear waste issue.

References

1. Beck U (1992) Risk society: towards a new modernity. Sage Publications, London, UK
2. Beck U (2009) What is globalization?. Polity Press, Cambridge, UK
3. Beck U (2002) The terrorist threat: world risk society revisited. Theory cult soc 19(4):39–55
4. Berkhout F (1991) Radioactive waste: politics and technology. Routledge, New York, NY
5. Brown PA, Letourneau C (2001) Nuclear waste fuel policy in Canada. In: Doern GB (ed) Canadian nuclear energy policy: changing ideas, institutions, and interests. University of Toronto Press, Toronto, pp 113–128
6. Brunnstorm D (2016, March 23) China still committed to nuclear reprocessing despite Asia stockpile fears. Reuters News Agency. http://www.reuters.com/article/us-nuclear-summit-china-reprocessing/china-still-committed-to-nuclear-reprocessing-despite-asia-stockpile-fears-idUSKCN0WX2KT
7. China Post (2016, October 22). Taipower mulling undersea site for disposal of nuclear waste. China Post. http://www.chinapost.com.tw/taiwan/national/national-news/2016/10/22/481802/taipower-mulling.htm

8. Cho M (2014, October 12) As nuclear waste piles up, South Korea faces storage crisis. Reuters News Agency. http://www.reuters.com/article/us-southkorea-nuclear-storage/as-nuclear-waste-piles-up-south-korea-faces-storage-crisis-idUSKCN0I10WR20141012
9. DIET Report Executive Summary (2013) https://www.nirs.org/fukushima/naiic_report.pdf
10. Dalton T, Cha M (2016, February 23) South Korea's nuclear energy future. Diplomat. http://thediplomat.com/2016/02/south-koreas-nuclear-energy-future/
11. Douglas M, Wildavsky A (1982) Risk and culture: an essay on the selection of technological and environmental dangers. U California P, Berkeley, CA
12. Dupont A (2001) East Asia imperilled: transnational challenges to security. Cambridge University Press, Cambridge
13. Flüeler T (2006) Decision making for complex socio-technical systems: Robustness from lessons learned in long-term radioactive waste governance. Springer, Amsterdam, ND
14. Glionna J (2011, December 27) Japan, utility at fault for response to nuclear disaster, report says. Los Angeles Times. http://articles.latimes.com/2011/dec/27/news/la-mobile-japan-tepco-nuclear-disaster
15. Global Post (2014, May 12) Taiwan in talks with China, France for nuclear waste deals. Global Post. http://www.globalpost.com/dispatch/news/kyodo-news-international/140512/taiwan-talks-china-france-nuclear-waste-deals (dead link)
16. Green J (2017, May 25) Is South Korea's nuclear industry a model for others to follow? Nuclear Monitor 844. https://www.wiseinternational.org/nuclear-monitor/844/south-koreas-nuclear-industry-model-others-follow
17. Guston DH, Sarewitz D (2002) Real-time technology assessment. Technol Soc 24(1–2):93–109
18. Guston DH, Fisher E, Grunwald A, Owen R, Swierstra T, van der Burg S (2014) Responsible innovation: motivations for a new journal. J Responsible Innov 1(1)
19. Han H (2009) Moved to action: motivation, participation, and inequality in American politics. Stanford University Press, Palo Alto, CA
20. Helbing D (2013) Globally networked risks and how to respond. Nature 497:51–59
21. Hibbs M (2009, June 1) Long-term spent fuel dilemma at issue in Taiwan-US renegotiation. Nuclear Fuels
22. Huiling C (2008) Strategy for high level waste management in China. J Nucl Sci Technol 45(Supplement 5):16–19
23. Hung C, Wu Y, King F (2017) Corrosion assessment of canister for the disposal of spent nuclear fuel in crystalline rock in Taiwan. Corros Eng, Sci Technol 52(S1):194–199
24. Japan Atomic Energy Agency [JAEA] (2015) Progress in the decommissioning of nuclear facilities and the treatment and disposal of radioactive waste. JAEA R&D Rev 8
25. Jia H (2016, October 31) New nuclear faces public suspicion in China. Chemistry World. https://www.chemistryworld.com/news/new-nuclear-faces-public-suspicion-in-china-/1017622.article
26. Jung-a S (2017, June 19) South Korea steps back from nuclear power. Financial Times. www.ft.com/content/40871236-45da-11e7-8519-9f94ee97d996
27. Kahan D (2015) What is the "science of science communication"? J Sci Commun 14(3):1–10
28. Kahan D, Braman D, Jenkins-Smith H (2011) Cultural cognition of scientific consensus. J Risk Res 14:147–174
29. Kunreuther H, Easterling D, Desvousges W, Slovic P (1990) Public attitudes toward siting a high-level nuclear waste repository in Nevada. Risk Anal 10(4):469–484
30. Lochbaum D (1996) Nuclear waste disposal crisis. Pennwell, Tulsa, OK
31. MacFarlane A (2011) It's 2050: do you know where your nuclear waste is? Bull At Sci 67(4):30–36
32. Masys AJ (2012) Black Swans to Grey Swans: revealing the uncertainty. Int J Dis Prev Manage 21(3):320–335
33. Masys AJ (2016) Counter-terrorism and design thinking: supporting strategic insights and influencing operations. In: Masys AJ (ed) Disaster forensics: understanding root cause and complex causality. Springer Publishing, Berlin

34. Masys AJ (ed) (2015) Applications of systems thinking and soft operations research in managing complexity. Springer Publishing, Berlin
35. Masys AJ, Ray-Bennett N, Shiroshita H, Jackson P (2014) High impact/low frequency extreme events: enabling reflection and resilience in a hyper-connected world. In: 4th international conference on building resilience, 8–11 September 2014, Salford Quays, United Kingdom. Procedia Economics and Finance, vol 18, pp 772 –779
36. McCurry J (2017, June 19) New South Korean president vows to end use of nuclear power. https://www.theguardian.com/world/2017/jun/19/new-south-korean-president-vows-to-end-use-of-nuclear-power
37. Metlay D (2013, July-August) Consent-based siting: what have we learned? Radwaste Solutions 28–36
38. Nikkei Asian Review (2017, July 28) Japan map shows possible permanent sites for nuclear waste. https://asia.nikkei.com/Politics-Economy/Economy/Japan-map-shows-possible-permanent-sites-for-nuclear-waste?n_cid=NARAN1507
39. Owens S (2000) Engaging the public: information and deliberation in environmental policy. Environ Plann A 32:1141–1148
40. Peachey C (2014, July 22) Korean repository realised. Nucl Eng Int. http://www.neimagazine.com/features/featurekorean-repository-realised-4323899/
41. Peters E, Slovic P (1996) The role of affect and worldviews as orienting dispositions in the perception and acceptance of nuclear power. J Appl Soc Psychol 26:1427–1453
42. Prizzia R, Levy J (2017) Towards climate security and sustainable security in the Asia-Pacific region. In: Masys AJ, Lin L (eds) Asia/Pacific security challenges: managing black swans and persistent threats. Springer Publishing, Berlin
43. Ramirez R, Wilkinson A (2016) Strategic reframing: The Oxford scenario planning approach. Oxford University Press, Oxford
44. Reason J (1999) Managing the risks of organizational accidents. Ashgate Publishing, Aldershot
45. Rich M (2017) Struggling with Japan's nuclear waste, six years after disaster. New York Times, March 11, 2017. https://www.nytimes.com/2017/03/11/world/asia/struggling-with-japans-nuclear-waste-six-years-after-disaster.html?mwrsm=Email
46. Roberts C (2011, October 7) Taiwan's paradise island fights to save its identity. BBC News. http://www.bbc.com/news/world-radio-and-tv-15182502
47. Rosa EA et al (2010) Nuclear waste: knowledge waste? Science 329:762–763
48. Ryall J (2017, May 5) Japan seeks final resting place for highly radioactive nuclear waste. Deutsche Welle. http://www.dw.com/en/japan-seeks-final-resting-place-for-highly-radioactive-nuclear-waste/a-38709488
49. Schon DA, Rein M (1994) Frame reflection: toward the resolution of intractable policy controversies. New York, NY: Basic Books
50. Slovic P (1993) Perceived risk, trust, and democracy. Risk Anal 13(6):675–682
51. Slovic P (2012, May 1) The perception gap: radiation and risk. Bull At Sci 68(3)
52. Sokolski H (2016, March 28) Can East Asia avoid a nuclear explosive materials arms race? Bull At Sci. http://thebulletin.org/can-east-asia-avoid-nuclear-explosive-materials-arms-race9295
53. Sudworth J (2013, July 13) Plans for Chinese nuclear plant cancelled. BBC News (2016). http://www.bbc.com/news/av/world-asia-china-23298472/plans-for-chinese-nuclear-plant-cancelled
54. United Nations Environmental Programme (1992) Rio declaration on environment and development. http://www.unep.org/documents.multilingual/default.asp?documentid=78&articleid=1163
55. Wang J (2010) High-level radioactive waste disposal in China: update 2010. J Rock Mech Geotech Eng 2(1):1–11
56. Wen LH (2017, January 18) Taiwan joins global anti-nuclear trend. Asia Times. http://www.atimes.com/article/taiwan-finally-joins-anti-nuclear-movement/

57. World Information Service on Energy [WISE] (2012, January 5) Taiwan, Ukraine, United Kingdom, USA. Nuclear Monitor (746, 747, 748) https://www.wiseinternational.org/nuclear-monitor/746-747-748/taiwan-ukraine-united-kingdom-usa
58. World Nuclear Association (2015) Issues paper 4: management, storage, disposal of nuclear and radioactive waste. Submission to the South Australian Nuclear Fuel Cycle Royal Commission
59. World Nuclear Association (2017) International nuclear waste disposal concepts. World-nuclear.org. http://www.world-nuclear.org/information-library/nuclear-fuel-cycle/nuclear-wastes/international-nuclear-waste-disposal-concepts.aspx
60. World Nuclear Association (2017) Nuclear power in the world today. World-nuclear.org, updated August 2017. http://www.world-nuclear.org/information-library/current-and-future-generation/nuclear-power-in-the-world-today.aspx
61. Zhu C (2013, July 18) Public trust crisis threatens China's nuclear power ambitions. Reuters News Service. http://www.reuters.com/article/us-china-nuclear-protests/public-trust-crisis-threatens-chinas-nuclear-power-ambitions-idUSBRE96H1BT20130718

Building Effective Emergency Management Public-Private Partnerships (PPP) for Information Sharing

Jason Levy and Ross Prizzia

Abstract Recent disasters have highlighted the need for effective and proactive partnerships between public emergency responders and the "Whole Community" to assist in all phases of disaster management. This paper examines the importance of reliable, credible, and accessible information that can support the private sector and the "Whole Community" in times of disaster. It is shown that community resilience is enhanced through a shared, real-time "common operating picture", explicit linkages, and enhanced information sharing. Business Emergency Operations Center (BEOC) designs are compared and contrasted and recommendations are put forth. Optimal Public Private Partnerships for information sharing identified. It is shown that achieving disaster resilience involves government and businesses sharing potential risks, as well as identifying disaster consequences, priorities, information, and capabilities. The conclusions provide guidance that can be used as dual-use roadmap for improving information sharing: on the one hand governments can use the insights of this paper to build new public-private sector programs (or to improve existing programs); on the other hand, the private sector can better understand how to become connected with the information sharing processes of governments.

Keywords Public-Private Partnerships (PPP) · Information sharing
Emergency management · Business Emergency Operation Centers (BEOC)

1 Introduction: The Whole Community Approach to Disasters

The Federal Emergency Management Agency (FEMA) "Whole Community" concept is a recent attempt to engage the whole spectrum of organizations found in a local community as part of the emergency management lifecycle. Recent disasters

J. Levy (✉) · R. Prizzia
Disaster Preparedness and Emergency Management Program,
Public Administration Division, University of Hawaii-West Oahu,
91-1001, Farrington Highway, Kapolei, HI 96707, USA
e-mail: jlevy@hawaii.edu

A. J. Masys (ed.), *Security by Design*, Advanced Sciences and Technologies
for Security Applications, https://doi.org/10.1007/978-3-319-78021-4_18

have highlighted the need for effective and proactive partnerships between public emergency responders and the "Whole Community" to assist in all phases of disaster management. Emergency preparedness information sharing is essential to ensure, comprehensive, holistic and integrated emergency management approach that incorporates the "Whole Community" in emergency planning and communications. The popularity of the "Whole Community" concept motivated states and local governments to develop a wide range of programs (such as the Safeguard Iowa Partnership and the Colorado Emergency Preparedness Partnership) designed to integrate individual, business and government preparedness with the entire community in preparedness, planning, response, recovery, and mitigation efforts. These comprehensive and integrated programs were put into place within localities, states and regions as a conduit for disaster management and continuity of operations partnerships between the public and private sector across the entire emergency management cycle.

Within this context, the Whole Community (including the private and non-governmental sectors) has become vital partners in the solution of homeland security and emergency management issues. Homeland Security Presidential Directive (HSPD-5) states, "The Federal Government recognizes the role that the private and nongovernmental sectors play in preventing, preparing for, responding to, and recovering from terrorist attacks, major disasters, and other emergencies. The Secretary will coordinate with the private and nongovernmental sectors to ensure adequate planning, equipment, training, and exercise activities and to promote partnerships to address incident management capabilities."

2 Information Sharing and Public-Private Sector Partnerships During Disasters

Improved access to information and data at all levels benefits both the public and private sectors. However, disaster resilience requires more than information access; it also demands a better understanding and definition of priority information requirements (based on what information the public and private sectors most critically need to report on, share and access). The sharing of disaster-related information among and between government and business partners provide a robust decision-making landscape that helps the public and private sectors to prepare for, mitigate against, respond to and recover from, critical incidents. Timely, accurate, and critical information about the disaster environment increase efficiency in response and recovery operations planning. Enhanced operational coordination between the private sector and jurisdictions is *sina qua non* for businesses and governments to support effective disaster management. The improved understanding of disaster processes and capabilities across the public and private sector can reduce the challenges caused by uneven information sharing linkages.

As described in Fig. 1, businesses need to understand the emergency management priorities and resources of the public sector so they can align their business operations. For example, prior to the onset of an incident (i.e. "pre-event"), the private sector requires the following from the public sector: information about frameworks, regulations and incentives; situational awareness (from EOCs and Fusion Centers); evacuation information; access and credentialing information; pre-disaster declarations; vulnerability and hazard information, etc. Conversely, governments can benefit from understanding the needs and capacities of the private sector. As shown in Fig. 1, the public sector requires the following pre-event information from the private sector: threat awareness; industry critical assets and key resources; corporate risk and vulnerability assessments; inventory levels of life support systems and industrial goods; facilities in hazard zones, etc.

In 2012, FEMA launched the National Business Emergency Operations Center (NBEOC) to serve as a valuable centralized information-sharing hub for the US business community during the response phase of a disaster. Soon after its inception the NBEOC showed its value by coordinating multiple Hurricane Sandy calls with the private sector in late 2012; by consolidating and sharing information from multiple federal agencies; and by highlighting the impact and needs of businesses so that the public sector could respond more effectively. The ability of the private sector to respond and recover from Hurricane Sandy's impact was in large part due to the ability of the NBEOC to improve information sharing processes between businesses and the public sector, including senior FEMA, DHS, and White House staff. For example, during the disaster response phase jurisdictions benefit from knowing whether businesses remain open and if/when they plan to close.

During the response phase, identifying shared public-private sector priorities in operational time periods can help public and private sectors to incorporate the most important mission priorities for each company and agency into incident action plans

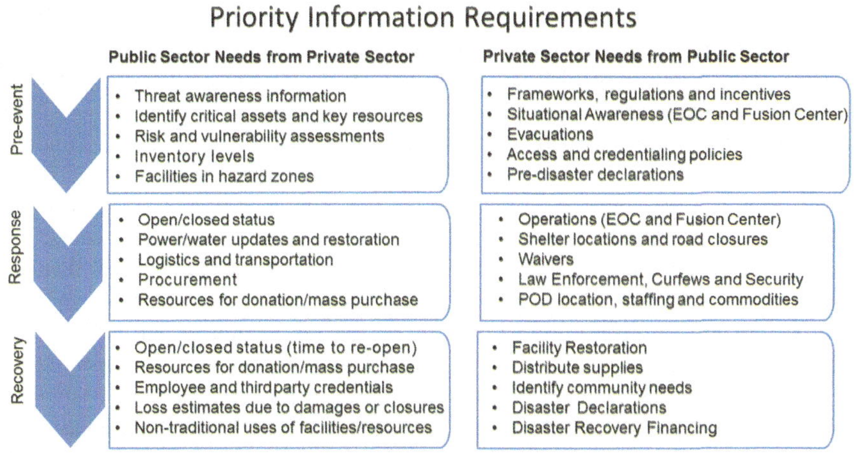

Fig. 1 Priority information requirements of the public and private sectors

(e.g. the need to re-supply stores and setup logistics). In this way, both governments and companies can develop a shared understanding of the most essential operational function for continuing functionality of an organization. In particular, during the response phase, the private sector needs public sector information about EOC and fusion center operations, shelter locations, road closures, emergency waivers to regulations, law enforcement information pertaining to curfews and security, Point of Delivery (POD) locations and staffing and commodity levels. The public sector can further improve disaster operations if they know the status of a business' critical infrastructure (power, water, water life support systems) as well as any logistical/transportation bottlenecks. Finally, governments are greatly aided by knowing private sector resources that are available for donation or mass purchase.

Business sector recovery is critical for the resilience of the entire community. After a major disaster, it is known that a quarter of small companies affected by business interruptions never reopen (Insurance Institute of Business and Home Safety, 2012). During the recovery stage, governments need to know when businesses plan to re-open, the credentials of business employees and third party contractors, disaster loss estimates (due to damage or closure) and non-traditional uses of facilities and resources. Finally, post-disaster the private sector require information about the public sector's plans for facility restoration, the distribution of supplies, community needs, disaster recovery financing and disaster declarations. Figure 1 provides more details about these issues.

However, effective information sharing processes between the public and private sectors are often lacking and companies face at least four major challenges in achieving this. First, the private sector is often burdened by a lack of coherent information sharing standards and there is often an inconsistent process for obtaining the priorities of the public sector. In general, public sector information sharing processes directed at businesses associated with the US National Infrastructure Protection Plan (NIPP) goals and supporting objectives is less mature than information sharing processes between fusion center partners and businesses. While this general gap exists (information sharing networks associated with NIPP private sector partners lag behind fusion center information sharing), more efforts is needed by all federal, state, local, territorial and tribal partners to improve information sharing processes and associated organizational structures. In the US, major challenges have been identified with information sharing processes and communication conduits used by the public sector in order to communicate with businesses: inadequate information sharing continues to hamper the resumption of normal business operations and the ability of governments mount an effective response to disasters.

Second, the private sector shares data and information much differently than the public sector; accordingly, governments will need to be more flexible in their approach to elicit disaster information from businesses. Moreover, businesses often lack a coherent, clear and official conduit of disaster information. Regardless of type or size of the private or public sector organization, governments and businesses require versatile, integrated, collaborative, effective and agile coordination of emergency response capacities to increase disaster resilience and reduce the associated costs of critical incidents for citizens, customers and communities.

Third, companies may face overlapping, time consuming and onerous information requests by multiple public sector entities. For example, multiple levels of government may request insights into the logistics aspects of private sector supply chain and operations. Hence it is important to minimize the resulting workload and frustrations faced by the private sector, particularly in times of crisis. Such delays in obtaining information can have major consequences since obtaining the right information in a timely manner is more valuable to responders and businesses than more complete, perfected information that arrives too late. Knowing the kinds of information needed before the event, helps improve speed, accuracy, and actionability in an event.

Fourth, information sharing is currently crisis-oriented. The crisis oriented nature of current information sharing arrangements often creates false assumptions and misunderstandings since the extent of the disaster and the resources and capabilities may not be fully known. While the intensity, quality, and effectiveness of information-sharing are often superior during the crisis phase of an emergency, bi-directional communication is equally important during the pre-disaster and post-disaster phases. An operational coordination framework is needed that aligns key decision thresholds, information requirements, data sources, and resource allocation decisions throughout the disaster management process.

3 Introduction to Private Public Sector Partnerships (PPPs) and Business Emergency Operations Centers (BEOCs)

The concept of Public-Private Partnerships (PPP) is increasing in importance in the emergency management community. PPPs have demonstrated their value due to their remarkable Return on Investment (ROI): they leverage modest investments into significant growth in emergency management capabilities. PPPs strengthen trust among partners as business and government partners work together at the local, regional and national levels to improve the disaster resilience of their communities on an ongoing basis across many initiatives. Such Partnerships can also help businesses to more effectively coordinate multiple and overlapping requests from federal, state and local government agencies. Partnerships also allow business and government leaders to creatively design and rank their own disaster management initiatives according to the needs and priorities of their respective regions. The principles of joint planning and operational coordination have been used to develop actionable, flexible, consistent, and scalable PPPs that promote the development and implementation of coherent emergency management plans and support practical and operational cross sector and cross industry relationships Most PPPs embrace the participation of the "whole community" (including all levels of government and all sizes of businesses from very small businesses to large corporate enterprises) and emphasize the development of robust channels of communication that can be used before, during and after a critical incident occurs. They can also include joint training, planning and exercises among the public and private sectors.

A business sector that is able to effectively share the status of their industry and to contact a resource during a disaster will be far better positioned for an effective response and recovery effort than one that is not. Desirable outcomes of the PPP approach include improved information sharing and a positive return on investment (ROI) for both public and private sectors due to the partnership. This requires the use of PPP models that capture business requirements and standards and can be widely adopted at the local, state, regional and national level.

Opportunities to actualize PPPs and information sharing guidance include expanding and revising FEMA's Emergency Management Institute (EMI) PPP Courses; developing technical assistance workshops to assist in developing the initial operating capability for governments and businesses; and partnering with non-governmental organizations to produce community-driven PPP guidance (rather than applying government-driven top down approaches). A range of partnerships, such as Safeguard Iowa, the Colorado Emergency Preparedness Partnership, and FEMA's NBEOC not only strive to implement effective information sharing but also serve to improve operational coordination efficiency, enhance community resilience and increase the speed of economic recovery by emphasizing the "Whole Community approach" to disaster management. For example, Minnesota's P2CAT program (Public/Private Coordination & Action Team)—constitutes an information sharing and response coordination program with the Minnesota Homeland Security and Emergency Management (HSEM). Led by the Minnesota FBI Infragard Chapter this public/private partnership effort is located within the State of Minnesota's EOC and supports the sharing of information; P2CAT also helps to coordinate response actions between the public and private sectors in Minnesota.

In general, a PPP deals with a broader range of emergency management activities than response and spans the full-cycle of emergency management from preparedness to long-term community coordination and recovery. There are many models for enabling efficient public-private operational coordination through improved information sharing. A PPP is a broad coalition of members from which BEOC participants are drawn when a RRCC/BEOC is activated to respond to an event. Hence BEOCs and PPPs differ in that the former typically focuses on life-saving, life-sustaining disaster operations in the critical 72-hour window of a response operation while the latter is broader than the disaster response phase, often spanning the full-cycle of emergency management from preparedness and mitigation to long-term community coordination and recovery.

Regardless the PPP program used, successful public/private industry engagement requires improving the intensity, quality, and effectiveness of information-sharing and understanding both the workflow of activities cutting across various business units and the standard operating procedures as they relate to an organizations' operational phases. Once critical business activities, procedures and phases are known, emergency support functional area annexes should be designed to specifically align (and provide definition) to them. As noted in *Lessons Learned Information Sharing, Best Practice, Public-Private Partnerships for Emergency Preparedness* (2017):

> Both the public and private sectors have a stake in disaster prevention, mitigation, response, and recovery, but they often lack the relationships necessary to coordinate preparedness efforts. In many communities, public safety agencies and private sector entities conduct planning and preparedness operations largely independently of each other. As a result, each often fails to fully understand or appreciate the other's role in emergency prevention, preparedness, mitigation, response, and recovery. Public safety agencies often underestimate the private sector's interest and involvement in emergency preparedness efforts. Private sector groups can overestimate the capabilities of public sector partners, failing to recognize the need for their own contributions to an incident response. In addition, the private sector often perceives cooperation with public safety agencies as risky because of concerns about the protection of proprietary information and public disclosure regulations.

4 BEOCs Begin to Emerge at All Levels

A Business Emergency Operations Center (BEOC) seeks to improve two-way communication between the private sector and public response officials within an Emergency Operations Center (EOC), Response Coordination Center (RCC) or Incident Command Center (ICC). Hence, a BEOC constitutes a common type of PPP program formed around emergency response principles which focuses on the principles of joint planning, operational coordination and disaster-resilience. Discussions on the BEOC concept began around 2007, and were followed up by efforts in a small number of states to create them. Consistent with NIMS and ICS, BEOCs help to better integrate private sector assets into response operations to improve government response effectiveness. In this way BEOCs have helped to reduce business interruptions and help companies restart critical operations more quickly.

However, in practice, the terms "BEOC" and "PPP" tend to be used interchangeably. Often times, the same entity engages businesses with preparedness, mitigation and continuity activities during steady state as a general PPP, and then converts to a more focused BEOC during a specific incident. In other words, the BEOC tends to be more directly integrated into a response center focusing on life safety issues (i.e. an EOC, RRCC, NRCC, etc.) during the incident, and then steps back out of that role once the response/recovery phases end. As shown in Table 1, the authors identified 30 PPPs in place across the USA including those at State level (Florida, Louisiana, Illinois, Missouri, New Jersey, Colorado, etc.) and a smaller number at the regional and city levels. There is also a National Business Emergency Operations Center (a federal BEOC), run by FEMA. The type of incident required to activate a state, regional and national level BEOC is provided in Fig. 2.

5 BEOC/PPP Case Studies

The US National BEOC is located in FEMA's National Response Coordination Center (NRCC) and works closely with FEMA's Private Sector Division in Washington, DC. FEMA's Regional Response Coordination Centers (RRCC) are

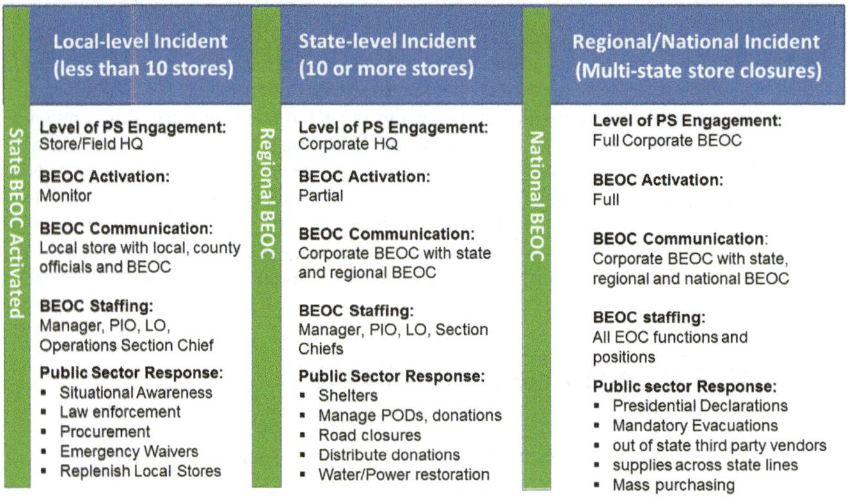

Fig. 2 Activation of state, regional and national BEOCs

Table 1 A sample of PPPs identified in the United States

Jurisdiction	Public-Private Sector Partnership	Type	Platform
1. Los Angeles	LA Business Operations Center	City	Physical
2. New York City	NYC EOC Private Sector (PS) Business ESF	City	Physical
3. Northeast	Nedrix	Interstate	
4. Chicago	ChicagoFIRST	Regional	
5. Cincinnati	Greater Cincinatti VBEOC	Regional	Virtual
6. San Diego	ReadySanDiego Business Alliance	Regional	Virtual
7. California	California Governor's Office of Emergency Services	Regional	
8. FEMA Region VII	RRCC/Regional BEOC	Regional	Virtual and physical
9. Washington DC	DC Business Emergency Management and Operations Center (DC BEMOC)	Regional	
10. Alaska	The Alaska Partnership for Infrastructure Protection (APIP)	State	
11. Arizona	Business Emergency Coordination Center	State	
12. Arkansas	ARBEOC	State	
13. Colorado	Connect Colorado/Emergency Preparedness Partnership	State	Physical
14. Florida	Florida Virtual BEOC	State	Virtual
15. Georgia	Business Operations Center	State	
16. Idaho	Idaho BEOC (WebEOC and Saber)	State	Virtual
17. Illinois	Illinois BEOC	State	Physical

(continued)

Table 1 (continued)

Jurisdiction	Public-Private Sector Partnership	Type	Platform
18. Indiana	Indiana Department of Homeland Security	State	
18. Iowa	Safeguard Iowa Partnership (SIP)	State	Independent, physical and integrated
19. Kentucky	KYVBEOC	State	
20. Louisiana	LA BEOC	State	Physical and virtual
21. Los Angeles	Los Angeles EOC (business breakout room)	City	Physical
22. Maryland	Maryland Emergency Management Agency	State	Physical and virtual
23. Minnesota	P2CAT (Public/Private Coordination and Action Team)	State	Physical
24. Missouri	Missouri Business Operations Center (BOC); Missouri Public Private Partnership (MOP3)	State	Physical and virtual
25. New Jersey	BEOC Alliance	State	
26. North Carolina	North Carolina BEOC	State	
27. Ohio	Ohio Public Private Partnership Program (OP3)	State	
28. Rhode Island	Rhode Island Business Alliance (RIBA)	State	Virtual
29. Texas	TX BEOC	State	
30. National	National BEOC	Federal	

located and operate within each of FEMA's ten regional offices across the US. These facilities provide support to each of the states within the regional boundaries and assist with disaster operations occurring in other FEMA regions during major multistate events. Each RRCC works directly with the NRCC in Washington, DC. Each Regional Private Sector Liaison support RRCC operations when their respective Regions are activated, though not all of them have a formal "RBEOC" function. In this manner the RBEOC serves as a liaison with and between the public and private response. For example in FEMA Region IX (where Hawaii is located), Michael Cummings is that Region's PS Liaison and staffs the RRCC's PS Specialist seat during an activation. The use of BEOCs for risk analysis are shown in Fig. 3.

5.1 FEMA Region VII's BEOC

FEMA Region VII's BEOC, the first Regional BEOC (RBEOC) in the country, is located in Kansas City, Missouri. The Region VII BEOC serves as a vital coordination linkage among state BEOCs (within the FEMA Region VII) and the FEMA National BEOC during the disaster response phase. The FEMA Region VII BEOC (RBEOC) is part of the FEMA-RRCC Emergency Support Function (ESF) 15 (i.e. the "External Affairs cell") and interfaces with FEMA Region VII

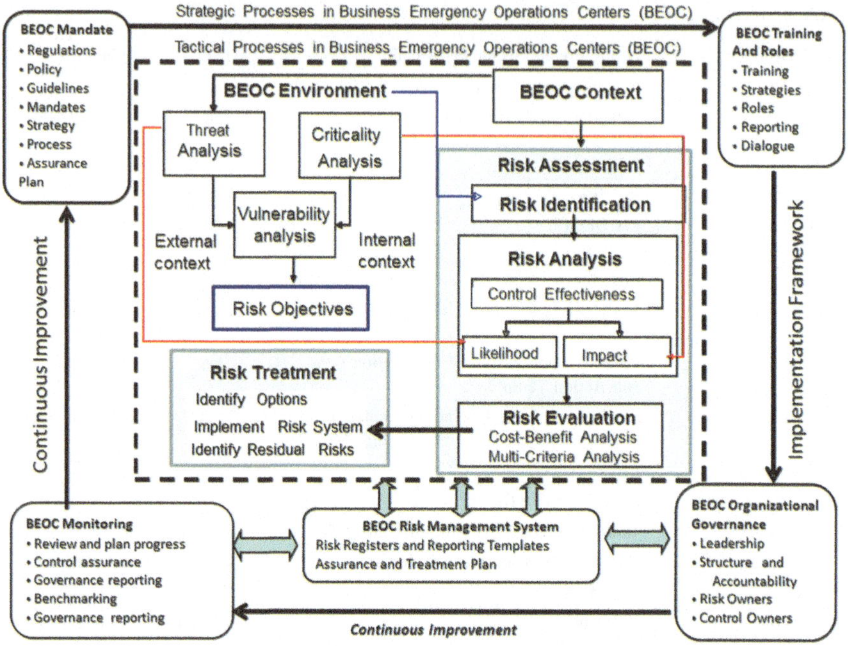

Fig. 3 BEOCs and risk analysis

Logistics and Operations. The primary mission of the FEMA Region VII BEOC (RBEOC) is to strengthen response operations by integrating private sector interests, assets and situational awareness into response operations. In contrast, the Regional BEOC (RBEOC) in Region III (comprised of DC, Virginia, and Maryland) is a more formalized process, but similar to Region VII in that the RBEOC is managed virtually by FEMA's Regional Private Sector Liaison (PSL) while they staff the RRCC.

Region VII established an RBEOC within the RRCC provide two-way communication between public and private sectors during a national exercise and a real response event. This has the potential to add significant value to FEMA's disaster response capabilities in support of disaster response efforts. The Regional BEOC is only activated during an event (not during steady state activities). When FEMA's Region VII Response Division activates its RRCC, the Region VII Response Chief and External Affairs Director may decide to activate the RBEOC physically (i.e. inside the FEMA Region VII RRCC) or virtually. Hence, in the event of an activation, representatives of regional businesses have the option of participating in response activities virtually or being physically present in the Region VII RBEOC (the number of private sector representatives to the RBEOC/RRCC recently expanded from three to ten). Public representatives from states outside of Missouri tend to interact virtually with the RBEOC.

Operations may be augmented to integrate private sector interests into the response operation, primarily through sharing information to enhance situational awareness and response among the four state regions of Iowa, Missouri, Nebraska, and Kansas. For example, following the Joplin tornado, the FEMA Region VII RBEOC and the Missouri BEOC partners coordinated to augment the response operation by sharing situational awareness about the critical need to expedite road clearances and debris removal so businesses could expedite the shipments of supplies to aid the response. This was coordinated from the federal to the local level as follows: first, the US National BEOC (NBEOC) worked with the Region VII (Regional) BEOC (through logistics and operations channels). Next the Regional BEOC coordinated with the state BEOC that in turn assisted with the local response operation. In this way, the coordination quickly opened roads so that private sector firms could bring in emergency supplies.

In addition to an RBEOC, FEMA Region VII also has a regional Public-Private Partnership (PPP) which includes multiple businesses and entities throughout the four state regions of Iowa, Missouri, Nebraska, and Kansas. When the Region VII RRCC is activated, one or more PPP members will form an RBEOC in the Region VII RRCC. The goal of the Region VII PPP is to strengthen emergency management operations during the disaster and promote economic development following disaster events. Region VII private sector partners include Walgreens, The Home Depot, Sprint, Cyber Sciences Corporation, Commerce Bank and the University of Kansas Medical Center. Region VII partners frequently meet in various regional forums and business partnership workshops. As FEMA's Regional Private Sector Liaison for Region VII, Jonathan Weinberg serves as the Point of Contact (POC) for both the Regional BEOC and PPP. Participation is voluntary for the Region VII external partners so there are no costs, except for very minor invitational travel on very rare occasion, which is part of FEMA's regular External Affairs operational budget.

5.2 *Missouri Public Private Partnership (MOP3)*

The Missouri Public Private Partnership (MOP3), authorized under the Missouri Homeland Security Advisory Council (HSAC), is a voluntary coalition of Missouri's private and public sector organizations which seeks to foster the direct involvement of the private sector to augment and support the capacity of Missouri to strengthen the to prevent, prepare for, respond to, and recover from disasters. The MOP3 has partnered with FEMA Region VII to credential Private Sector Representatives (PSRs) within the Regional Business Emergency Operations Center (RBEOC) in the Regional Response Coordination Center (RRCC). MOP3 currently has a number of PSRs that report to FEMA's Kansas City office to represent private sector interests during a disaster response when the Regional BEOC is activated. During steady state activities the Regional PPP is still active (although the Region VII RBEOC is not).

The state of Missouri Business Emergency Operations (BEOC) is a voluntary affiliation of dozens of businesses and associations committed to a private-public partnership with State of Missouri Emergency Management Agency (SEMA). Embedded within the State EOC, the state of Missouri BEOC utilizes training, exercises, and activations to enhance the state's disaster response capabilities in an incremental and iterative process. The Missouri BEOC ensures "continuity of community" and helps companies to maintain the continuity of their own private sector operations, thereby relieving pressure on State resources. The Missouri BEOC enhances the bidirectional flow of information pertaining to activities, policies, and other disaster management efforts affecting key assets and critical infrastructure. In this way, private sector representatives can inform the public sector about industry's needs and how the government can meet those needs. The private sector can also identify industry strengths and ways to increase private sector support to the government: there are numerous ways in which industry can provide supplies, services, and assistance to the public sector, thereby promoting community resilience. Working side-by-side with government, BEOC private representatives benefit from regular state updates, direct access to critical information and enhanced situational awareness in order to support more informed emergency management decisions. Through the BEOC, both the private sector and the government have the opportunity to develop trusted relationships and become true partners in the emergency response effort. These relationships will benefit both government and the private sector in addressing a range of issues. For example it may be mutually advantageous to increase the protection of companies responding to disasters against lawsuits arising from their disaster response efforts (i.e. updated state liability protection for businesses).

5.3 *Safeguard Iowa Partnership (SIP)*

The Safeguard Iowa Partnership (SIP) seeks to strengthen the capacity of Iowa to prevent, prepare for, respond to and recover from disasters through public-private collaboration. The Safeguard Iowa Partnership (SIP) sends LNOs (Liaison Officers) to the state of Iowa EOC but neither Iowa nor SIP has a BEOC. The SIP recently trained 10 new LNO to the State EOC according to its 2014 annual report. During an "activation" the SIP LNOs go to the state of Iowa EOC at the Joint Forces Headquarters in Johnston, Iowa where they have a seat at the table at the Emergency Support Function (ESF) for business industry and commerce. In this sense the SIP is physical and "independent" but also integrated within state EOC.

5.4 *North Carolina BEOC*

According to the North Carolina BEOC representative, the state of North Carolina has both a physical and virtual BEOC. The State of North Carolina BEOC operates inside the State EOC which is organized using the Incident Command System (ICS) with Emergency Support Functions. The Private Sector emergency management partners are represented through representatives from the various private sectors (referred to as 'Chairs') carry out the following important tasks: monitor EOC operations, examine situation reports, provide status updates and carry out sector analyses on possible and actual disaster impacts. These private sector representatives also are involved in face to face coordination, support BEOC Resource Requests, and assist with the donations process. The North Carolina BEOC is extremely versatile as it contains both physical and virtual components. N. Carolina has created uniform guidelines across the entire state for private sector disaster re-entry which have also been implemented in S. Carolina and Florida.

In the North Carolina physical BEOC, the Private Sector emergency management representatives are located inside the State EOC (SEOC). The presence of a physical BEOC allows the sector 'Chairs' inside the SEOC to identify disaster information that is critical to their particular industry. They can also work with the State Emergency Response Team (SERT) on sector analyses. The private sector representatives also become involved in coordination calls and BEOC resource requests. Other critical activities involve supporting the donations management process, monitoring the task board, engaging in face to face coordination with the SERT and examining disaster briefs pertaining to the changing environment.

In the North Carolina virtual BEOC, the private sector representatives are connected virtually through WebEOC which enables real time updates. The representatives also have the opportunity to obtain information on the operational status and to add information about potentially significant incidents to enhance situational awareness and create a comprehensive common operating picture. Specific North Carolina initiatives include "Eyes on the Ground". The private sector organizations have a number of options to access to WebEOC so they can be fully engaged in the disaster response process: the private sector can use not only County Emergency Management Boards but also BEOC Boards and the Duke RSS feed.

5.5 Maryland Emergency Management Agency's (MEMA) Private Sector Integration Program (PSIP)

The Maryland Emergency Management Agency's (MEMA) Private Sector Integration Program (PSIP) seeks to facilitate open communication lines between MEMA and the private sector. The PSIP is used to foster disaster management partnerships, to enhance situational awareness, and to improve information sharing without undercutting existing vendor contracts or replacing existing procurement procedures. Members of the PSIP can access the Maryland Virtual Business Operations Center (VBEOC) at any time during an emergency activation for access to the latest situational awareness and other disaster-specific resources. MEMA operates on a 24/7 schedule during large scale emergency events and requests that each PSIP member organization provide at least one emergency representative who can serve as the business contact person during disaster periods.

Through the state of Maryland VBEOC, PSIP members can also contact State Emergency Operations Center officials. During emergency events, the Business Operations Center (BOC), the operational component of the Private Sector Integration Program (PSIP) is activated. The BOC provides PSIP members with disaster information to assist the private sector in making business operations and continuity decisions. In this way, the Business Operations Center, housed within the State of Maryland's Emergency Operations Center, seeks to ensure effective communication with the private sector during emergencies and supports OSPREY Business. Publication of a PSIP member organization's association with the PSIP/BOC occurs at the discretion of the PSIP Liaison at MEMA.

Both the Maryland Emergency Management Agency (MEMA) and the North Carolina Emergency Management agency collaborate on a BEOC Concept of Operations format known as the Business 'B' (Fig. 4): the left side (stem) of the 'B' shows private sector daily operations: partnership building, outreach, WebEOC training, exercises, re-entry certification & renewal, the application of corrective action, review and updates to the State Emergency Operations Plan (SEOP). Other important daily private sector activities include the implementation and use of Private Sector Working Groups (PSWG) and the continued development of private-public sector programs and initiatives.

The upper curve in the B depicts BEOC activities in the response phase. The multi-pointed star at the top of the B (start of the response phase) highlights that a critical incident has occurred or is expected. This event is immediately followed by the activation of the SEOC which in turn causes the BEOC to begin continuous operations. In its operational period the BEOC performs the following activities: enhances situational awareness and information sharing to the private sector through a number of activities, including coordination calls, private sector briefings and sector analyses, real-time private sector credentialing for re-entry to the disaster zone, fostering private sector linkages to enhance business to business mutual aid, and the coordination of private sector donations (by including county EM Directors and private sector partners). The lower curve of the B shows the BEOC moving into the recovery phase of operations. In the recovery phase, private sector chairs continue private sector disaster analyses and examine issues regarding critical infrastructure and sector interdependencies that may pose a challenge to business recovery. The private sector chairs also carry out preliminary estimates of disaster losses and explore local, state and federal Business Recovery resources and tools, links and contacts. The recovery phase is carried out in close collaboration between the private sector representatives and the Joint Field Office (JFO). Key recovery activities may include monitoring re-entry operations, liaising with local law enforcement, developing recovery strategies and promoting business to business mutual aid.

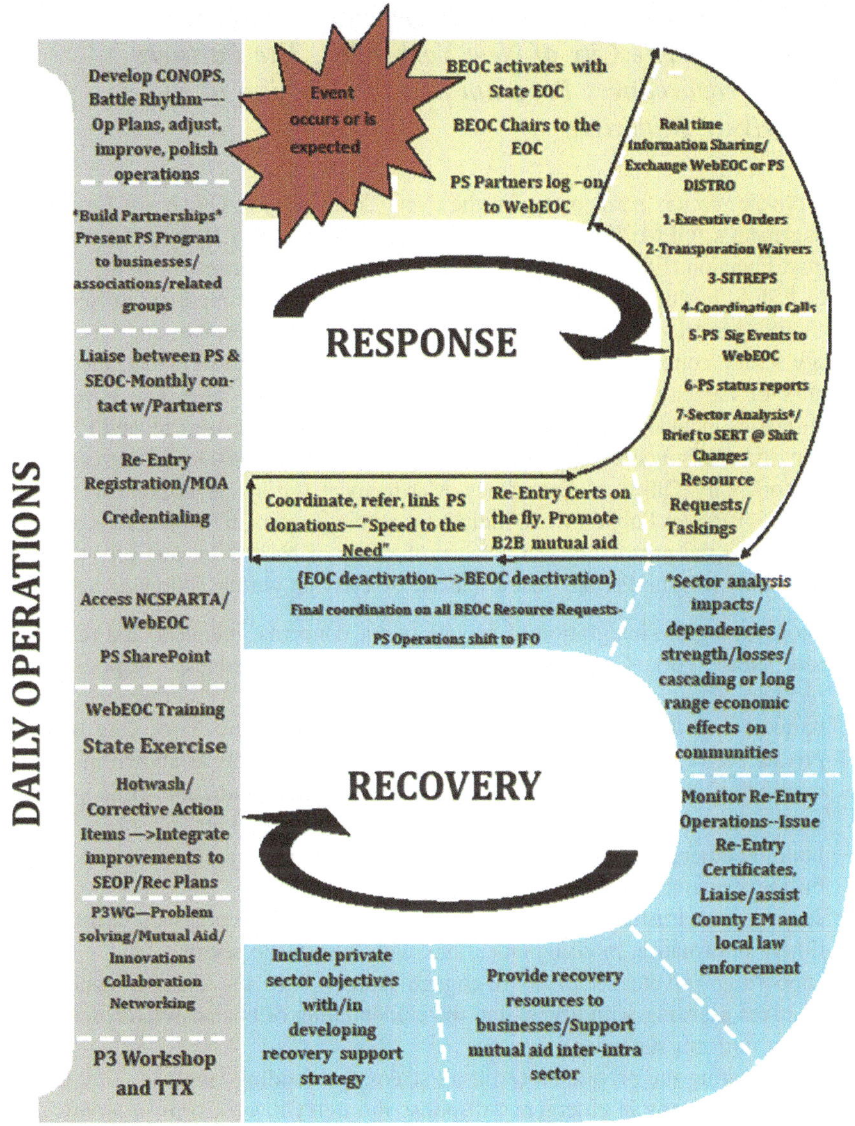

Fig. 4 Daily operations, response and recovery in the Maryland and North Carolina BEOC

5.6 *New York City Office of Emergency Management (OEM): The City of New York EOC, The Partners in Preparedness Program and the Division of Public Private Initiatives*

The private sector participates in the New York City's Office of Emergency Management's (OEM) EOC through the "Private Sector (PS) Business" Emergency Support Function (ESF) which is physically located as a separate row in the 145 seat EOC. ESF 8 (health) and ESF 2 (utilities/power) are critically important in New York City's ESF Model since there are 58 New York City public hospitals and several vital energy sector companies (e.g. ConEdison, Hudson and Madison) that affect large swathes of the city when the power system de-energizes. Other key government organizations include Army Corps of Engineers, the Parks Department, and Planning, Design and Construction agencies. The Director of Public/Private Initiatives for the New York City Office of Emergency Management (OEM) is responsible for the OEM's Division for Public Private Initiatives and the Partners in Preparedness program and serves as the primary liaison between the City of New York and private sector organizations (i.e. the Private Sector liaison) by carrying out the following activities:

- coordinating the integration of private sector concerns, interests, and resources into New York City's emergency planning, preparedness, response, and recovery activities (e.g. include the private sector in the City's emergency plans)
- harmonizing the exchange of information between New York City and the private sector, using pre-existing communication channels as shown in Fig. 5.
- overseeing the Private Sector Emergency Support Function (ESF), which incorporates private sector capabilities into OEM operations.
- building emergency management relationships and integrating them into normal business interactions
- serving as a trusted communication source: providing consistent, trusted and timely information in communications with the private sector
- supporting private sector to re-engage and re-enter the disaster zone. This involves assisting with access and the credentialing of businesses (credentialing is not uniform statewise)
- incorporating the private sector at all scales and funding levels
- engaging citizens in emergency response through Citizen Corps programs (e.g. Community Emergency Response Teams) and other initiatives

6 Types of BEOCs: Physical, Virtual, Integrated, Mobile and Hybrid

There are a number of BEOC/PPP formats and platforms in use by emergency management organizations in the US and internationally.

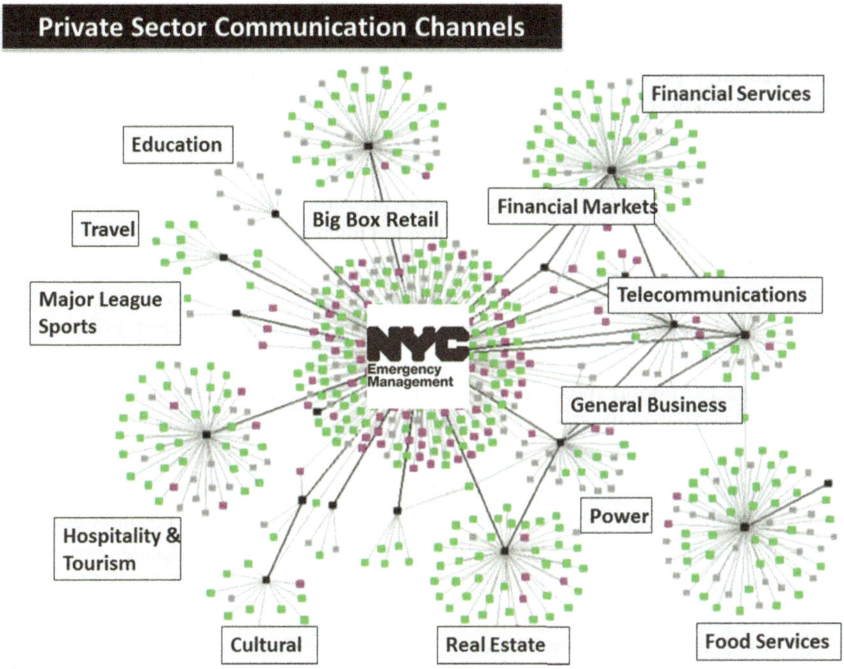

Fig. 5 Private sector communication channels in New York City's Office of Emergency Management (OEM)

6.1 Virtual BEOCs

A virtual platform represents the most common BEOC/PPP formats. Over a dozen US BEOCs/PPPs use virtual platforms to enhance communication and collaboration between the business community and the public sector during disasters. A virtual component allows for private partners to become involved in disaster management activities without co-locating in a shared physical space, thereby dramatically increasing the potential for information-sharing. WebEOC is one of the most popular virtual crisis information systems for information sharing. A comprehensive Virtual Business Emergency Operations Center (VBEOC) provides a single stage for emergency management communication, collaboration, community, and response actions. A VBEOC can serve as a valuable online clearinghouse for information that the "Whole Community" can use in disaster management. This constitutes a best practice in use in the state of Florida.

Once registered, businesses in Florida may use the tools on the state of Florida VBEOC site to report their status, complete a business damage assessment, or even offer or request resources. All of this information can feed directly into the BEOC to provide emergency management professionals with the status and needs of the business community during a disaster including road closures, utility service status, ongoing response actions, weather, analysis, etc. A virtual public/private sector portal can allow for-profit and not-for-profit private entities to receive and provide critical situational awareness information before, during, or after disasters. This facilitates private sector and government responder cooperation to promote mutual resiliency. Once registered, businesses may use the tools on a VBEOC site to report their status, develop a business continuity plan or access the "resource system" to work directly with each other to help fill resource gaps. When the resource system is made active, businesses can enter in the resources they can offer or, if overwhelmed, the resources they need. This allows the private sector to work directly with one another and ensure that the needs of the whole community are met during a disaster. All of this information can be used to feed directly into the State Emergency Operations Center to provide emergency management professionals with the status and needs of the business community during a disaster. The VBEOC can be used to support the delivery of State and Federal disaster assistance funds.

Communication with the Private Sector at the state level in Idaho is achieved through a virtual BEOC using WebEOC: The state of Idaho's main goal is information sharing and Idaho is now attempting to integrate SABER with WebEOC to ensure a bi-directional flow information between private sector partners and government officials (Idaho used SABER during the Cascadia Rising exercise to interact with private sector partners that were using Google Earth). Idaho also produced and electronically distributes its "Plan to Stay in Business guide" to help small businesses prepare for disasters and recover (businesses are offered a certificate upon completion). There are a large number of competing options to WebEOC from Ning to Yammer.

In order to minimize barriers to accessing information, the US National BEOC (NBEOC) in Washington DC uses Adobe Portal to create free, customized Adobe Connect dashboard links for business partners that include national level disaster resources (from general information to targeted information dealing with transportation, business continuity and donations management):

NBEOC Adobe Connect Dashboard Links

- Gotham Shield Exercise: https://fema.connectsolutions.com/nbeocgs17
- Hurricane Matthew: https://fema.connectsolutions.com/hurricanematthew
- Tropical Storm Cindy: https://fema.connectsolutions.com/june2017tropical
- Great Plains Tornadoes: https://fema.connectsolutions.com/greatplains
- Oroville Dam Spillway: https://fema.connectsolutions.com/r9dams.

National Level Disaster Resources
General Info

- FEMA: www.fema.gov
- (Hurricane Matthew: https://www.fema.gov/node/292516?utm_source=hp_promo&utm_medium=web&utm_campaign=femagov_hp)
- National Weather Service: http://www.weather.gov
- National Hurricane Center: http://www.nhc.noaa.gov
- Pacific Hurricane Center: http://www.prh.noaa.gov/cphc.

Transportation

- FHWA Traffic & Road Closures: https://www.fhwa.dot.gov/trafficinfo/
- FMCSA Decs, Waivers, Permits: https://www.fmcsa.dot.gov/emergency
- Federal Highway Administration (511): http://www.ops.fhwa.dot.gov/511
- USCG Port & Directory Status: https://homeport.uscg.mil/mycg/portal/ep/home.do
- All Hazards Consortium: http://www.ahcusa.org/mathew-1032016.html
- FAA Flight Delays: http://www.fly.faa.gov/flyfaa/usmap.jsp

Business Resources

- Google Crisis Map: https://www.google.org/crisismap/
- SABER Status Map: http://www.saberspace.org/status-map.html
- Healthcare Ready: https://www.healthcareready.org/hurricane-matthew
- Red Cross Shelters: http://www.redcross.org/get-help/disaster-relief-and-recovery/find-an-open-shelter
- Chamber Foundation - Disaster Help Desk: https://www.uschamberfoundation.org/site-page/disaster-help-desk-business
- Chamber Foundation - Corporate Aid Tracker: https://www.uschamberfoundation.org/corporate
- Small Business Administration - Disaster Press Releases: https://www.sba.gov/about-sba/sba-newsroom/disaster-press-releases

Volunteers and Donations

- Red Cross Volunteers: http://www.redcross.org/volunteer/become-a-volunteer#step1
- NVOAD Resource Center: http://www.nvoad.org/resource-center
- Center for Disaster Philanthropy: http://disasterphilanthropy.org
- "Donate Responsibly" Campaign: https://www.youtube.com/watch?v=UsUUAVmu6Og&feature=youtu.be
- Donations: www.fema.gov/donate | www.redcross.org/find-your-local-chapter

During an activation, the NBEOC also has developed specialized WebEOC boards with Adobe Portal. As information technology becomes more ubiquitous, increasing numbers of community members, responders and businesses have easy access to smartphones or computers. Virtual EOCs have a number of limitations. First, face-to-face collaboration may foster unique solutions when the government's capabilities are exceeded. Second, proprietary crisis information software may require specific training that limits accessibility. Third, the reliability of telecommunication networks during a disaster is a concern both because they are prone to damage and face a surge in demand.

6.2 Physical BEOCs

Physical BEOCs constitute a physical location where private and public sector representatives can co-locate in a shared dedicated space for direct person-to-person interactions. For example, the LA BEOC has a dedicated 44-seat command center (i.e. a physical facility) hosted at the University of Louisiana at Lafayette which also serves as an alternate/backup to the state EOC. The LA BEOC is also the control point for the virtual, on-line BEOC (www.labeoc.org) that ties into the State WebEOC system. The LA BEOC is also activated to support the Louisiana Governor's Office of Homeland Security & Emergency Preparedness (GOHSEP) in disaster response and recovery activities and is being developing as a resource for coordinating community recovery. A major incident may be required to activate a physical BEOC (as is the case in Illinois, Missouri, New Jersey, Colorado and Los Angeles) since it may be difficult for community members and businesses to co-locate to a remote location. As well, the capacity of a BEOC may be limited. For example, the City of Los Angeles's EOC has a business breakout room for only 12 private sector representatives.

6.3 Other BEOCs

A BEOC may also be fully integrated within the state EOC, or located off-site: Many private sector representatives interact with the state-level EOCs through a designated row in the EOC (i.e. New York City EOC), as part of a logistics team (Missouri) or designated business operations desk or breakout room (Los Angeles). Those representatives may be tasked with passing information along to and from the sectors they represent or to and from the BEOC.

There are also mobile BEOCs and traditional BEOC platforms such as call centers for businesses to call into during activation, where they can get specific information, troubleshoot problems, and make and offer resource requests. Both the National US BEOC and the New York City EOC has a similar telephone hotline system. All platforms present various opportunities and challenges, where strengths and synergies can be leveraged according to need. There are also hybrid versions of the above.

7 Selecting the Optimal BEOC

The following six BEOC alternatives are now evaluated according to the criteria of effort, transferability, costs, operations, collaboration and communication. Emergency management government officials, private sector continuity planners and non-governmental disaster management organizations from across the US were interviewed to determine the relative weights of criteria and the importance of each BEOC alternative on each criteria.

- Hybrid (Virtual-Physical Integrated)
- Hybrid (Virtual-Physical Independent)
- Physical Integrated
- Virtual (only)
- Mobile BEOC

As shown in Fig. 6, the general problem of deriving priorities from a matrix of pairwise comparison judgments is to solve for the derived priorities from the matrix $A = (a_{ij})$ where the a_{ij} are judgments from the fundamental 1–9 scale as discussed by Saaty [2, 3] in the Analytic Hierarchy Process (AHP) and Analytic Network Process (ANP). If $a_{ij} = a_{ik}/a_{kj}$ for all i, j, and k, the matrix in Fig. 6 is consistent. A priority vector derived from paired comparisons represents the impact of a given set of elements in a component on another element in the system. When an element has no influence on another element, its influence priority is assigned a zero.

The priority vectors derived from pairwise comparison matrices are each entered as a part of a column of a "supermatrix", which represents the influence priority of an element on the left of the matrix on an element at the top of the matrix [1, 5]. In order to model complexity (dependence, feedback and non-linear interactions) our BEOC research will use a nonlinear hierarchical structure and its corresponding supermatrix (Fig. 6b) rather than a linear hierarchy (Fig. 6a, which constitutes a simple and degenerate case of a network).

A node represents a component (or cluster) with elements inside it; a straight line/ or an arc denotes the interactions between two components; and a loop indicates the inner dependence of elements within a component [1, 4]. Figure 6 illustrates the supermatrix of a hierarchy and a network with three levels of clusters: where $w21$ is a

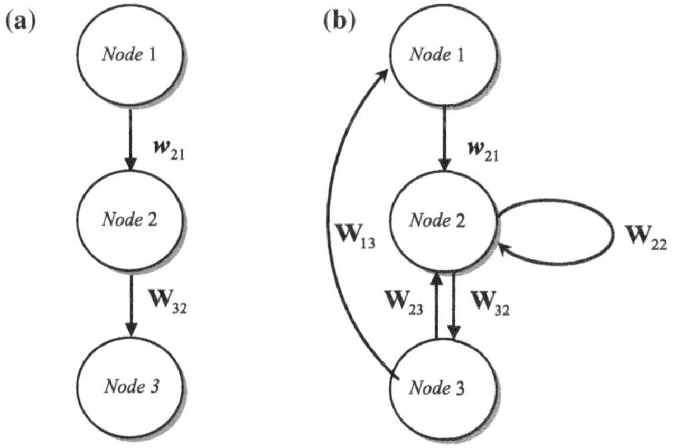

$$\begin{array}{c} \\ \mathbf{W}_h = \end{array}\begin{array}{r} \\ Node1 \\ Node2 \\ Node3 \end{array}\begin{array}{c} \begin{array}{ccc} Node1 & Node2 & Node3 \end{array} \\ \begin{bmatrix} 0 & 0 & 0 \\ w_{21} & 0 & 0 \\ 0 & \mathbf{W}_{32} & \mathbf{I} \end{bmatrix} \end{array} \qquad \begin{array}{c} \\ \mathbf{W}_n = \end{array}\begin{array}{r} \\ Node1 \\ Node2 \\ Node3 \end{array}\begin{array}{c} \begin{array}{ccc} Node1 & Node2 & Node3 \end{array} \\ \begin{bmatrix} 0 & 0 & \mathbf{W}_{13} \\ w_{21} & \mathbf{W}_{22} & \mathbf{W}_{23} \\ 0 & \mathbf{W}_{32} & \mathbf{I} \end{bmatrix} \end{array}$$

Fig. 6 **a** Linear Hierarchy and **b** Nonlinear Network [4] vector that represents the impact of *Node1* on *Node2*; **W**13 is a matrix that represents the impact of *Node 3* on each element of *Node 1*; **W**22 is a matrix that represents the impact of *Node2* on itself (inner loop); **W**23 is a matrix that represents the impact of *Node 3* on each element of *Node 2*; **W**32 is a matrix that represents the impact of *Node2* on each element of *Node3*; and **I** is the identity matrix.

The scores of various BEOC alternatives are now provided in Fig. 7 and Table 2 which shows that a hybrid (virtual-physical-integrated) BEOC obtains the highest

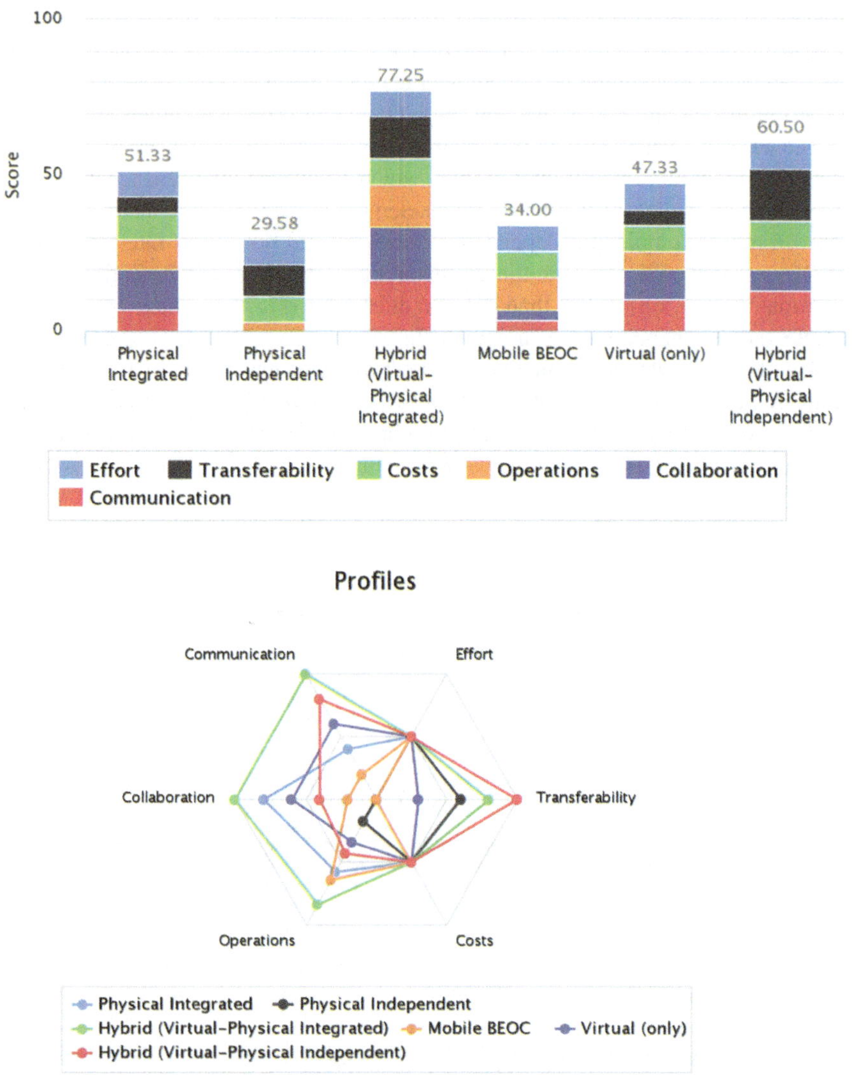

Fig. 7 Multiple criteria evaluations of six BEOC alternatives

Table 2 BEOC evaluation scores

BEOC number	BEOC description	Evaluation score
1	Hybrid (Virtual-Physical Integrated)	77.248
2	Hybrid (Virtual-Physical Independent)	60.499
3	Physical Integrated	51.334
4	Virtual (only)	47.333
5	Mobile BEOC	34.001
6	Physical Independent	29.585

evaluation followed by a hybrid (virtual-physical-independent) BEOC. Sensitivity analysis for each criteria are provided in Appendices 1 through 6.

8 Conclusions

As the emergency management expanded in scope throughout the twentieth century the growth of public-private sector partnerships became increasingly important. It is shown that the design of PPPs and BEOCs can help to understand the origins of complex disasters and security challenges and ways to prevent them. In-depth interviews with emergency managers, security professionals, continuity planners and non-government officials are undertaken to investigate how private sector organizations, government agencies and non-governmental organizations perceive BEOCs and how they are using EOCs to manage disasters. The scores of various BEOC alternatives show that a hybrid (virtual-physical-integrated) BEOC obtains the highest evaluation followed by a hybrid (virtual-physical-independent) BEOC. Sensitivity analysis for each criteria are provided in Appendices A through F. Rounded to one decimal point for ease of presentation, the Hybrid (Virtual-Physical Integrated) BEOC has an overall score of 77.2, followed by the Hybrid (Virtual-Physical Independent) with a score of 60.5 and the Physical Integrated BEOC yields a value of 51.2. The bottom three BEOC alternatives are now provided: Virtual (only), Mobile BEOC, and Physical Independent. In particular, the Virtual (only) alternative has a score of 47.3 while the Mobile BEOC yields a value of 34.0 and the Physical Independent BEOC has the lowest overall value of 29.6.

Appendix 1: Sensitivity Analysis for Costs

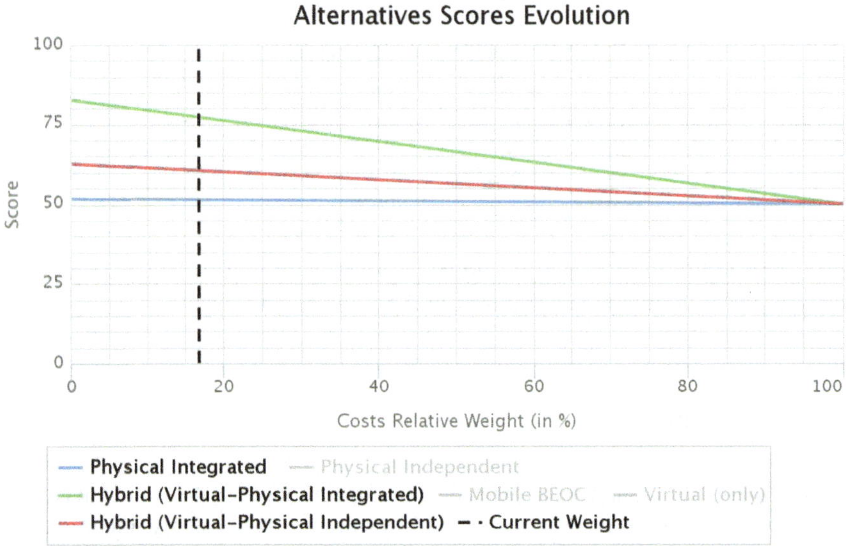

Appendix 2: Sensitivity Analysis for Effort

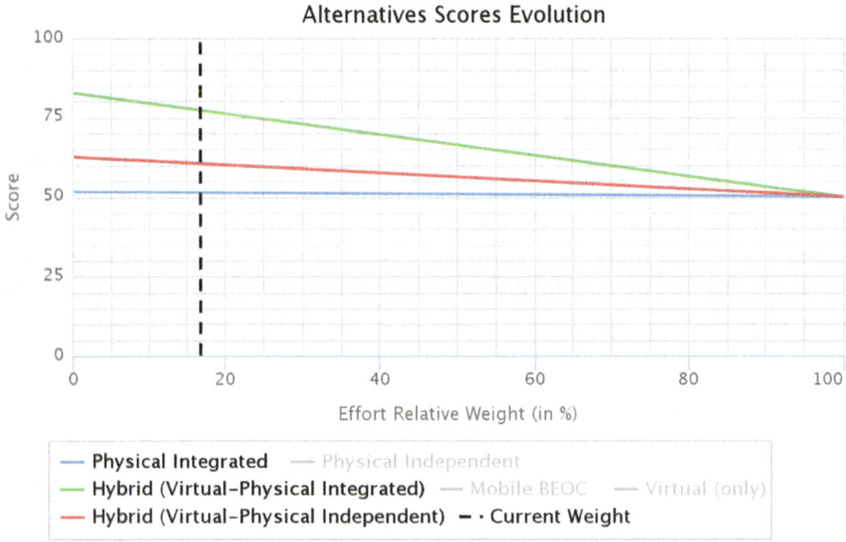

Appendix 3: Sensitivity Analysis for Transferability

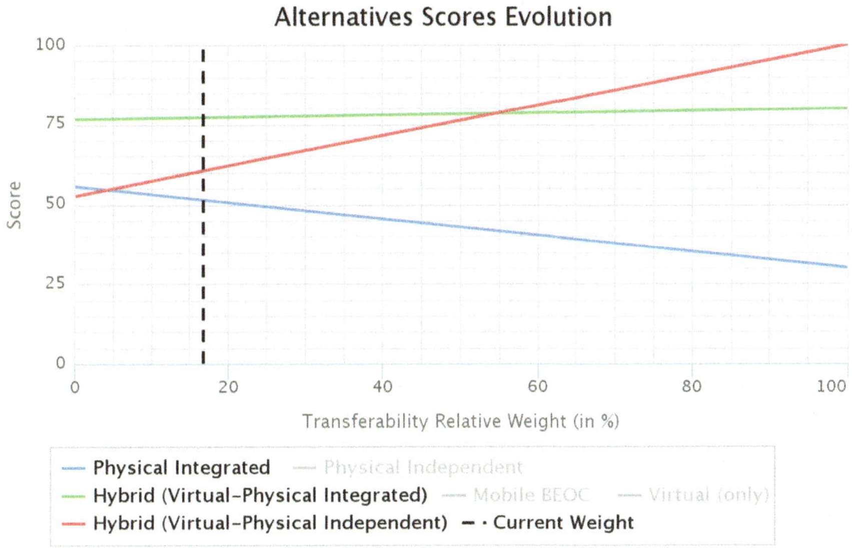

Appendix 4: Sensitivity Analysis for Collaboration

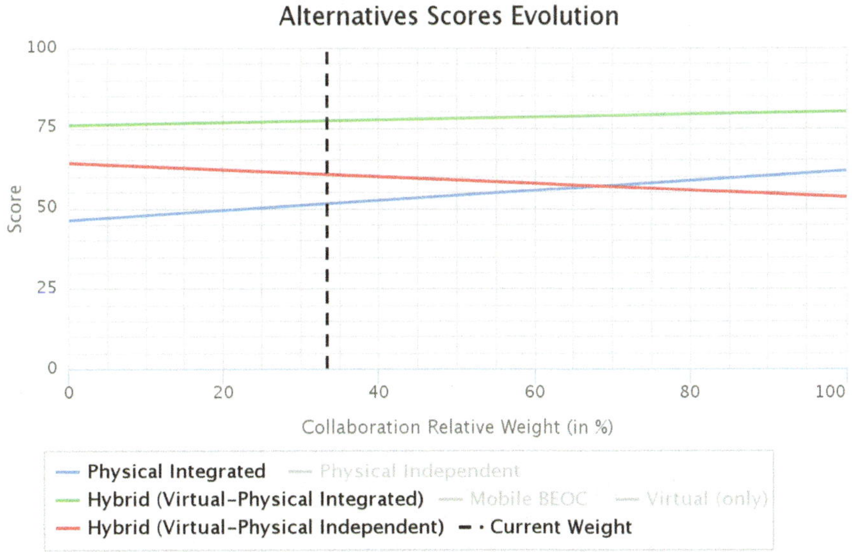

Appendix 5: Sensitivity Analysis for Operations

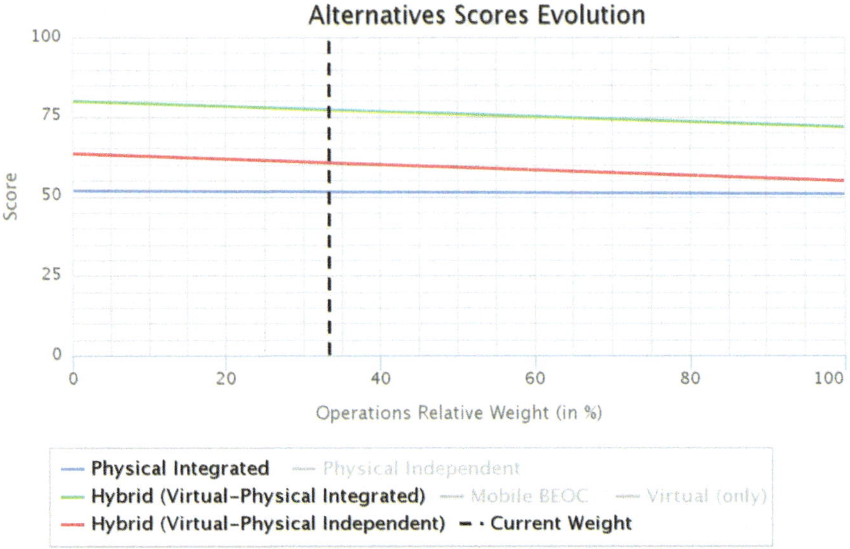

Appendix 6: Sensitivity Analysis for Communications

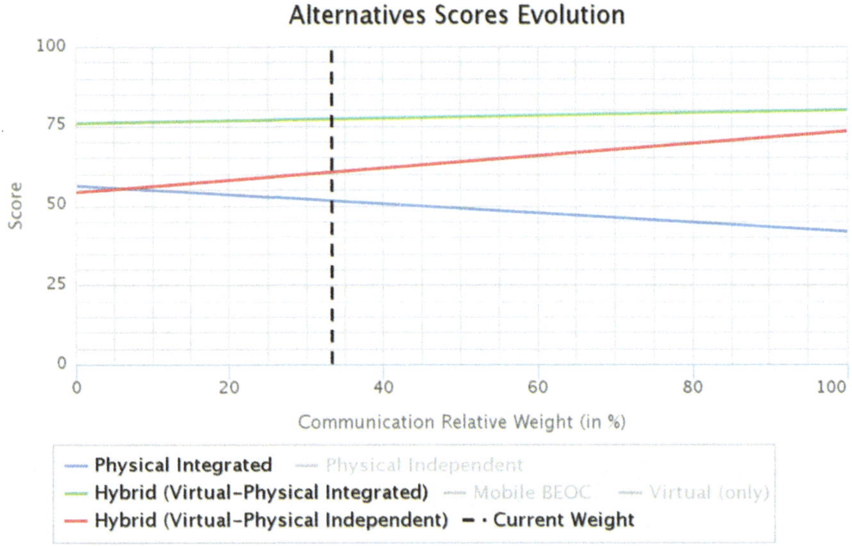

References

1. Levy JK, Taji K (2007) Group decision support for hazards planning and emergency management: a Group Analytic Network Process (GANP) approach. Math Comput Model: Int J 46(7–8):906–917
2. Saaty TL (2001) The analytic network process: decision making with dependence and feedback. RWS Publications, Pittsburgh, PA
3. Saaty TL (2004) Decision making—the analytic hierarchy and network processes (AHP/ANP). J Syst Sci Syst Eng 13:1–35
4. Saaty TL, Takizawa M (1986) Dependence and independence: from linear hierarchies to nonlinear networks. Eur J Oper Res 26(1986):229–237
5. Saaty TL, Vargas L (2006) Decision making with the analytic network process: economic, political, social and technological applications with benefits, opportunities, costs and risks. Springer, Dordrecht

The manufacturer's authorised representative in the EU is Springer Nature Customer Service Centre GmbH, Europaplatz 3, 69115 Heidelberg, Germany. If you have any concerns regarding our products, please contact ProductSafety@springernature.com

Printed and bound by CPI Group (UK) Ltd, Croydon, CR0 4YY
30/09/2026
02225655-0011